Social Inequality as a Global Challenge

RIVER PUBLISHERS SERIES IN CHEMICAL, ENVIRONMENTAL, AND ENERGY ENGINEERING

Indexing: all books published in this series are submitted to the Web of Science Book Citation Index (BkCI), to SCOPUS, to CrossRef and to Google Scholar for evaluation and indexing

The "River Publishers Series in Chemical, Environmental, and Energy Engineering" is a series of comprehensive academic and professional books which focus on Environmental and Energy Engineering subjects. The series focuses on topics ranging from theory to policy and technology to applications.

Books published in the series include research monographs, edited volumes, handbooks and textbooks. The books provide professionals, researchers, educators, and advanced students in the field with an invaluable insight into the latest research and developments.

Topics covered in the series include, but are by no means restricted to the following:

- Energy and Energy Policy
- Chemical Engineering
- Water Management
- Sustainable Development
- Climate Change Mitigation
- Environmental Engineering
- Environmental System Monitoring and Analysis
- Sustainability: Greening the World Economy

For a list of other books in this series, visit www.riverpublishers.com

Social Inequality as a Global Challenge

Editors

Medani P. Bhandari and Shvindina Hanna

River Publishers

Published, sold and distributed by:
River Publishers
Alsbjergvej 10
9260 Gistrup
Denmark

www.riverpublishers.com

ISBN: 9788770225991 (Hardback)
9788770225984 (Ebook)

Contents

Dedication

Firstly, we dedicate this book to all loved one Worldwide who lost their lives due to Corona pandemic from the beginning to date; we acknowledge that some of you were victimized due to inequality in healthcare systems. In your memory—in this book, we have raised the questions of why the world was not able to save you. The world will remember and salute you all, and we hope that future science will be able to control such a pandemic.

Secondly, we dedicate this book to Prof. Bishnu Poudel, (January 3, 1936–April 14, 2020) who always insisted us (and every one of us to do the right thing in right time), we feel that your departure time was not right. We never forget your encouragements and support to all of us. Your forward for the Inequality book is included, as token of your blessings to us.

To Dr. Poudel, each person and each event in a day of life had meaning. He was one of those unique souls that actually lived a life. There were no expectations, no barriers, and no complaints, just the joy and humility of being (Peter Poudel—son of Dr. Poudel).

I have known late Prof. Bishnu Poudel since 2012 here in Fairfax, Virginia. He was one of the very few individuals of Nepali origin who settled in Fairfax, Virginia in early 1960s. He was a towering and inspiring intellectual personality in Nepali-American community. Ever since, I knew him, I found him very actively involved in the community activities here in Virginia. He was very knowledgeable about wide ranging subjects including politics, history, religion, and Eastern philosophy. He was a scholar of eastern Vedantic philosophy and Sanskrit, the oldest language of human civilization. Last time, we travelled together in 2019 from Virginia to New York during which I have had a very interestingly long conversation on some intellectually interesting subjects such as Eastern philosophy, crisis in human civilization, and climate change, etc. His views on such complex subjects were enriched with profound knowledge, understanding, and wisdom. Anybody who interacted with him cannot remain positively unmotivated and uninspired. I personally missed an inspiring "guru." He will be greatly missed by Nepali-American community here in Fairfax, Virginia.

Prof. Gopi Upreti, Fairfax, Virginia, USA

"It was my great privilege to know, work, and be mentored by Dr. Bishunu Poudel. He had a unique personality of philosophy, spirituality, and international relations among people, countries, geo politics, and diversities. Both of us were bothered by digital divide, social divide, and all sorts of inequalities. Therefore, we established the Equality Foundation in 2015 and waged the war against digital divide from Nepal. He served as the founding president. We learned so much from him that his influence in our lives will remain intact and we felt that we must carry his sacred legacy especially being the founder and current president of the Equality Foundation. Overall, Dr. Poudel was a complete and content human soul, who was able to make friends and inspired people before all kinds of barriers."

Medini Adhikari, President Equality Foundation, Professorship at, American Military University, American Public University System (APUS), USA

Arriving in the United States 50 years ago, after completing all his studies here, Prof. Dr. Vishnu Poudel was always standing on the equality issues on behalf of South Asian arena. A lover of Nepali language, literature, society, and Hindu religion and culture, Dr. Poudel was at the forefront not only in the Nepali community but also in the South Asian and Indian communities in the United States for their rights. He has contributed to the minority communities when he served as the White House Asia Advisor during President Clinton's tenure.

Representing Nepalis in the establishment, development and various intellectual conferences of Nepali and American institutions, he had profound philosophical views. As an expert writer of Eastern philosophy, I am also a person who has been very impressed by his many books, discourses, and synthetic remarks. He spoke on behalf of minority groups and raised the equality issues in the international platforms.

Gunaraj Luitel, President, INJA Network, USA

Prof. Bishnu Poudel was a "Leading Figure" in our community. He encouraged to serve honestly to USA and help to Nepal to many young Nepalese, who are learning and earning in USA. Because of his leadership we are standing together today for good cause. Our team will remember him forever.

Niranjan Regmi, Equality Foundation, USA

Preface

The book "Inequality a Global Social Challenge" covers the various factors of inequality embedded within the social, economic, and political systems. Inequality can be seen as a global communicable disease of human civilization. Social inequalities are seen in the service sectors—access to healthcare, access to education, social protection, access to housing systems, childcare, elderly cares, etc. Another neglected but deep-rooted issue is cultural inequality, which segregates people from the mainstream. The minority groups face recognition problems of their social status, language, religion, customs, norms, etc. This book tries to unveil the real picture of these issues with exemplary cases studies from various countries (Nepal, USA, India, Mexico, Ukraine, European Union, Russia, Pakistan, Serbia, Iran, Nigeria, Middle East, Latin America, North America, Asia, Pacific, Turkey, Bulgaria, Lebanon, Jordan, Egypt, Jordan, Syria, etc.).

This volume begins with why inequality is the global problem, and why and how society is divided dude to economic, political, religious, and due to traditional framework, which still divides society on the basis of gender, age, origin, ethnicity, disability, sexual orientation, class, refugee status and religion, etc. It is a common challenge for both developed and developing countries, however, the developing world has been severely victimized throughout the human history. Book begins with the theories, history, and jumps on the current scenarios, whereas mostly chapters unveil the stories of discrimination on the basis of personal and social identity of a person or a group. Gender inequality: inequality due to color, ethnicity, country of origin, social discrimination due to haves and have-nots group (which is created by the social elites of all time); Discrimination Against LGBTQ+ (in education, job, social gathering, and political system) is well described in the book, whereas authors have provided the history of LGBTQ+ movements in the world with the chronological data and evidence. Social Inequality can be seen in every sphere of social systems; therefore, chapters are presented from the Behavioral View on Inequality; Inequality in Public Health; Inequalities in Education and also unveils the truth of Social Exclusion due to gender, age, origin, ethnicity, disability, sexual orientation, class, refugee status and religion, etc.

This book is a combined effort of several scholars around the world who have considerable experience in research, teaching, and social development. There are sixteen chapters in the book, and each chapter unveils the real picture of how societies are divided, demoralized, marginalized, and dominated mostly through the grounded culture, social atmosphere, religions, economic and political systems dominated by the elite of society throughout human civilizations. Once again, mostly, when we talk about inequality, the focus remains on economic inequality; however, the major inequality persists due to discrimination to gender, age, origin, ethnicity, disability, sexual orientation, class, and religion within the society. It is true that primarily the world is divided between haves and have-nots; and social strata are created as comfort zones to those who dominate the social, economic, and political systems. The rules and regulations, social norms and values are always created and implemented by the social elite, therefore, the marginalized groups have no or very minimal stake in the society, and they are always the victims. There is a need for a complete social, economic, and political reform to minimize the designed division. Until or unless the marginalized group is empowered, the inequality issue cannot be solved or even minimized. This edited book tries to capture these issues through case studies from various part of the world and tries to unveil the real picture of the societies. The book examines the historical perspectives of inequality which is grounded within the society, politics, and the economy, the causes of inequality, and how inequality has been persistent for centuries given multidisciplinary factors challenge with the development economic, social justice, dominance and through the sustainability discourses. Inequality is one of the major divisive factors of human civilization and the evolutionary history of development. Inequality can be seen in every sphere of social, political, and economical structure of community, national and international level. Such unequal strata can be grounded on innumerable factors—such as social, cultural, political, geographical, or due to environmental (anthropogenic or natural) catastrophe. On the basis of various case studies, this book encourages to rethink societal development through the lens of growing inequalities and disparities. The book presents new insights for evaluating the progress on social development, with the coverage of major inequality due to gender, age, origin, ethnicity, disability, sexual orientation, class, and religion. Book unveils the main challenges of social inequality. Every chapter is different and yet in combination, they give an integrated understanding of the social phenomenon of "why society is inequal and what is the major social inequality". This book will be useful for those stakeholders, who want to make or contribute to the change and see an undivided, socially inclusive society, and to those who want to contribute to empowering society, so no one left behind to enjoy the beauty of human lives.

Prof. Medani P. Bhandari, PhD.

Akamai University, Hawaii, USA;
Sumy State University, Ukraine

The Final Words of Prof. Bishnu Poudel (January 3, 1936–April 14, 2020) on Inequality

It is my pleasure to write a few words about Dr. Medani **P.** Bhandari's work. I have known Dr. Bhandari since 2017. His scholarship, research, and in-depth analysis of subjects he handles are remarkably profound and timely for major shifts in policy decisions by all the concerned societies and countries equally. I have read his previous work "Green Web-II: Standards and Perspectives from the IUCN" and the manuscript of "Getting the Climate Science Facts Right: The Role of the IPCC." It is a very profound analysis of justice, equity, and sustainable development. In this edited work "Social Inequality as a Major Challenge" his analysis is very crucial to fathom the depth of globe marching toward its own destruction. All scholar's contribution in the volumes are His write-up in "The Problems and remarkably maintain very high standards, which adds new knowledge in inequality and Sustainable Development." Every scholar succinctly lays out the fundamental issues of the world today. They are all divides, viz economic, income, race, religion and, human relations as well as geopolitical intolerance. This has given the rise to further dictatorial attitudes to have in all levels of social strata including governance. Book outlines statistical analysis of the issues in the world today. I admire Dr. Medani P. Bhandari and Shvindina Hanna, and all authors frank and timely pointers that humanity needs to wake up.

Otherwise, we may be doomed before we see the twenty second century. I congratulate Dr. Medani Bhandari and his team for their attempt to help mankind rise to the occasion to bring in sanity to its own existence.

Thank you.

Bishnu P. Poudel, Ph. D

Forewords

Inequality stems from different factors operating within a given geographical area, country and at global level. Primarily few factors such as political stability of country, whether there are ongoing conflicts, violence's, application of new technology in manufacturing and its adoption, the level of education, sharing and caring of economic for the benefits of the people, endowment of Natural resources and their optimum use, saving and investment culture of a country for public infrastructure and uniform health services. These factors are influenced by economic system and working and collaborating with different culture. In the current economy, these are the influencing factors and directly impacting for global inequality. The effects are immense including social unrest, increasing conflicts, frustration in people leading drug abuse, crime, poverty, and eventually terrorism.

In order to address those issues, countries in the world have progressively implemented policies for access to education, financial subsidies, improving workers' rights, increasing political stability, promotion of free enterprises and overall adopted the concept of welfare state. Despite this however there is an increase in globally organized crime, exploitation of poor developing countries, land speculation, exploitation on social security, increase in environmental and atmospheric pollution and dependency.

This edited version of the book chapters **"Social Inequality as a Global Challenge"** has tried to address the issues raised earlier. In the contemporary world inequality has become a pandemic and threat to Human civilization. Which has segregated people **between have and have not**, promoted racial and cultural discrimination and social unrest.

This book is the result of collective work of scholars around the world. At the present time when we talk of inequality, we immediately focus our vision to economic inequality. But there are other major areas where human has discriminated their fellow members in terms of culture, languages, ethnicity, sex, social status, and others. This compilation indeed will contribute to raise innovative debate among policy makers and bureaucrats and private sector decision makers in different parts of the world. I hope this will generate robust discussions among academia and young generation leaders of public and private domain. I am thankful to all the contributing authors and editors for undertaking this arduous job.

Dr. Krishna Prasad Oli, Ph.D.
Member National Planning Commission
Government of Nepal
Kathmandu

Social Inequality as a Global Challenge

The issues related to social inequality and related matters of discrimination and injustice to the marginalized group and societyare at the forefront of cultural crises worldwide. This book, Social Inequality as a Global Challenge edited by Prof. Medani P. Bhandari and Dr. Hanna Shvindina, addresses the importance of universal social equality for effective addressing the vital global challenges worldwide.

Important to state here is the reality that the United States of America was founded upon the premise that." *All men are created equal, that they are endowed by their Creator with certain unalienable rights, that among these are life, liberty and the pursuit of happiness.* (Thomas Jefferson, 1776)

The present-day deterioration of this foundation worldwide discloses the importance of this book in support of our understanding of the path forward away from inequality in race, religion, ethnicity, gender, LGBTQ+ and cultural inequalities. It is known that the social patterns of poverty and deprivation are closely related to deprivation and lack of opportunity that emerge from broad-based inequalities and severe lack of sustainability of cultural advances across the globe.

Consequently, Bhandari and Shvindina and their team (the individual chapter authors) have successfully endeavored to clarify the importance and the path toward reducing pervasive inequalities with the purpose of moving toward sustainable development goals across the global community. Our cultures are judged by how we treat the least among us, how we support those who are most abused and what help we provide to those who hunger and suffer. I am encouraged by the depth and courage of this book.

Douglass Lee Capogrossi, PhD
President Emeritus
Akamai University

Social Inequality as a Global Challenge

Now is the time. No longer can we ignore the obvious disparities in the social structure of our countries. This is a worldwide problem that is not going to be solved by the tunnel vision and cultural norm expected to be followed.

Courage is needed, demanded to look inside within each of us our thoughts, emotions and actions that drive our inner being into leadership out of the darkness or compliance of the status quo. The range of inequalities are staggering. Where to start? What to focus on? Then what to do with what I uncover through my research and discernment to follow truth and innovation to being into light the discrepancies worldwide.

Can you stand up to the truth of your prejudices and short sightedness? Do you turn your back once you even take the risk to become informed? Or do you settle into complacency like so many have done over the centuries and throughout many countries?

Can you risk becoming informed? Being inspired? Being hopeful and excited into action? I throw you this challenge as you investigate what the authors have gathered as evidence to shine the light in the darkness. I dare you to take the risk and carry the torch to your destiny as motivated by taking on the challenge to be an innovator practitioner, educator and / or researcher.

The exemplary scholars have gathered the evidence for you to review, ponder, reflect, and empower to act on behalf of the injustices you see.

Stand up.
Stand out.
Stand firm.
Stand tall.
Stand strong.
Stand resilience.
Stand firm.
Stand unwavering.
Stand facing the Light ... and know ... you are not alone.

Change can happen. Will you be part of the Light beams illustrating the path to a new height for all? Not for a "chosen" few, rather a leveled playing field all can operate on and win the game of life well spent living your dream. You have done wonderful job Prof. Medani P. Bhandari and Dr. Shivindina Hanna. The inequality is a great challenge of all time. Your team contribution means a lot to academia as well as to all stakeholders, who thrive for the positive change in the society and certainly help to minimize inequality in the world.

Dr. Mary Jo Bulbrook, President,
Akamai University, USA

Social Inequality as a Global Challenge

It is my pleasure to write a few words about Professor Medani P. Bhandari and Associate Professor Hanna Shvindina's book. "Social Inequality as a Global Challenge" is another outcome of fruitful cooperation between the US and Ukrainian academic communities. This book is devoted to Professor Balatskyi who was a well-known environmentalist, and this is another proof of scale and scope of the presented work as it unites efforts of scholars in the field from many countries, many areas of research in their intentions to change the world for a better place, where inequality is not a challenge anymore. The contributions of scholars bring new knowledge into the field, establish new interlinks between complicated and complex phenomena related to social inequality and enable to rethink the strategy in the long run for future developments. The editors did a great job, as well as all authors of the chapters. All collaborators of this project brought the light to different aspects of social inequality which is timely and relevant especially in times of crises. I appreciate the contributions of my colleagues, Dr. Medani P. Bhandari, Dr. Hanna Shvindina, and the team of authors in the further development of social sciences.

Prof. Vasyl Karpusha, Rector,
Sumy State University, Sumy, Ukraine

Social Inequality as a Global Challenge

The problem of social inequality has dominated public consciousness over the past few years being the theme of discussion on the strong social stratification which is increasingly present in the era of rapid economic changes. The discussion is intensified by waves of global recessions, the effects of which could be strongly felt for example during the Great Recession of 2008–2010. This particular downturn resulted in a sharp increase in the level of economic inequality which, in turn, gave rise to social inequalities. While in case of the economy, the recovery processes are relatively efficient, a considerable number of people remain insecure about their economic future and social situation. The insecurity has been further deepened by the current crisis of the SARS-COV-2 virus pandemic as many people have lost employment and found themselves on the verge of social exclusion.

The concept of social inequalities, understood as a socio-economic phenomenon, referring to discrepancies in distribution of resources, power, income, housing and welfare status among social groups, has been evoking extreme emotions and heated discussions. A related concept is that of stratification which refers to a model of social inequality that specifies the relationship between wealth and social standing.

Social inequality strongly contrasts with the sense of justice, causing a series of controversies, at the root of which lies economic inequality. Inequality means that those at the bottom of the economic and social ladder experience disadvantage, often extreme poverty, which significantly influences their social development. According to Forbes magazine, on 18 March 2020 there were 2.095 billionaires around the globe, while the data from a report published by the World Bank in 2015 quote that in that year about 736 million people were living on less than $1.90 per day. However, it should be emphasized that when compared with 1990, when 1.9 billion people lived in extreme poverty and the world had only 269 billionaires, the situation has improved considerably (Forbes, *The World's Billionaires: 25th Anniversary Timeline,* 2012).

Looking at the effects produced by the social debate, one can get the impression that no country in the world has yet found a perfect tool that would be fully effective in overcoming social inequalities. Understanding social inequalities requires considering not only the very fact of the existence of differences but also the causes rooted much deeper in human consciousness. Henry Ford said: "*If money is you hope for independence, you will never have it. The only real security that a man can have in this world is a reserve of knowledge, experience and ability.*" These words make one realize the importance of the role that factors such as upbringing, mentality, and lifestyle play in the life of a society. One of the main tools to combat social inequalities magnified by financial crises or global pandemics would be building strong partnerships to achieve the sustainable development goals.

The book by Professor Medani P. Bhandari and Associate Prof. Shvindina Hanna refers to the timeless issues related to economic and social inequality. The authors of the book attempt to answer the question of what needs to be done to reduce the scale of social inequality in the world. The book deals with issues related to, among others, historical aspects of inequality, religious and gender inequality, public health policy communication, employment discrimination, and social exclusion. The examples of various known and hidden meanings of social inequalities refer to the situation in many countries of the world. Each chapter presents a new perspective on issues that readers may also apply to their personal and professional life, depending on their cultural realities. The authors also illustrate in a comprehensive way how the world faces economic and social challenges. This book can be recommended for those readers who are interested in the subject of social inequality and/or for those involved in the elimination of strong economic and social imbalances.

Dr. Jacek P. Binda Recotr and a researcher at Bielsko-Biała School of Finance and Law, Poland (pol. Wyższa Szkoła Finansów i Prawa w Bielsku-Białej, Polska).

Social Inequality as a Global Challenge

It is a pleasure to be writing foreword for this timely and important book "Social Inequality as a Global Challenge" edited by Prof. Medani Bhandari and Prof. Shvindina Hanna. This is a topic of current interest to many across the world.

I grew up in the remote hills of Nepal at a time when almost everyone lived on subsistence agriculture. Looking back, poverty was the norm in my village and the surrounding localities that I knew of. The prevalent poverty and lack of any opportunities to improve once's life was common for everyone except for a handful of individuals in the entire village. The existence was basically hand-to-mouth, and every generation was stuck in the same level of standard of living. Except in a few urban areas with a limited level of prosperity or a few, I believe the socio-economic situation I described above was the norm in Nepal.

For many, the impact of the pervasive poverty was exacerbated by the various forms of discrimination prevalent in the Nepali society then. The women toiled in the fields and faced drudgery at home. Caste-based discrimination made it difficult for the families of the lower castes to experience any upward mobility. Nepal had officially embarked on creating a modern nation-state and the national laws had vowed to end discrimination and create a fair society. However, elimination of discriminatory practices would take a long time, and even today, many forms of discrimination continue to exist in the country.

Achieving social equity is an important issue everywhere. Despite the vast improvements in human well-being the world has seen in the past, social inequality remains a stubborn problem in the human society. Whether it is about access to positions of power, employment opportunities, health and education services, availability of decent housing or other amenities, there is a gap between what is available for the more privileged and marginalized groups in societies.

The social inequalities can be driven by various factors such as economic, racial, and ethnic class, and sexual orientation or gender differences. For example, in most societies, minorities and women often face discrimination in workplace, political arena, and social life in general. To preserve and protect their power and advantageous positions, the elites and groups belonging to the higher echelon in the society may create policies, laws, regulations, and protocols to secure their position of privilege in the system. The legal and policy regimes could seem fair, but there might be a tacit agreement among many in the advantageous situation to help preserve the status quo.

Global situation on equality remains challenging. Even when humanity continues to gain unprecedented improvement in the overall well-being, many parts of the world still practice regressive policies and programs related to dignity and equity for all. For example, some countries allow lesser rights to women, and other limit the rights of religious minorities. The nationalist and isolationists trends in the domestic politics of several countries remains a worrying trend. Within the borders of many countries, social intolerance seems to be growing, and marginalized groups are suffering the brunt of the nationalistic and xenophobic politics. Even in developed, secular, and human rights practicing societies such as in the United States and United Kingdom, the minorities and marginalized groups can face an uphill struggle in realizing their aspirations for equality and better life. For example, in the aftermath of the killings of many African American individuals in the hands of mostly white police, racial tensions have recently flared up in the United States as can been seen in the confrontation between the supporters of the "Black Lives Matter" and their opponents.

Social discrimination can have many dimensions. In addition to the common forms of discrimination based on race, color, sex, and national origin, even speech accents can be a basis for discrimination. Several studies done in UK and USA have concluded that persons who have accents different than the language spoken by the mainstream society can face tacit discrimination. When the accent can be identified with national regions, immigrant categories, minority groups or working class, it can be a cause for lower pay, diminished business opportunities, and impediments to upward mobility.

As the world continues to benefit from the cultural evolution where the intrinsic worth of every individual and social group is respected, improvement in social mobility and offering respect for every individual becomes an even more important task. The ideal situation will be where there is no discrimination of any individual or social group, and everyone gets to live a life that is dignified and respected and full of opportunities. When everyone lives in a fair and just society and is able to realize his/her full potential as an individual, the society, country and the world all benefit. The world becomes a morally just place.

This book is designed to fill the knowledge gap especially in comparative analysis of social inequalities. The book editors Prof. Medani Bhandari and Prof. Shvindina Hanna have extensive background in this field as educators and practitioners in social and environmental development. Bhandari started his education and a career in the environmental and development field in the rural and small-town environment in Nepal. He later moved to the USA to obtain his doctorate degree and now teaches in several countries. His experience in different countries have provided him with a comparative perspective on how societies function and how social inequalities can impact in their abilities to realize the full potential as members of the society. Hanna grew up in Ukraine and was educated in Ukraine and later in France and USA. Her education and training in economics, management and development has helped her to teach and write in support of social and economic development internationally. Her vast education and experience have provided her with important perspectives to co-edit this book.

The editors have gathered an impressive constellation of authors from various countries and disciplines to provide a wide overview of social inequality issues. This has helped make the book rich in perspectives, insights and provocative analysis that can greatly benefits the reader.

This book will be an important reading for students, academics, researchers, policymakers, and laypeople who are interested to understand social inequalities and learn about possible options to reduce inequality.

Dr. Ambika P. Adhikari, PhD
Sr. Sustainability Scientist,
Arizona State University, Arizona, USA
Former Country Representative,
International Union for Conservation of Nature (IUCN),
Nepal

Social Inequality as a Global Challenge

Profs. Medani P. Bhandari, Shvindina Hanna edited the book "Social Inequality as a Global Challenge" covers a wide range of issues. The discussion starts from historical perspectives of inequality with reference to society, politics, economy, religion, and gender using examples from rural and urban areas of South Asian, European, and African countries. It also addresses inequality issues related to public health by investigating discrimination on employment and social exclusion from Eastern European countries. It not only analyzes the impacts of subsidies on social equality but also explains hidden inequalities in education and its long-term consequences on society. The book also looks at how LGBTQ+ people are experiencing cultural discriminations. To substantiate the discrimination against the LGBT+ groups, the editor /author adds examples of the addition of unprecedented number of refugees from conflict ridden countries to the European realm to provide further explanation on inequalities. This book also discusses the household food security taking examples from South Asia. Using examples from rural and urban areas of South Asia, Eastern European, and African countries, the Prof. Bhandari tries to engrain regions, countries, and local specific these cases within the theoretical frameworks postulated by Karl Marx, Emile Durkheim, Max Weber, Henri Fayol, Frederick Taylor, Luther Halsey Gulick, Herbert A. Simon, and Berton H. Kaplan.

Contributors of various chapters in this book connect inequality issues associated with economic, social, political, cultural, and religious rights and opportunities, and living conditions of people. Prof. Bhandari explains how inequality stratifies a society. He explains how economic inequalities further ramify skewed ownership of human, natural resource, social assets, political opportunities, deconstrue institutions, distort military structures, and how dominant group opportunistically empower themselves in every sector of a society. These developments keep marginalized group at the lower strata of political, social, and economic main streams. Such a skewed concentration of resources empowers a wealthy few to monopolize languages, religions, customs, and social norms. Taking examples from South Asia, this book examines how inequalities in resource distribution affect gender and caste orientations, creates fear among low community strata, gender, and minority groups. The author explains how such stratification has added gender-based domestic violence, where mostly women are victimized. Almost 30 percent of income inequality resulting at the household levels further fuels such violence against women. This book also presents cases where men tend to earn more than women and women are often underrepresented in senior positions, and relegation of women to fewer and less important responsibilities.

Citing examples from various countries, this book analyzes how social inequalities are grounded within the religious and cultural covers from self-made stories [like Q'Anon]. Despite the **United Nations attempting to address inequalities under its 1, 3, 4, 5, 6, 7, 8, 9, 10, 11, 16 and 17 goals, inequalities have further increased today. The** United Nations has been publishing voluminous reports periodically on inequality including the recent report on *Inequality— Bridging the Divide—Shaping Our Future Together* (2020). Yet, inequalities exist everywhere, for examples:

a. In 2018, the world experienced 12th consecutive year of decline in global freedom, with 71 countries suffering from net declines in political and civil liberties resulting into multidimensional inequalities.

b. An average income of North American people is 16 times higher than the people in Sub-Saharan Africa.

c. According to an Oxfam report, since the financial crisis of 2008 and 2012, global resources are controlled by a wealthy few, for example, in 2018, the 26 richest people in the world held as much wealth as half of the global population (the 3.8 billion poorest people), which is down from 43 people in 2017.

d. Inequalities of opportunity affect a person's life expectancy and access to basic services such as healthcare, education, water, and sanitation.

These inequalities have led to nativism and extreme forms of nationalism in the wake of globalization. The agony is that if issues like climate change are unaddressed and nativism and ultra-nationalist slogans remain strong pillars, they will increase further inequality not only at the global, regional, and national levels but also at the local levels because climate change has ramification in every sector of the global society. Without appropriate policies and institutions, and the use of 21st century technology, inequalities and social divisions will continue. Theoretically, it is hard to fully pinpoint the root cause of inequality. Left leaning politicians fabricate equality among all communities is possible only under socialism, but without strong economic growth, living conditions of people cannot be improved by strong slogans. As of today, social justice theory, social dominance theory, and development economics theory have not been able to address the nature of inequality. This book tries to fill the knowledge gap between theoretical and practical aspects to alleviate inequalities. However, so far, none of the published literature have been able to provide a workable model to end inequality. Profs. Bhandari and Hanna provide their personal experiences and tries to fill the knowledge gap to make this book a useful tool for planners, policy makers, and politicians (P-actors) to address inequalities. Illustrating case examples within the framework of social science theories, qualitative and quantitative methodologies, and organizational sociology, this book attempts to find a meeting point between theories and practice to find a sustainable solution to social inequality. This book will be an important resource for students, academicians, researchers, P-actors, and ordinary people to understand inequalities from the sociological theories.

Keshav Bhattarai, Ph.D.
March 16, 2021
Professor of Geography
School of Geoscience, Physical & Safety
University of Central Missouri, Humphreys 223C
Warrensburg, MO 64093
bhattarai@ucmo.edu
Phone: 660-543-8805

Social inequality is a dangerous societal disease worldwide. It has multitudes of causes including, race, caste, gender, language, income, job status, education, and culture. Social hierarchy leads to discrimination compromising social justice. People falling in the higher strata of social structure systematically enjoy basic services such as healthcare, education, employment opportunities, and empowerment, while people falling in lower strata struggle throughout their lives for these basic services. Social discrimination leaves permanent scars on people's lives. Social discrimination impedes individual's personal and intellectual growth and changes social behavior. Communities that are historically left behind on education and economic wellbeing are trapped into these societal maladies of social inequality unless massive measures are taken to lift them out of this vicious cycle. It is important to know and understand clearly the causes and consequences of social inequalities and identify appropriate measures to solve this global problem effectively.

This book "Social Inequality as a Global Challenge" is a treatise of facts and knowledge on social inequality. This book presents causes and consequences of social inequality and suggests appropriate mitigation measures for solving this societal problem. Book chapters written by scholars from different parts of the world are full of insights into the theoretical and practical aspects of the problem. This book presents analyses on social inequality from historical perspective and discusses various dimension of human society including migration, intercultural fusion, health and educational inequality, and socio-economic divisions. This book establishes the fact that social inequality is the major obstacle for the progress of our global societies. Social inequality is spreading all over the world and it must be brought under control in order to create an equitable and just society.

It is clear that unless there are serious initiatives and undertakings from the governments globally, the social inequality is not going away anytime soon. In order to wipe out social inequality from the world, it requires a concerted effort from all stakeholders including government authorities, international organizations, academic institutions, mass media, community organizations, and whole population. Relatively a small number of people enjoy the most resources whether it is in the form of education, healthcare, living standards, disposable income, or leisurely vacations in almost every country in the world. As income gaps are widening in most countries in recent decades, social inequalities are also increasing. This book will be a very valuable resource to those who like to understand and learn about social inequality and want to be a change agent for creating a socially just global society.

Professor Durga D. Poudel
University of Louisiana at Lafayette, Lafayette, Louisiana, USA

Social Inequality—Pervasive social inequality is a pivotal debate of our time. Generally, political discourse oscillates around different interpretations of equality across societies and approaches associated in tackling the problem with various dimensions and degree. Definition elaborating equality equilibrium purportedly differentiates between the North and the South on account of geography, culture, race, cast, religion, gender, disabilities, level of economic prosperity, income, opportunities, education, employment, health care and system of governance. It is not uncommon to witness social inequalities occurring deliberately due to unfair treatments, biases and/or prejudices toward community, social groups, individual, gender, ethnicity, and origin. In precise, inequalities are the primary product of socially recognized differences and adopted systems. Equality has become a possibility only if one recognizes the virtues of ***equity***. This is, in practice, absolute ***equality*** is like dreaming of a utopian welfare philosophy of the past mirrored against the challenges of the 21st century, dominated by science and technology, AI, innovation, and their impact on human behavior, society, and systems, unseen hitherto.

Ambitious Millennium Development Goal (MDG) of the United Nations was a step forward in right direction, to a certain extent. But, the successor initiative, the Sustainable Development Goals, is no less ambitious and likely falter as well. One must recognize the stark reality that this would not be attained in post-pandemic era when the world has been witnessing states' failing strategies in containing the spread of a contagious disease, simply called COVID-19. This situation is heavily impacting on peoples' lives and psychic in an eroding environment of public health care system and rapidly deteriorating quality of life, coupled with vanishing small and medium scale economic enterprises and loss of employment opportunities. Recovery and transformation from this dual-edge shocks (health and economy), created by the global pandemic, is unknown and faltering, and future is uncertain. Multilateralism, in a multi-polar world tension, faces serious challenges in addressing growing socio-economic inequalities. We need transformative re-thinking whether traditional measures are still relevant for new challenges if we are serious about finding pragmatic approaches to equitable equilibrium level of wellbeing for all, and thereby reducing deep disparities and inter-social inequalities. Although this may not sound new, but the landscape has altered significantly warranting re-definition of practical approaches and realistic measures for reducing inequalities for creating a semblance of equity equilibrium.

Time to contemplate differently and deep than in the past. I leave this question to scholars, academics, economists, sociologists, and political pundits to drill down for a workable answer. I am optimistic that this book will excite all enthusiasts, who are interested in social equity and advancement because this is a lingering debate of the century. Arguments presented in this book would prove invaluable guide in this perspective. Professor Medani Bhandari and Associate Professor Shvindina Hanna have jointly taken a bold step forward in this journey at enhancing knowledge of the subject matter while clarifying general understanding of continuing socially embedded phenomenon of inequalities.

Kedar Neupane, Executive Board Member, Nepal Policy Institute, and retired staff member of United Nations High Commissioner for Refugees. February 2, 2021.

Endorsements

Social inequalities grounded through discriminative behavior on the basis of race, sex, gender, ethnicity, country of origin, language, refugee status, color, and sexual orientations persist globally. Despite the tremendous efforts and progress made, large disparities and discriminations remain worldwide in terms of access to food, health, education, judiciary services and right to vote, and especially in least developed and developing countries. Such inequalities and discriminations have resulted in cultural, social, and economic unrest and conflicts in society and widening gap between haves and have-nots. This is a comprehensive book dealing with all aspects of social inequalities. Prof. Medani P. Bhandari and Dr. Hanna Shvindina have done an excellent job of authoring and editing this book that includes chapters by eminent authors from several countries highlighting the extent, problems, and consequences of social inequalities. This book attempts to close the social inequality gap in the society and achieve the SDGs related to social inequality, especially SDG 10. I found it rich in facts and easy to read and understand different aspects of social inequalities prevailing and is a helpful resource for ideas on how social inequality can be reduced in the society. I endorse the book as there are hardly such books available in the market that capture the depth of information that this book has provided.

Dr Jagadish Timsina,
Editor-Agricultural Systems,
Melbourne, Australia

Inequality has become a major challenge facing the world. Its causes are simultaneously political, economic, and social. The escalating level of inequality has led to social instability, crisis, and conflicts around the world.

While national governments and United Nations recognize this as a major challenge, too little is being done on the ground to overcome this. This book, edited by Prof. Medani P. Bhandari and Dr. Shvindina Hanna, offers a timely and critical analysis of socio-economic foundations of inequality, covering all key parameters such as income and wealth distribution, equity & equality of outcome, and equality of opportunities. With chapters written by eminent scholars using latest research and incisive analysis, the book is an important primer for understanding the nuances of economic inequality in the contemporary world.

Dr. Hemant Ojha, Associate professor,
University of Canberra, Australia.

Acknowledgements

First of all, I would like to thank my colleague, co-editor of this volume Associate Professor Shvindina Hanna and all members of her family (parents, Olexandr and Valentina), Prof. Volodymyr Boronos, Prof. Tetyana Vasilyeva, Prof. Oleg Balatskyi (her mentor 1999–2012). I should acknowledge that, she is the key manager of this project —making initial book plan to establishing contacts with authors —collecting manuscripts, and other editorial tasks as needed.

Equally, I would also like to thank all authors who contributed to this volume —Drs./ Profs. G.B. Sahqani ; Salvin Paul and Maheema Rai; Tetyana Vasilyeva, Anna Buriak, Yana Kryvych and Anna Lasukova; Oleksii Demikhov; Mirjana Radovic-Markovic, Milos Vucekovic and Aidin Salamzadeh; Sotnyk Iryna, Oleksandr Kubatko, Tetiana Kurbatova, Leonid Melnyk, Yevhen Kovalenko, Almashaqbeh Ismail Yousef Ali; Oksana Zamora, Svitlana Lutsenko; Tetiana Semenenko, Volodymyr Domrachev, Vita Hordiienko; Alla Krasulia, Yevhen Plotnikov, and Liudmyla Zagoruiko; Shapoval Vladyslav, and Shvindina Hanna; Sergij Lyeonov, Tatiana Vasylieva, Inna Tiutiunyk and Iana Kobushko; Alla Krasulia1, Yevhen Plotnikov, Liudmyla Zagoruiko, Prem B. Bhandari, Madhu Sudhan Atteraya; Teletov Aleksandr Sergeevych, and Teletova Svetlana Grigorievna. Each of you have done an excellent job, your scholarly work bridges the knowledge gap in the complex field—inequality with reference to sustainable development goals.

I would like to thank to Prof. Bishnu Paudel (Guru of all of us, who passed away due to heart attack in early 2020, when COVID-19 pandemic was just at the beginning stage. Still, I have a great pain, because, we were not even allowed to say a final goodbye to our beloved Guru, due to restrictions of Coronavirus). Prof. Poudel was the motivator, encourager, and supporter for me to continue my collaborative work to address the inequality challenges around the world. I have kept his brief encouraging note in this book as well as another book on economic inequality. Prof. Poudel used to make his suggestions in every book I wrote, every paper I wrote. His understanding of Bashudhiva Kutumbakam (all living being are our relatives and neighbors) was slightly different than ordinary meaning. He used to say that the essence of Bashudhiva Kutumbakam is to see everyone as you see yourself, love everyone as you love yourself. He always used to say that any contribution toward addressing the challenges like inequalities, climate change, sustainability, pandemic, and other social, economic, political, and natural challenges are the pathway to change human behavior toward Bashudhiva Kutumbakam (all living being are our relatives and neighbors).

I would also like to thank Prof. Douglas Capogrossi (my mentor, Emeritus, President of Akamai University, Hawaii), Dr. Mary Jo Bulbrook, (President, Akamai University), Dr. Ambika Adhikari (mentor for environment conservation), Dr. Krishna Prasad Oli (Member of National Planning Commission, Government of Nepal); Prof. Keshav Bhattarai (Arizona State University), Mr. Kedar Neupane (former staff of UNHCR), Prof. Durga D. Poudel, University of Louisiana at Lafayette, Lafayette, Louisiana, USA, , Dr. Jacek Piotr Binda (Poland); Dr Jagadish Timsina, and Dr. Hemant Ojha, (Australia) and Prof. Vasyl Karpusha, PhD (Ukraine) for their reviews, forewords, and endorsements. Your togetherness with us adds value in this book project.

Special thanks to Prof. Boronos Volodymyr, Head of the Department, of Finance and Entrepreneurship, I never forget your encouragements during my stay in Ukraine. You have given me new insights specially to dig on Marxist philosophy. I would also like to thank all colleagues of Finance and Entrepreneurship. You all are amazing, your knowledge sharing expertise is commendable. Thank you to you all.

I would also like to thank my wife Prajita Bhandari for her encouragements and for giving insightful information on how women are victims of inequality—at home to work environment. Without your support, this project was impossible. I would also like to thank our son Prameya, daughter Manaslu, and daughter-in-law Kelsey for helping me to find relevant resources in the field of inequality, social division, stratifications, and gender issues. I would also like to thank our granddaughter for giving us joyful environment, which always helps to concentrate in the work I am tuned in. I would also like to thank our son-in-law Abhimanyu Iyer, and his parent Mahesh and Uma Iyer for insightful thoughts on inequality issues. I would also like to thank my mother Hema Devi, brothers Krishna, Hari, sisters Kali, Bhakti, Radha, Bindu, Sita, and their families for encouraging me by providing peaceful environment. I would also like to thank my friends Kshitij Prasai, Prem Bhandari, Man Bahadur Khari, Rajan Adhikari, Govinda Luitel, Medini Adhikari, Prof. Gopi Uprety, Tirtha Koirala, Prof. Sanjay Mishra, Bijay Kattel , Dr. Aleksander Sapinski (Poland), Dhir Prasad Bhandari and all my Facebook and LinkedIn friends, who have been always encouraging to us to give back to the society through knowledge sharing. Thank you to Rajeev Prasad, Junko, and all friends of River Publishers, for encouraging and empowering us to complete this book project on time. Thank you to you all, who have given their inputs for this book project directly or indirectly.

Thank you,

Medani P. Bhandari

Chapter 1

Social Inequality as a Global Challenge: Scenario, Impacts, and Consequences

Medani P. Bhandari, PhD[1]

1. Introduction

> *"Inequality—the state of not being equal, especially in status, rights, and opportunities—is a concept very much at the heart of social justice theories. However, it is prone to confusion in public debate as it tends to mean different things to different people. Some distinctions are common though. Many authors distinguish "economic inequality", mostly meaning "income inequality", "monetary inequality" or, more broadly, inequality in "living conditions". Others further distinguish a rights-based, legalistic approach to inequality—inequality of rights and associated obligations (e.g. when people are not equal before the law, or when people have unequal political power*)" (United Nation 2018:1).

Inequality is one of the major human to human divisive factor since the evolutionary history of human development and civilizations. Inequalities present in every sphere of social, political, and economical structure of community, national and international level throughout social, economic, religious, and political histories. The strata created by inequalities are grounded

[1] Professor of Sustainability, Akamai University, Hilo, Hawaii, USA; Prof. of Innovation and Finance, Sumy State University, Ukraine. Medani.bhandari@gmail.com

on innumerable factors—such as social, cultural, political, geographical, or due to environmental (anthropogenic or natural) catastrophe.

Inequality can be seen as a communicable global disease. Inequalities can be understood in many forms, however, mostly known, and discussed forms are economic, social, political, cultural, and religious inequalities.

As Stewart (2010) notes:

> "Economic inequalities include access to and ownership of financial, human, natural resource-based and social assets. They also include inequalities in income levels and employment opportunities.
>
> Political inequalities include the distribution of political opportunities and power among groups, such as control over local, regional, and national institutions of governance, the army and the police. They also include inequalities in people's capabilities to participate politically and express their needs. The dominant group capture the political opportunities and power of coherence, so that marginalized group always remain at the lower strata of political, social, and economic system.
>
> Cultural inequalities include disparities in the recognition and standing of the language, religion, customs, norms, and practices of different groups".

Religious inequalities can be seen in the inequal access to the religious activities (Prayer, Worship, Bhajan Chanting, Procession, Meditation Camps, Religious Discourses—Story Telling or Gatherings to Listen, Travel for the Pilgrimage, Washing and Bathing in and around the Pilgrim Places etc.)—due to religious beliefs (Hindus and other religious believers are not allowed to Islam's places like Mecca and Medina; similarly, many Hindus temples in India and Nepal do not allow entrance to the non-Hindus; and even in most of the Hindu Temples women are not allowed to conduct religious or spiritual rituals during the mensuration period; utmost of the Hindu temples; only high caste male can be the priest. Women are not allowed to take the major role of in the religious conduct even at the home). Religious houses discriminate people based on race, cast, wealth, power, color, ethnic identity, gender, age, sexual orientations, and physical structure of the people. Religion and culture combinedly bound the society to continue the discriminatory practices in the society. Mostly, in the developing world, religion still creates fear among the low strata community and help the elite community to rule them. So called male Gurus create a fearful environment through made up stories, so the innocent poor people fearing those stories make donations even if they have no food for themselves or for their children. They follow the instructions of those male elites which are on the basis of religious books, stories, and tradition. The traditional cultural practices are totally male centric; therefore, male members of the society get more favorable treatment than women. The discrimination between male and female begins right after the birth and remains until death. Gender inequality can be seen in families/households in most part of the world (especially the developing world, where access to education, healthcare and resources for the livelihood is limited). Women always have multiple workloads than men. They need to be involved in earning, cooking, cleaning, maintaining house, take care of babies and elderly people and along with all these they have to be prepared to be attractive for male partner's satisfaction. There were numerous incidences of domestic violence, where the victims are mostly women.

Women are the main workforce to maintain the house (shown in pictorial form- Nepal examples).

Photo by Medani Bhandari (2018)

The picture (cover page picture of this book) provides an example of women labor of semiurban area of Nepal (Morang, District—on the road to Babiya Birta from Karsiya Bazaar—Summer of 2018). They live very close to the Karsiya Bazaar (local market emerging city). They have semi concrete--build standard houses. There are four women; two of whom are eight months pregnant are hiding behind, the other two in front have seven-, eight-, and nine-month-old babies at home. They both do breast feeding. These women represent four different middle-income households. Each family has a small farm (subsistence farming), few goats, at least one milking cow, some ducks, chickens, and few pairs of doves. In their family, they have husbands, husbands' brothers, fathers-in-law, mothers- in- law, and other relatives. Mothers at home have plenty of work—cooking, taking care of grandchildren, cows, ducks, chicken, and even preparing tea almost every hour to their husbands or sons or other family members. The women members of the family have no time for themselves. However, they are not considered as earners in the family. These four women in the picture are carrying loads of grass (fodder) in their bicycles for the cows at home. None of the milking cows normally get fresh grass; however, they get hay (the dry grasses). The male members of the family do the external work which mainly involves farming. However, these women also equally involve in the farming activities as well as other labor-intensive work. Men can have leisure time, their work is considered as hard work, but women's workload is always undermined. This is the real picture of gender inequality. Women get less access to food, education, health, and even in the process of socialization. In general, girls are to remain inside their homes while boys can go and play with other boys. Until they got married, they will be under father's or brother's control and after marriage they will be under the control of the husband, father-in-law, mother-in-law, or other male members of the new family (this is a type of inequality). This applies to all families, poor or rich. The causes of such inequality are due to male-centric attitude, culture, social system, customs, and traditions.
Even baby girls have responsibility, but they complete their tasks with no complaint.

Girls of Parsa Harpur Nepal are returning home with dried leaves in sacks. The dried leaves are burned around the house in the evening to repel mosquitoes and insects.

"certain cultural practices that could make gender inequality in today's poor countries persist even in the face of economic growth, such as patrilocality and male-centered funeral rituals. These cultural norms help explain the extremely male-skewed sex ratio in India and China, for example. Similarly, the anomalously low FLFP rate in India, the Middle East, and North Africa is likely rooted in the high value these cultures place on women's "purity." (Jayachandran 2015:84)

Photo Source: Kantipur News, Kathmandu, February 20, by Angad Dhakal https://ekantipur.com/photo_feature/2021/02/20/161382924654279507.html

A global scenario of gender inequality

- *All over the world men tend to earn more than women.*
- *Women are often underrepresented in senior positions within firms.*
- *Women are often overrepresented in low-paying jobs.*
- *In many countries men are more likely to own land and control productive assets than women.*
- *Women often have limited influence over important household decisions, including how their own personal earned income is spent.*
- *In most countries the gender pay gap has decreased in the last couple of decades.*
- *Gender-equal inheritance systems, which were rare until recently, are now common across the world.*
- *Composite indices that cover multiple dimensions show that on the whole gender inequalities have been shrinking substantially over the last century. Source: Esteban Ortiz-Ospina and Max Roser (2018)-Our World in Data-https://ourworldindata.org/economic-inequality-by-gender*

Source: As provided by Dr. Man Biswakarma—Kantipur publication and other media—The burden of survival (women with baby, heavy load on the back and food on a bag and women carrying 10–12 empty containers to find drinking water, also other women are following her with same load). Why only women?

Source: As provided by Dr. Man Biswakarma–and other media (Raute huts and Raute women in western Nepal—the burden on women—she is carrying six pots on her head; she might be going to collect drinking water).

Source: Hidden Apartheid—As provided by Dr. Man Biswakarma and Medani Bhandari.

(Girl in the mud hut is in mensuration period.) In western Nepal, still during mensuration period, girls have to stay out from the family, they are considered completely untouchable (picture 1); and a middle-aged woman, Kalika Devi has three educated brothers, five well settled sisters, all belong to a middle class family; however, Kalika Devi knows nothing but continuous work. Her responsibility is to take care of every needs of the cows (cleaning dung, collecting grass, etc.) and nothing else matters. She begins her day by cleaning cow dung and finishes her day by giving fodder to the cow for the night, giving no notice to the weather or her own health condition . Women like Kalika Devi, begin their life with housework and continue it over their lifetime until death. They do not know or care whether they are treated unequally.

This is the situation of low-income middle-class family. What is the situation of low income- family who are living in poverty? Pictures speak

Needless to explain, no need of evidence; the women of the poor families have been bearing a double burden—male centric social practices and at the same time they have to maintain life support system without resources for survival.

For the poor nothing is easy

Source: Kantipur—schooling for poor kids

Crossing river to reach school

These pictures are from Nepal; however, kids of many developing countries are still facing similar difficulties. Despite the difficulties, the new generation is getting stronger as these two girls are sailing on a tube to cross the river for school.

Why this book?

Informally, this book holds relatively a long educational trajectory from the social events and from my primary schooling days to date. As usual, practice of societal practice, at home, my parents, grandparents, relatives, siblings, neighbors, my playmates, and the surroundings primarily tried to teach me how to live in the stratified, classified, caste, race, ethnically divided inequal society, and how to maintain the social harmony with each other, where societal norms and values were imposed by the elite high caste males. I have always wondered, why all my friends' parents are not equally participating in the social events and why even an old person of low caste origin used to salute me or used to say JADAU to me, though I was below the age of his youngest grandchildren. I have also seen a big problem in every month for about four days during the mensuration period of my maternal aunty. Firstly, there was problem of water collection, and other household activities, because during the period she was not supposed to touch anything which are in liquid form. In fact, she was one of the major labor forces to maintain the house, when she had no access to the regular work, house environment used to be abnormal. Another problem was seen and unseen quarrels in the family indirectly created due to regular period cycle. Underlining factor was related to fear of physical and mental preparedness to have a baby. My uncle and aunty had no children for a long time and there was a clear blame on my aunty. I think, my grandparents might hold the hope of conceiving a baby by their daughter-in-law and when there was regular period cycle in her, I think, they felt hopelessness and my aunty was blamed. That conflict remained in the family until my uncle and aunty had a baby boy after decades of marriage.

Primarily, my young brain was disturbed due to unseemly, divided society, due to haves and have nots situation of the society, and the divisions created on the name of tradition, culture, religion, and other social systems—norms, values, regulations, and rules etc. In one hand my surroundings were enforcing me to mingle or merge within the existing social systems; on the other hand, my inner self was questioning "why we all are not equal, why we all do not have the same access to social events, education, labor, food, and even in shelters" and "also why society is giving more priority and love to me and why not to girls of my age (like Indira, Yogi, Kali, Bhunti etc.)." Why my aunty was not equally treated at home and my parents repeatedly called her as PARAYA DHAN (the property of someone else), since my uncle and she had the same parents (my grandparents). I was also shocked that why there are no answers for questions—why there was unequal treatment between brothers and sisters? why sisters were not given priority for education? As soon as I was able to have some courage to ask questions, I first asked my grandfather, why they blame my aunty (his daughter-in-law) without any mistakes, why my uncle had the opportunity for education and why not my aunty have the same (though both were his children). My grandfather never responded to my questions, always just praised me without any reason. I was not happy with such praise at all. However, no body answered my questions. My uncle was a schoolteacher, so I thought he will answer, but he also simply replied that, I was too young to discuss such matters. I asked those questions also to the primary and secondary school teachers, as I grew up, however, no one gave me a clear answer on why there was inequality, what is society and why social system is not equal for everybody. Time never stops, it was the time to join high school; however, the closest high school was in three hours walk from my home. I wanted to continue my study and decided to join the school. My family members did not oppose my decision. However, I was the only student from my village who chose the six hours long walk. There were several boys of my age and several girls of my age, but no one preferred to take the risk of six hours a day walk for school. I still wonder why none of boys and girls of my age joined that educational journey to school. Who makes them weak? During the three years walking from village to the school, I saw such inequalities in every steps of social, economic, political, and religious life. Even I saw such inequality in school because there were no female teachers in middle and high school. I was very surprised, and I raised that question to some in the school. But no one answered. There was a hostel in the school but only for boys. I wanted to know the reason of this problem, once I reached grade ten, I asked the headmaster, about this situation of inequality in the society, no answer. I reached college, university, became professor of social sciences and still I am not able to provide the factual reasons of inequalities prevailing in local communities, nations, and in international level. The inequalities due to race, cast, wealth, politics, power, color, ethnic identity, gender, age, sexual orientations, and physical condition still exist and economic inequality is growing. This book tries to search the answers of why social inequalities are growing (from Bhandari 2021, Theories, and Methods application in Organizational Sociology-*forthcoming*).

Social inequalities are grounded and boasted by the religious and cultural systems, therefore, social phenomena automatically create the socially biased inequal circumstances particularly to the women. In another word, inequality in the society is created by the elite groups for their benefit, it is present in each society; however, most dangerously, the gender inequality is maintained by the males of each houses. The social ethics and cultures are prepared by the male; therefore, it is omnipresent. Another neglected but deep-rooted issue is cultural inequality, which segregates people from mainstream. The minority groups face recognition problems of their social status, language, religion, customs, norms etc.

Social, cultural, and religious discriminations often shadowed, in the society, because the show-up face is made up by the elite males, who hold the knowledge of made stories; therefore, actual discriminated face of society is rarely understood. Society functions through the economic, social, religious, cultural, and political (the pillars of the society) systems. Even if one pillar is dysfunctional, it can disbalance the entire social order.

> "To speak of a social inequality is to describe some valued attribute which can be distributed across the relevant units of a society

> in different quantities, where 'inequality' therefore implies that different units possess different amounts of this attribute. The units can be individuals, social groups, communities, nations; the attributes include such things as income, wealth, status, knowledge, and power" (Wright 1994: 21, as in Soares *et. al.* 2014:1). Iinequality is multidimensional—and can be seen in complex forms—including assets, access to basic services, infrastructure, and knowledge, race, gender, ethnic and geographic dimensions (Soares *et. al.* 2014). The concept of inequality also has human attitudinal aspect (happiness) (Brockmann and Delhey 2010).

It is important to restate that, when we talk about inequality, the focus remains on economic inequality; however, the major inequality persists due to discrimination based on gender, age, origin, ethnicity, disability, sexual orientation, class, and religion within the society. It is true that primarily the world is divided between haves and have not; and social strata are designed to create the comfort zones to those who dominate the social, economic, and political systems. The rules and regulations, social norms and values are also created and implemented always by the social elite, therefore, the marginalized groups have no or very minimal stake in the society, and they are always victims. There is a need of complete social, economic, and political reform, so that the designed division can be minimized. Until or unless the marginalized group is empowered, the inequality issue cannot be solved or even minimized.

As inequality has been a black shadow, a dividing factor of human civilization; it has been an important discourse of discussion in the academic as well as development domains (Barker 2004; Beall, Guha-Khasnobis and Kanbur, 2012; Bhandari 2018; 2019, 2020; Bhandari and Shivindina 2019; Brockmann, and Delhey 2010; Carson, 1962; Cheng, 2018; Colyvas, and. Powell 2006; Crossan, and Bedrow, 2003; Daly, 2007; DESA 2013; Freistein, and. Mahlert, 2015; Holvino, and Merrill-Sands 2004; IISD 2005; IUCN, UNDP & WWF 1991; Jackson, 2009; Jan-Peter et al 2007; Harris 2000; Lélé 1991; Marshall 1961; Marx 1953ff; Meadows, 1998; Meadows, et al 1972; Norgaard, 1988; North, 1981; Purvis et al 2018; Rockström 2009; Scott, 2003; Soares, et al 2014; Sterling, 2010; United Nations 1972;1997; 2002; 2010; 2012, 2015, 2015 a, b, c, d; UNEP 2010; UNESCO 2013; Vander-Merwe and Van-der-Merwe 1999; Vos 2007; Wals 2009; World Bank 2015; WCED 1987; WID 2018; Wright 1994; World Economic Outlook 2007).

In recent history, the major international agencies such as all development banks, especially the World Bank and Regional Development Banks have been implementing various programs to reduce inequalities (World

Bank 2011; 2020; United Nations 2020; World Economic Forum 2020; ILO 2020; UNDP 2020). All of the United Nations line agencies have special focus to reduce inequalities (United Nations 1972; 1997; 2002; 2010; 2012, 2015, 2015 a, b, c, d; UNDP 2019).

The academic scholars have published thousands and thousands of papers and books about the problems and consequences of various types of inequalities in the societal level to global level (Blau, and Kahn 2017; Goldin and Rouse 2000; Neumark, Bank, and Nort, 1996; Olivetti, and Petrongolo, 2008; Ortiz-Ospina 2018; Jayachandran 2015). However, instead of reducing, inequalities continue to increase globally (Christiansen and Jensen 2019; Bhandari and Shvindina 2019; United Nations 2020a, b; United Nations 2020; Dehm, 2018; Chhabria 2016).

All concerned stakeholders of the world—the academia, media, development agencies accept the fact that, inequality has been a major problem of social development (World Bank 2011; 2020; United Nations 2020; World Economic Forum 2020; ILO 2020; UNDP 2020). Among them, the United Nations has been insisting its member countries to implement the various programs to reduce inequality. As the acknowledgement of inequality, a major challenge of contemporary world, United Nations (2015) has given a priority in its sustainable development goals. As an evidence "Out of the 17 goals, eleven goals address various forms of inequities, (Goals 1, 3, 4, 5, 6, 7, 8, 9, 10, 11, 16 and 17), and one goal (Goal 10) explicitly proposes to reduce various forms of inequalities" (Freistein and Mahlert 2015:7).

Further, United Nations has been publishing various reports to reveal the factual challenges of inequality. Among them, recent publication, titled: Inequality—Bridging the Divide- Shaping Our Future Together (2020), states:

- "Inequalities are not only driven and measured by income, but are determined by other factors—gender, age, origin, ethnicity, disability, sexual orientation, class, and religion. These factors determine inequalities of opportunity which continue to persist, within and between countries. In some parts of the world, these divides are becoming more pronounced.
- The average income of people living in North America is 16 times higher than that of people in sub-Saharan Africa.
- Today, 71 percent of the world's population live in countries where inequality has grown. This is especially important because inequalities within countries are the inequalities people feel day to day, month to month, year to year.

- An Oxfam report shows that in the 10 years since the financial crisis, the number of billionaires has nearly doubled, and the fortunes of the world's super-rich have reached record levels. In 2018, the 26 richest people in the world held as much wealth as half of the global population (the 3.8 billion poorest people), down from 43 people the year before ... This matters because rapid rises in incomes at the top are driving and exacerbating within country income inequality.
- There are also inequalities within communities—and even families. Up to 30 per cent of income inequality is due to inequality within households. When it comes to women and girls, progress is uneven.
- Groups such as indigenous peoples, migrants and refugees, and ethnic and other minorities continue to suffer from discrimination, marginalization, and lack of legal rights. This is pervasive across developing and developed countries alike and is not tied to income.
- The measurements and impacts of inequality go far beyond income and purchasing power. Inequalities of opportunity affect a person's life expectancy and access to basic services such as healthcare, education, water, and sanitation. They can curtail a person's human rights, through discrimination, abuse and lack of access to justice. In 2018, we saw the world's 12th consecutive year of decline in global freedom, with 71 countries suffering net declines in political and civil liberties.
- High levels of inequality of opportunity discourage skills accumulation, choke economic and social mobility, and human development and, consequently, depress economic growth. It also entrenches uncertainty, vulnerability and insecurity, undermines trust in institutions and government, increases social discord and tensions and trigger violence and conflicts. There are growing evidence that high level of income and wealth inequality is propelling the rise of nativism and extreme forms of nationalism.
- the evolution of issues such as climate change, technology, and urbanization raise urgent policy challenges. For example, climate change is exacerbating environmental degradation, increasing the frequency and intensity of extreme weather events, and by no means impacting people uniformly. If climate change continues unaddressed it will increase inequality within countries and may even reverse current progress in reducing inequality between countries.
- With a global trend toward urbanization, cities are becoming a growing site for inequalities. They find high levels of wealth and modern

infrastructure coexist with pockets of severe deprivation, often side by side" (United Nations 2020a-website).

- "Inequality within countries is very high ... Since 1990, income inequality has increased in most developed countries.
- The world is far from the goal of equal opportunity for all: circumstances beyond an individual's control, such as gender, race, ethnicity, migrant status and, for children, the socioeconomic status of their parents, continue to affect one's chances of succeeding in life.
- High or growing inequality not only harms people living in poverty and other disadvantaged groups. It affects the well-being of society at large.
- Without appropriate policies and institutions, inequalities in outcomes create or preserve unequal opportunities and perpetuate social divisions.
- Rising inequality has created discontent, deepened political divides and can lead to violent conflict" (United Nations 2020:20b).

There is a clear need to pursue inclusive, equitable, and sustainable growth, ensuring a balance among economic, social, and environmental dimensions of sustainable development (United Nations 2020a website).

The pictorial depictions of Nepal case and above-mentioned summary statements listed in "*Inequality—Bridging the Divide—Shaping Our Future Together*" provide the factual truth of the unequal society.

The major goal of any society is sustained harmony in the society; however, it rarely exists in human sociopolitical history. The divisive element of inequality—is oppression to weak according to strata which can be economic, social, political, and cultural or others forms-grounded and influenced by societal circumstances. Theoretically, it is hard to fully pinpoint the root cause of inequality—why oppressed are being oppressed and oppressors have been maintaining the domination. However, social justice theory, social dominance theory and development economics theory provide some basis for understanding the nature of inequality.

Having this general theoretical basis, this book, unveils societal inequalities through regional and country specific cases, mostly with the social principles. Each chapter use and strictly follow the prescribed scholarly standards of scientific writing and provide new insights, therefore each chapters are unique.

Using quantitative approach, this book covers social aspects of inequality (in terms of gender; religion, migration, labor etc.).

> *"We must all realize that inequality reduction does not occur by decree; neither does it automatically arise through economic growth, nor through policies that equalize incomes downward via blind taxing and spending. Inequality reduction involves a collaborative effort that must motivate all concerned parties, one that constitutes a genuine political and social innovation, and one that often runs counter to prevailing political and economic forces" (Genevey et.al. 2013:6).*

Whereas, the chapters in this book provide the in-depth problems and consequences of social inequality, with the case studies of various countries of the world. Each chapter is unique, complete, theoretically grounded and supported by the factual data. The chapter—Inequality a Historical Perspectives—With Reference to Society, Politics, and the Economy—provides the theoretical perspective of inequality; chapter on Religion and Gender Inequality in India, reveals how women are marginalized in the South Asia and India; another chapter provides the Behavioral View on Inequality Issue with the Bibliometric Analysis; chapter on shows the problems of Public Health Policy Communication with Other Policies in The Context of Inequality; the case study chapter on Serbia Investigates Employment Discrimination and Social Exclusion; three chapters on Ukraine cases analyze Impact of Subsidies on Social Equality and Energy Efficiency Development; Combating Inequality Via an Intercultural Strategy of The Ukrainian City. Similarly, other two chapters unveil the Causes and Ways to Overcome Socio-Economic Inequality and Hidden Inequalities in Education: Case of In-Service English Language Teachers Training in Ukraine. Social exclusion of LGBTQ+ People common problem. The chapter on Education as A Tool for Prevention Discrimination Against LGBTQ+ People outlines the general history of such struggles and the current situation of LGBTQ+ people. In general, inequality can be seen in every spheres of political, social, economic, cultural, religious systems; however, LGBTQ+ people's struggles are relatively new and painful. Chapter on the Effect of Shadow Economy on Social Inequality: Evidence from Transition and Emerging Countries, unveils the interconnectedness of social and economic inequalities. The world has been witnessing discrimination due to color, country of origin, religious beliefs, ethnicity, language etc. Chapter on Cultural Dynamics and Inequality in the Turkish Sphere Concerning Refugees provides a true picture of war victims and war generated refugee's crisis in the Middle East. And the chapter presents the Socio-Cultural Dimensions of Socio-Economic Inequality in Household Food

Security in Nepal. Likewise, chapter on "Genesis and Reasons for Social Inequality in the Globalization Era" demonstrates that globalization has not reduced social differentiation between people. Moreover, globalization deepened it in some way since the scope of global inequality, reflecting the difference in welfare between the richest and poorest countries in the world, exceeds the inequality within a particular country. At the same time, inequality indices in individual countries continue to increase.

This edited book, follow listed theoretical frameworks, and tries to fulfill the knowledge gap of complex social discourse in reducing the global, regional, national and local inequality. The chapters' result shows that even though United Nations and many other international organizations, civil society, scholars, governments have been trying to reduce the inequality—it is increasing and there are no symptoms of minimizing inequality. Almost all governments of the world, every development agency, each of the United Nations agencies and academicians as well as general stakeholders have been advocating, documenting the problems of inequalities; also implementing various programs to reduce inequalities, however, in so far, the result has been always negative—and in fact, inequality in increasing. Why? This book tries to provide some of the basic answers. As noted, in our earlier book on Inequality, the main cause of inequality is exclusion of others from me or us territory. We do not value to them who differs with us, by color, religion, ethnicity, language, social status, economy, political ideology, and race or by gender. Elite group creates the boundary and continuously practices the divide and rule tactics, so the marginalized group remains in low strata and elite continuously rule. In fact, in one hand, there is lack of awareness about the equity, social inclusion, equality, among the socially, economically, politically and culturally marginalized group; and another hand the elite group do not want undivided equal society. There is a need of factual knowledge and practices which can bridge the gap in the divided society. Awareness needs to come from individual to individuals, society to societies, and nation to nations. To overcome the inequality problems the concept of "BashudaivaKutumbakkam"—The entire world is our home, and all living beings are our relatives" and "Live and let other live"—the harmony within, community, nation and the globe is needed. There is a need of trust, love, and respect to each other.

> *"Trust is at a breaking point. Trust in national institutions. Trust among states. Trust in the rules-based global order... We must repair broken trust. We must reinvigorate our multilateral project" (United Nations Secretary-General António Guterres, September 2018).*

References / Bibliography

AASHE (2018), Stars Technical Manual. (The Association for the Advancement of Sustainability in Higher Education). Version 2.1. Available online: https://stars.aashe.org/pages/about/technical-manual.html

Adomßent, M., Fischer, D., Godemann, J., Herzig, C., Otte, I., Rieckmann, M., Timm, J. (2014), Emerging Areas in Research on Higher Education for Sustainable Development – Management Education, Sustainable Consumption and Perspectives from Central and Eastern Europe, Volume 62, 1 January 2014, p. 1–7. http://dx.doi.org/10.1016/j.jclepro.2013.09.045

Adomßent, M., Godemann, J., Leicht, A., Busch, A. (Eds.), 2006. Higher Education for Sustainability: New Challenges from a Global Perspective. VAS –Verlag für Akademische Schriften, Frankfurt am Main.

Alghamdi, N.; den Heijer, A.; de Jonge, H. (2017), Assessment Tools' Indicators for Sustainability in Universities: An Analytical Overview. Int. J. Sustain. High. Educ., 18, 84–115

Arima A (2009), A plea for more education for sustainable development. Sustain Sci 4(1):3–5

Barbier, E. B., (1987), 'The concept of sustainable economic development,' Environmental Conservation, Vol. 14, No. 2 (1987), pp. 101-110.

Barker, Chris (2004), The SAGE Dictionary of Cultural Studies, SAGE Publications, London / Thousand Oaks / New Delhi https://zodml.org/sites/default/files/%5BDr_Chris_Barker%5D_The_SAGE_Dictionary_of_Cultural__0.pdf

Barnosky, Anthony, and others (2012). Approaching a state shift in Earth's biosphere. Nature, vol. 486, No. 7401 (7 June), pp. 52-58.

Barth M, Godemann J (2006), Study program sustainability—a way to impart competencies for handling sustainability? In: Adomßent M, Godemann J, Leicht A, Busch A (eds) Higher education for sustainability: new challenges from a global perspective. VAS, Frankfurt, pp 198–207

Barth, M., Rieckmann, M., Sanusi, Z.A. (Eds.), 2011. Higher Education for Sustainable Development. Looking Back and Moving Forward. VSA – Verlag für Akademische Schriften, Bad Homburg.

Baumert, Kevin A., Timothy Herzog and Jonathan Pershing (2005). Navigating the Numbers: Greenhouse Gas Data and International Climate Policy. Washington, D.C.: World Resources Institute.

Beall, Jo, Basudeb Guha-Khasnobis and Ravi Kanbur, eds. (2012). Urbanization and Development in Asia. Multidimensional Perspectives. New York: Oxford University Press.

Berg, Andrew, and Jonathan Ostry (2011). Inequality and unsustainable growth: two sides of the same coin? IMF Staff Discussion Note. SDN/11/08. Washington, D.C.: International Monetary Fund. 8 April.

Bertelsmann Stiftung and Sustainable Development Solutions Network (2018), SDG Index and Dashboards Report 2018-Global Responsibilities, Implementing the

Goals, G20 and Large Countries Edition. www.pica-publishing.com, http://www.sdgindex.org/assets/files/2018/00%20SDGS%202018%20G20%20EDITION%20WEB%20V7%20180718.pdf

Berzosa, A.; Bernaldo, M.O.; Fernández-Sanchez, G. (2017), Sustainability Assessment Tools for Higher Education: An Empirical Comparative Analysis. J. Clean. Prod. 161, 812–820

Bhandari, Medani P. (2019), Inequalities with reference to Sustainable Development Goals in Bhandari, Medani P. and Shvindina Hanna (edits) Reducing Inequalities Towards Sustainable Development Goals: Multilevel Approach, River Publishers, Denmark / the Netherlands- ISBN: Print: 978-87-7022-126-9 E-book: 978-87-7022-125-2

Bhandari Medani P. and Shvindina Hanna (2019), The Problems and consequences of Sustainable Development Goals, in Bhandari, Medani P. and Shvindina Hanna (edits) Reducing Inequalities Towards Sustainable Development Goals: Multilevel Approach, River Publishers, Denmark / the Netherlands- ISBN: Print: 978-87-7022-126-9 E-book: 978-87-7022-125-2

Bhandari, Medani P (2017), Climate change science: a historical outline. Adv Agr Environ Sci. 1(1) 1-8: 00002. http://ologyjournals.com/aaeoa/aaeoa_00002.pdf

Bhandari, Medani P. (2018), Green Web-II: Standards and Perspectives from the IUCN, Published, sold and distributed by: River Publishers, Denmark / the Netherlands ISBN: 978-87-70220-12-5 (Hardback) 978-87-70220-11-8 (eBook),

Bhattacharyya, Subhes C. (2012), Energy access programs and sustainable development: A critical review and analysis, Energy for Sustainable Development, Volume 16, Issue 3, September 2012, Pages 260-271

Blau, Francine D., and Lawrence M. Kahn. (2017). "The Gender Wage Gap: Extent, Trends, and Explanations." Journal of Economic Literature, 55(3): 789-865.

Brockmann, Hilke, and Jan Delhey (2010), "The Dynamics of Happiness". Social Indicators Research 97, no.1 (2010): 387-405.

Caeiro, S.; Jabbour, C.; Leal Filho, W. (2013), Sustainability Assessment Tools in Higher Education Institutions Mapping Trends and Good Practices around the World; Springer: Cham, Gemany, p. 432.

Callanan, Laura and Anders Ferguson (2015), A New Pilar of Sustainability, Philantopic-Creativity, Foundation Center, New York, https://pndblog.typepad.com/pndblog/2015/10/creativity-a-new-pillar-of-sustainability.html

Carson, Rachel (1962), Silent Spring, A Mariner Book, Houghton M1fflin Company, Boston, New York

Cheng, V. (2018) Views on Creativity, Environmental Sustainability and Their Integrated Development. Creative Education, 9, 719-743. doi: 10.4236/ce.2018.95054.

Chhabria, Sheetal (2016), Inequality in an Era of Convergence: Using Global, Histories to Challenge Globalization Discourse, World History Connected Vol. 13, Issue 2.

Christiansen C.O., Jensen S.L.B. (2019) Histories of Global Inequality: Introduction. In: Christiansen C., Jensen S. (eds) Histories of Global Inequality. Palgrave Macmillan, Cham. https://doi.org/10.1007/978-3-030-19163-4_1

Clark, W. C., and R. E. Munn (Eds.) (1986), Sustainable Development of the Biosphere, Cambridge: Cambridge University Press

Colyvas, Jeannette A. and Walter W.Powell (2006), Roads to Institutionalization: The Remaking of Boundaries between Public and Private Science, Research in Organizational Behavior, Volume 27, Pages 305-353

Crossan, M. & Bedrow, I. (2003), 'Organizational learning and strategic renewal'. Strategic Management Journal, 24, 1087-1105.

Daly, H. E (2007), Ecological Economics and Sustainable Development, Selected Essays of Herman Daly, Advances in Ecological Economics, MPG Books Ltd, Bodmin, Cornwall http://library.uniteddiversity.coop/Measuring_Progress_and_Eco_Footprinting/Ecological_Economics_and_Sustainable_Development-Selected_Essays_of_Herman_Daly.pdf

DESA (2013), World Economic and Social Survey 2013, Sustainable Development Challenges, Department of Economic and Social Affairs, The Department of Economic and Social Affairs of the United Nations Secretariat, NY https://sustainabledevelopment.un.org/content/documents/2843WESS2013.pdf

Esteban Ortiz-Ospina (2018). - "Economic inequality by gender". Published online at OurWorldInData.org. Retrieved from: 'https://ourworldindata.org/economic-inequality-by-gender' [Online Resource]

Esty, K., R. Griffin, and M. Schorr-Hirsh. (1995), Workplace diversity. A manager's guide to solving problems and turning diversity into a competitive advantage. Avon, MA: Adams Media Corporation.

Fischer, D.; Jenssen, S.; Tappeser, V. Getting (2015) an Empirical Hold of Thesustainable University: A Comparative Analysis of Evaluation Frameworks across 12 Contemporary Sustainability Assessment Tools. Assess. Eval. High. Educ. 40, 785–800

Fobes (2019), America's Wealth Inequality Is At Roaring Twenties Levels, Contributor-Jesse Colombo, Forbes (https://www.forbes.com/sites/jessecolombo/2019/02/28/americas-wealth-inequality-is-at-roaring-twenties-levels/#62f244642a9c).

Freistein, K., and. Mahlert, B. (2015), The Role of Inequality in the Sustainable Development Goals, Conference Paper, University of Duisburg-Essen

See discussions, stats, and author profiles for this publication at: https://www.researchgate.net/publication/301675130

Galbraith, James K. (2012). Inequality and Instability: The Study of the World Economy Just before the Great Crisis. Oxford: Oxford University Press

Girard, Luigi Fusco (2010), Sustainability, creativity, resilience: toward new development strategies of port areas through evaluation processes, Int. J. Sustainable Development, Vol. 13, Nos. 1/2, 2010 161

Goldin, C., & Rouse, C. (2000). Orchestrating impartiality: The impact of" blind" auditions on female musicians. American Economic Review, 90(4), 715-741

Guidetti, Rehbein (2014), Theoretical Approaches to Inequality in Economics and Sociology, Transcience, Vol. 5, Issue 1 ISSN 2191-1150 https://www2.hu-berlin.de/transcience/Vol5_No1_2014_1_15.pdf

Håvard Mokleiv Nygård (2017), Achieving the sustainable development agenda: The governance – conflict nexus, International Area Studies Review, Vol. 20(1) 3–18

Holvino, E., Ferdman, B. M., & Merrill-Sands, D. (2004), Creating and sustaining diversity and inclusion in organizations: Strategies and approaches. In M. S. Stockdale & F. J. Crosby (Eds.), The psychology and management of workplace diversity (pp. 245-276). Malden, Blackwell Publishing. https://wir2018.wid.world/files/download/wir2018-full-report-english.pdf

IISD, (2005), Indicators. Proposals for a way forward. Prepared L. Pinter, P. Hardi & P. Bartelmus. International Institute for Sustainable Development Sustainable Development, Canada Retrieved January 8, 2015, from https://www.iisd.org/pdf/2005/measure_indicators_sd_way_forward.pdf.

IISD, (2013), The Future of Sustainable Development: Rethinking sustainable development after Rio+20 and implications for UNEP. International Institute for Sustainable Development Retrieved November 5, 2015, from http://www.iisd.org/pdf/2013/future_rethinking_sd.pdf

ILO (2020). Gender Statistics, International Labor Organization (ILO), 1990-2016

IUCN (1980), World Conservation Strategy: Living Resource Conservation for Sustainable Development. Retrieved November 7, 2015, from https://portals.iucn.org/library/efiles/documents/WCS-004.pdf.

IUCN, UNDP & WWF, (1991), Caring for the Earth. A Strategy for Sustainable Living. International Union for Conservation of Nature and Natural Resources, United Nations Environmental Program & World Wildlife Fund Retrieved November 8, 2015, from https://portals.iucn.org/library/efiles/documents/CFE-003.pdf

IUCN, UNDP & WWF, (1991), Caring for the Earth. International Union for Conservation of Nature and Natural Resources, United Nations Environmental Program & World Wildlife Fund

Jackson, Tim (2009). Prosperity without Growth: Economics for a Finite planet. Abingdon, United Kingdom: Earthscan.

Jan-Peter Voß, Jens Newig, Britta Kastens, Jochen Monstadt† & Benjamin No¨ Lting (2007), Steering for Sustainable Development: a Typology of Problems and Strategies with respect to Ambivalence, Uncertainty and Distributed Power, Journal of Environmental Policy & Planning Vol. 9, Nos. 3-4, September–December 2007, 193–212 https://www.researchgate.net/profile/Jochen_Monstadt/publication/233049753_Steering_for_Sustainable_Development_A_Typology_of_Problems_and_Strategies_with_Respect_to_Ambivalence_Uncertainty_and_Distributed_Power/links/577ff29608ae5f367d370a97/Steering-for-Sustainable-Development-A-Typology-of-Problems-and-Strategies-with-Respect-to-Ambivalence-Uncertainty-and-Distributed-Power.pdf

Jayachandran S. (2015), The Roots of Gender Inequality in Developing Countries, Annu. Rev. Econ. 2015.7:63-88. Https://faculty.wcas.northwestern.edu/~sjv340/roots_of_gender_inequality.pdf Downloaded from www.annualreviews.org

Jonathan M. Harris (2000), Basic Principles of Sustainable Development, GLOBAL DevelopmentAndEnvironmentInstitute,WorkingPaper00-04,GlobalDevelopment and Environment Institute, Tufts University, https://tind-customer-agecon.s3.amazonaws.com/11dc38b4-a3e2-44d0-b8c8-3265a796a4cf?response-content-disposition=inline%3B%20filename%2A%3DUTF-8%27%27wp000004.pdf&response-content-type=application%2Fpdf&AWSAccessKeyId=AKIAXL7W7Q3XHXDVDQYS&Expires=1560578358&Signature=T%2BpMgFFZjQvmVL8EHsy74Ds%2FKAM%3D

Julia Dehm, (2018), 'Highlighting Inequalities in the Histories of Human Rights: Contestations Over Justice, Needs and Rights in the 1970s' https://doi.org/10.1017/S0922156518000456 (published online 19 September 2018).

Karlsson-Vinkhuyzen, Sylvia; Arthur L Dahl and Asa Persson (2018), The emerging accountability regimes for the Sustainable Development Goals and policy integration: Friend or foe? Environment and Planning C: Politics and Space, Vol. 36(8) 1371–1390

Klarin, Tomislav (2018), The Concept of Sustainable Development: From its Beginning to the Contemporary Issues, Zagreb International Review of Economics & Business, Vol. 21, No. 1, pp. 67-94, DOI: https://doi.org/10.2478/zireb-2018-0005 https://content.sciendo.com/view/journals/zireb/21/1/article-p67.xml

Lang, D.J., Wiek, A., Bergmann, M., Stauffacher, M., Martens, P., Moll, P., Swilling, M., Thomas, C.J., (2012), Transdisciplinary research in sustainability science: practice, principles, and challenges. Sustainability Science 7 (1), 25–43.

Lélé, Sharachchandra M. (1991), Sustainable development: A critical review. World Development, Vol 19, No 6, 607-621 https://edisciplinas.usp.br/pluginfile.php/209043/mod_resource/content/1/Texto_1_lele.pdf

Mair, Simon, Aled Jones, Jonathan Ward, Ian Christie, Angela Druckman, and Fergus Lyon (2017), A Critical Review of the Role of Indicators in Implementing the Sustainable Development Goals in the Handbook of Sustainability Science in Leal, Walter (Edit.) https://www.researchgate.net/publication/313444041_A_Critical_Review_of_the_Role_of_Indicators_in_Implementing_the_Sustainable_Development_Goals

Marin, C., Dorobanțu, R., Codreanu, D. & Mihaela, R. (2012), The Fruit of Collaboration between Local Government and Private Partners in the Sustainable Development Community Case Study: County Valcea. Economy Transdisciplinarity Cognition, 2, 93–98. In Duran, C.D.,

Marshall, Alfred (1961) (originally 1920), Principles of Economics, 9th edn, New York: Macmillan.

Marx, Karl (1953ff), Marx-Engels-Werke (MEW). Berlin: Dietz

Meadows, D.H. (1998), Indicators and Information Systems for Sustainable Development. A report to the Balaton Group 1998. The Sustainability Institute.

Meadows, D.H., Meadows, D.L., Randers, J. & Behrens III, W.W. (1972), The Limits of Growth. A report for the Club of Rome's project on the predicament of mankind. Retrieved September 20, 2015, from http://collections.dartmouth.edu/published-derivatives/meadows/pdf/meadows_ltg-001.pdf.

Neumark, D., Bank, R. J., & Van Nort, K. D. (1996). Sex discrimination in restaurant hiring: An audit study. The Quarterly Journal of Economics, 111(3), 915-941.

Nicolau, Melanie and Rudi W Pretorius (2016), University of South Africa (UNISA): Geography at Africa's largest open distance learning institution, (in book), The Origin and Growth of Geography as a Discipline at South African Universities, Gustav Visser, Ronnie Donaldson, and Cecil Seethal, eds. Stellenbosch, South Africa: Sun Press

Norgaard, R. B., (1988), 'Sustainable development: A coevolutionary view,' Futures, Vol. 20, No. 6 pp. 606-620.

North, Douglass C. (1981), Structure and Change in Economic History. New York: W.W. Norton & Co.

Olivetti, C., & Petrongolo, B. (2008). Unequal pay or unequal employment? A cross-country analysis of gender gaps. Journal of Labor Economics, 26(4), 621-654.

PAP/RAC, (1999), Carrying capacity assessment for tourism development, Priority Actions Program, in framework of Regional Activity Centre Mediterranean Action Plan Coastal Area Management Program (CAMP) Fuka-Matrouh – Egypt, Split: Regional Activity Centre

Purvis, Ben, Yong Mao and Darren Robinson (2018), Three pillars of sustainability: in search of conceptual origins, Sustainability Science, Springer, 23https://doi.org/10.1007/s11625-018-0627-5r15%20-low%20res%2020100615%20-.pdf

Rockström, J., Steffen, W., Noone, K., Persson, Å, Chapin, S., Lambin, E., Lenton, T., Scheffer, M., Folke, C., Schellnhuber, H., Nykvist, B., de Wit, C., Hughes, T., van der Leeuw, S., Rodhe, H., Sörlin, S., Snyder, P., Costanza, R., Svedin, U., Falkenmark, M., Karlberg, L., Corell, R., Fabry, V., Hansen, J.,Walker, B., Liverman, D., Richardson, K., Crutzen,P., Foley, J., (2009), A safe operating space for humanity. Nature 461, 472–475.

Sayed, A.; Asmuss, M. (2013), Benchmarking Tools for Assessing and Tracking Sustainability in Higher Educational Institutions. Int. J. Sustain. High. Educ. 14, 449–465

Scott, W. R. (2003). 'Institutional carriers: reviewing models of transporting ideas over time and space and considering their consequences'. Industrial and Corporate Change, 12, 879-894.

Soares, Maria Clara Couto, Mario Scerri and Rasigan Maharajh -edits (2014), Inequality and Development Challenges, Routledge, https://prd-idrc.azureedge.net/sites/default/files/openebooks/032-9/

ST/ESA/372, United Nations publication, Sales No. E.20.IV.1, ISBN 978-92-1-130392-6, eISBN 978-92-1-004367-0 https://www.un.org/development/

desa/dspd/wp-content/uploads/sites/22/2020/01/World-Social-Report-2020-FullReport.pdf

Sterling, S. (2010), Learning for resilience, or the resilient learner? Towards a necessary reconciliation in a paradigm of sustainable education. Environmental Education Research, 16, 511-528. DOI: 10.1080/13504622.2010.505427.

Tomislav Klarin (2018), The Concept of Sustainable Development: From its Beginning to the Contemporary Issues, Zagreb International Review of Economics & Business, Vol. 21, No. 1, pp. 67-94, DOI: https://doi.org/10.2478/zireb-2018-0005 https://content.sciendo.com/view/journals/zireb/21/1/article-p67.xml

UN, United Nations (1972), Report of the United Nations Conference on the Human Environment. Stockholm. Retrieved September 20, 2015, from http://www.un-documents.net/aconf48-14r1.pdf.

UN, United Nations (1997), Earth Summit: Resolution adopted by the General Assembly at its nineteenth special session. Retrieved November 4, 2015, from http://www.un.org/esa/earthsummit/index.html.

UN, United Nations (2002), Report of the World Summit on Sustainable Development, Johannesburg; Rio +10. Retrieved November 4, 2015, from http://www.unmillenniumproject.org/documents/131302_wssd_report_reissued.pdf.

UN, United Nations (2010), The Millennium Development Goals Report. Retrieved September 20, 2015, from http://www.un.org/millenniumgoals/pdf/MDG%20 Report%202010%20En%20

UN, United Nations (2012). Resolution „The future we want ". Retrieved November 5, 2015, from http://daccess-dds-ny.un.org/doc/UNDOC/GEN/N11/476/10/PDF/N1147610.pdf? .

UN, United Nations (2015). Retrieved September 21, 2015, from http://www.un.org/en/index.html.

UN, United Nations (2015b), 70 years, 70 documents. Retrieved September 21, 2015, from http://research.un.org/en/UN70/about.

UN, United Nations (2015c), Resolution „Transforming our world: the 2030 Agenda for Sustainable Development. Retrieved November 5, 2015, from http://www.un.org/ga/search/view_doc.asp?symbol=A/RES/70/1&Lang=E.

UN, United Nations (2015d), The Millennium Development Goals Report 2015. Retrieved November 5, 2015, from http://www.un.org/millenniumgoals/2015_MDG_Report/pdf/MDG%20

UNDESA-DSD –(2002). United Nations Department of Economic and Social Affairs Division for Sustainable Development, 2002. Plan of Implementation of the World Summit on Sustainable Development: The Johannesburg Conference. New York. UNESCO – United Nations Educational, Scientific and Cultural Organization, 2005. International Implementation Scheme. United Nations Decade of Education for Sustainable Development (2005-2014), Paris.

UNDP (2020). UN Human Development Report, United Nations Development Program http://hdr.undp.org/en/data# http://data.worldbank.org/data-catalog/world-development-indicators

http://reports.weforum.org/global-gender-gap-report-2020/dataexplorer/

http://www.ilo.org/ilostat/

UNEP (2010), Background paper for XVII Meeting of the Forum of Ministers of Environment of Latin America and the Caribbean, Panamá City, Panamá, 26 -30 April 2010, UNEP/LAC-IG.XVII/4, UNEP, Nairobi, Kenya http://www.unep.org/greeneconomy/AboutGEI/WhatisGEI/tabid/29784/Default.aspx.

UNESCO (2013), UNESCO's Medium-The Contribution of Creativity to Sustainable Development Term Strategy for 2014-2021, http://www.unesco.org/new/fileadmin/MULTIMEDIA/HQ/CLT/images/CreativityFinalENG.pdf

United Nations (2015), Transforming our world: the 2030 agenda for sustainable development. New York (NY): United Nations; 2015 (https://sustainabledevelopment.un.org/post2015/transformingourworld, accessed 5 October 2015).

United Nations (2020), World Social Report 2020, Inequality in A Rapidly Changing World,

United Nations (2020). – Gender Statistics, United Nations (Minimum Set of Gender Indicators, as agreed by the United Nations Statistical Commission in its 44th Session in 2013) https://genderstats.un.org

United Nations (2020a), Inequality – Bridging the Divide- Shaping our future together- United Nations 75-2020 and Beyond, United Nations, New York https://www.un.org/en/un75/inequality-bridging-divide

United Nations General Assembly. (1987), Report of the world commission on environment and development: Our common future. Oslo, Norway: United Nations General Assembly, Development and International Co-operation: Environment.

Vander-Merwe, I. & Van-der-Merwe, J. (1999), Sustainable development at the local level: An introduction to local agenda 21. Pretoria: Department of environmental affairs and

Vos, Robert O. (2007), Perspective Defining sustainability: a conceptual orientation, Journal of Chemical Technology and Biotechnology, l 82:334–339 (2007)

Wals, A., (2009), United Nations Decade of Education for Sustainable Development (DESD, 2005-2014): Review of Contexts and Structures for Education for Sustainable Development Learning for a sustainable world 2009. Paris.

WB, The World Bank (2015), World Development Indicators. Retrieved September 2, 2015, from http://data.worldbank.org/data-catalog/world-development-indicators.

WCED (1987), Our Common Future World Commission on Environment and Development New York: Oxford University Press

WID (2018), World Inequality Report 2018, The Paris School of Economics, Inequality Lab, WID.world,

Wiseman, Erica (2007), The Institutionalization of Organizational Learning: A Noninstitutional Perspective, Proceedings of OLKC 2007 – "Learning Fusion", UK https://warwick.ac.uk/fac/soc/wbs/conf/olkc/archive/olkc2/papers/wiseman.pdf

World Bank (2020). – World Development Indicators, World Bank

World Bank (2020). Gender Statistics- World Bank – https://datacatalog.worldbank.org/dataset/gender-statistics

World Bank. (2011). Gender Equality and Development, World Bank http://siteresources.worldbank.org/INTWDR2012/Resources/7778105-1299699968583/7786210-1315936222006/Complete-Report.pdf

World Economic Forum (2020). – Global Gender Gap Report, World Economic Forum

World Economic Outlook (2007), Globalization and Inequality, World Economic Forum, DC

Wright, E. O., (1994), Interrogating Inequalities. New York: Verso

Wu,SOS, Jianguo (Jingle) (2012), Sustainability Indicators Sustainability Measures: Local-Level SDIs494/598–http://leml.asu.edu/Wu-SIs2015F/LECTURES+READINGS/Topic_08-Pyramid%20Method/Lecture-The%20Pyramid.pdf

Yarime, M.; Tanaka, Y. (2012), The Issues and Methodologies in Sustainability Assessment Tools for Higher Education Institutions—A Review of Recent Trends and Future Challenges. J. Educ. Sustain. Dev. 6, 63–77.

Chapter 2

Inequality, a Historical Perspective: With Reference to Society, Politics, and the Economy

G.B. Sahqani

ABSTRACT

This chapter consists of four parts. The first part looks at the origin of inequality using historical perspectives that look at society, politics, and the economy. Inequality is defined in generalized terms and according to specific discipline in the light of economic and political theories. In the second part, we look at what causes inequality, and how inequality has been persistent for centuries given multidisciplinary factors. In the third part, we discuss the relation of inequality with various sectors of the economy and factors which determine the strength of inequality. We look at stakeholders, the role of governments, and behavior of the free market for not realizing the problems caused by inequality in the society. In the last part, we look at how fast-paced technological development will shape inequality and our future socio-economic life.

1. Introduction

Inequality is the harsh reality of our socio-economic life, which we start realizing through the beginning of our social life. Socio-economic history tells us—from the days of social Darwinism during the gilded age to the present idea of a caring economy—equality and justice are the core objectives for the development of society. A long history of human struggle on social, political, and economic fronts appears to be a never-ending movement. Even during the evolutionary stages, the structures of society and state institutions remained under constant change for the improvement of the standard of justice and equality, but they failed to detach the legacy of inequality.

Every industrial revolution is followed by development in new sectors of political and economic fields, and by virtue of that, a new class of billionaires emerges every time. Similarly, the shape of inequality changes, but inflicted inequality never ends. It is strange that historic mass movements followed by revolutions were in fact the reasons for high inequality and injustice in society; accordingly, political systems changed, governments toppled, and new faces came to power to run the governments, but global inequality prevails till today. We have to look into the reality, whether or not inequality is a permanent feature of our socio-economic life, does it relate to our mindset, nature, or general human psychology?

Our economic society has not yet reached a level where it can think judiciously, behave fairly, and live humanly; therefore, informalities which actually provide basis for the establishment of inequalities need to be realized in real time. Policy makers of the past century have struggled between implementing a centrally controlled economy and a free market school system based on the theories of Adam Smith, Thomas Malthus, and David Ricardo. In both cases, the social element has been largely ignored. The capitalist school of thought wants to keep the purity of the economic discipline and serves the needy and poor through a charity-based system or social security systems, some of which are borrowed from socialist ideologies. On the other hand, centrally controlled economies have their own system run by the government where definition of rights and privileges are completely different from capitalist systems. The mixed economy system is prevalent in many countries, where education and health are free, and the states also run charitable and social security programs to look after the people. In these governments, the spending in welfare or caring economies is more than 50% of the GDP (OECD, 2020). It is evident that some mixed economies have absorbed certain changes with time, which were necessary for the improvement of people at the time and those arrangements gradually became part of their system. This is one of the reasons that income inequality in many EU countries is less than in the USA and United Kingdom based on OECD inequality data. Those economies which practically remained under the influence of theories of minimum regulatory interference in market affairs by the government, have established inequality as a regular feature of their economic system. "*It is obvious that economic development and increase in economic activities under a free market economy makes the rich richer—widening the inequality gap*" (Molander, 2016). A study on disposable income finds that from the period of 2006 to 2014, inequality remained high in the USA as compared to EU and other European area countries (Filauro, 2018).

Psychologically, the human race is carrying an inherited odd complexion of socio-economic imbalances. This is due to the genomic building blocks representing unequal ideological foundations. Its balancing can be made through change in social framework to remove all odds whether they are intrinsic or inflicted.

Development is a constant process but unavoidably it carries forward inequality to the next generations. If we claim that human ideological essence has reached the prime of its civilization—after going through three industrial revolutions with the fourth one at our doorstep, to take us into the digital and beyond the age of AI, but we lack the human inclusive approach. If at the prime of our development, we cannot bring justice and equality in society starting from social and ending at economic aspects, it will be a disaster as "*without institutional transformation, economic elites will move further away from the rest of humanity, putting world social structure in danger of collapse*" (Cozens, 2019, para 3). History of being rich is full of odds; it surely costs others their rights, lives, and shares. Man has gained power and wealth through thick and thin whereas natural justice demands that wealth should be distributed equally. We cannot shower the poor with money, but rather should trace out their difficulties and fix them through better economic policies and a perfect justice system.

1.1. Industrial Revolutions and Inequality

After every industrial revolution, a new class of rich capitalists emerged by making unequal gains from trade. In the first revolution, between 1760 and 1820, three main developments were visible: first, a shift from a traditional way of manufacturing to a mechanized version, second, the establishment of factories that formed three social classes, instead of two, creating a middle class, then, in the late 19th century and early 20th century, rapid standardization and industrialization began, leading to the Second Industrial Revolution. Through this, a new class of rich emerged with more wealth and capital i.e., bankers and traders.

The third revolution began in 1980 and is present till now in the Information Technology (IT) sector. The third revolution is unique in the sense that it introduced new trends in the IT sector or IT oriented services sectors. Collaboration, convergence, and mergers developed two or more sectors into giant multinational companies with multiple functions i.e., sale, supply, transportation, and manufacturing sector under one roof. As a result, transformation in every sector started with new technology adoption i.e., IT Postal service developed into private courier services to e-commerce and now authorized economic operators are in the field doing door to door multiple services ensuring fast growing global trade and supply chain. The fourth industrial revolution is ready to install a new class of billionaires. This will be a class of rich who converge the different sectors successfully. The development theory of private business under the free market system has not provided safeguards for the widening of the income inequality gap. Additionally, it does not look possible to delimit the development of AI and modern technology-oriented economy in the future. Even with development in the recent and coming decades, many companies will be laying off employees—replacing manpower with AI-aided machines with state-of-the-art systems. Moreover, new business models that are already in the development phase, to be installed in the fourth industrial revolution, have built-in online, globally controlled features: relocation of intangible assets, user's activities and business operations. "*It thus concentrates power and resources in new ways. New institutions will be needed to redefine, redistribute, and re-employ. Without institutional transformation, economic elites will move further away from the rest of humanity, putting world social structure in danger of collapse*" (Cozzens, 2019, para 3). Those sectors chasing technology on a real time basis will attract global resources through the fourth revolution in the development phase of global capitalism (Butcher, 2019).

Global capitalism is the next phase of free market capitalism, wherein giant companies will prevail through their big volume of capital, managing capacity and global business models with efficient private governance systems. They will take up some public affairs departments, local governments will work in collaboration with the companies—will give finances as aid, loan and also provide technical support through international consulting companies. Just like the IMF or World Bank is supporting the economic and financial work in the world. The world will change to "actual global," in view of the presence of private companies in each country. Some companies are working along these lines, however more companies will take part in more activities in the future and the growing role of privately-owned companies having unlimited capital will ultimately make this world a global village in reality. The quality of capitalism is that private companies dictate—by virtue of having business and marketing tools—expertise and capital. They influence the minds of consumers and in the next phase they will dictate the states.

In the next phase, governments will spend more on public welfare, health and education, and very little will be left to finance the huge budget for government employees' salaries and offices as well as defense expenditures. Therefore, it will not remain feasible to keep certain departments and a big volume of defense forces' arms and ammunition. Rather, a public good paradigm will prevail.

1.2. Literature in Review

We will be looking at inequality as a socially originated phenomenon, which involves other disciplines through human interaction, interdependence and material needs that result in economic inequality, which is being measured as an outcome. The idea that the inequality of opportunities is related to social or political circumstances has not been explored deeply by economists. However, we argue that this is the basic issue, and inequality is evolving into a multidimensional affair of society.

The fact is that the idea of inequality of opportunity remains mostly theoretical, and policy implications for the reduction of poverty—which indirectly impacts inequality. However, the inequality of outcomes has a lot of data readily available to support the researchers and economists to prove its valuation in terms of the GINI index or Kuznet Curves.

In the beginning the theoretical approach has been adopted to prove the origin of inequality and what the classical economists and social scientists were thinking at that time. Theories of Marx, Rousseau, and Keynes have been seen in the light of the present scenario. At the same time proponents of the free market system—Milton Friedman, Herbert Spencer, Krugman, and Joseph Stiglitz—are very strong in their opinion. Research regarding the impact of inequality of opportunities and outcomes has been consulted in this chapter. Most of the literature which is consulted is regarding the hidden factors which create inequality within the capitalist economies. Research papers on the growing impacts of inequality individually and as a society i.e., suicides and other causations have been found very interesting and strong in their analysis. This topic has rich material for analyses in socio-political inequality which plays an active role in exacerbating economic inequality and vice versa—in a circular movement pushing and pulling each other. Amartya Sen also suggested government regulations to check the free market at the same time, which supports the ideas of Karl Marx regarding wages and capacity of labor to work. Research work done by the IMF, EU, and OECD are also very helpful in making an analysis on the subject.

2. Origin of Inequality

In fact, social inequality is the mother of all inequalities. Its impact is referred to other disciplines through a socially created mindset, wherein psychology of self-interest, self-projection, and self-benefits is developed. This makes some people distinctive and superior and this is reflected in the form of inequality and discrimination in all other disciplines (Bhandari 2020; Bhandari and Shvindina 2019). Systematically and gradually, humanity and business have been separated for the purity of the economic discipline. "*While inequality in terms of opportunity is defined on an ex-ante basis and is concerned with ensuring a common starting place*" (United Nations, 2015, para 10). In fact, inequality starts from self-interest, business, and a profit-driven mindset, further developed through various political systems. "*Natural inequality of mankind must be increased by the inequalities of social institutions*" (Rousseau, 1755, p40). Today's inequality is inflicted through the system and its minimization is possible through the development of morality standards, reforming the system, policies, safeguards against inequality and the due role of regulators. The states or world community cannot afford to take further risk simply relying on operators with blind trust. In 2008, the USA injected billions of dollars to support its financial sector. This was not the government's fault but rather heavy reliance on free market operators for the crisis.

Inequality lies in the concept of individuality; it is in fact the mindset based on natural competition among individuals—such instincts are even found in animals. Therefore, this ideal of social contract was in fact to avoid unregulated competition. Justice and equality could be regulated, and rivalries and unhealthy competition may be curtailed. The concept of state is to run its institutions to regulate the competition under certain standard laws.

In many states minimum wages per month, per day, or per hour have been fixed. The irony is that these wages do not at all cover the normal expenses of a family; therefore, one thing becomes clearer that some level of inequality in the system is by design and some extent of inequality is by default. Inequality can only be minimized through government policies, structural reforms and by design is to be addressed through fiscal measures. It is necessary to change the government's regulatory framework and in the economic justice system.

2.1. Human Advancement

Human needs and desires played an important role for the development of a typical mindset to acquire and accumulate personal wealth and earn maximum profit. It fueled the need for creating a sense of financial security and to get the maximum comfort of life as desired by human nature. These factors grew up gradually with the development of community, society and state as Rousseau states, "*all the inequality which now prevails owes its strength and growth to the development of our faculties and the advance of the human mind, and becomes at last permanent and legitimate by the establishment of property and laws*" (Rousseau, 1755, p39).

Furthermore, Rousseau emphasizes as competition increases with people having conflicting interests, the need to profit while others suffered was consequential. This selfish nature was a direct result of establishing the concept of property and the reasons for inequality growing. Rousseau's analysis is even true today, and modern business is still relying on these principles.

2.2. Reasons for Inequality Growing

For the past more than four decades, economic inequality has been an important issue for economists and social scientists. Perhaps, inequality has acquired a lot of substance—reinforced with many multidisciplinary factors, directly or indirectly involved in creating and widening inequality gaps—and by virtue of that it has become the core issue of the 21st century. As with the development of global concepts, almost every sector has to embrace changes to catch the pace with time, adopting new technology. In that manner, inequality is now becoming more visible almost everywhere in different shapes, in both developed and developing countries. The way of thinking about inequality is becoming more realistic. For decades, global leaders, national as well as international agencies ignored volumetric gray transactions, millions of offshore accounts and huge tax avoidance. Rather, it was kept out of the ambit of tax evasion for saving the tax avoiders from criminal proceedings by giving it a soft name instead of categorizing it as evasion.

There is a deliberate ignorance on the part of the whole society that inequality is visible indirectly in the form of poverty, injustice, and corruption, everywhere. It needs to be realized that economic inequality causes so many social problems in society but despite having plenty of literature on these issues, comprehensive regulations are not introduced. OECD's efforts for tackling tax evasion, tax avoidance, profit shifting, and transfer pricing are slow in its pace. The Convention on Mutual Assistance on Tax Matters was developed in 1988, but its effectiveness has not been proved yet as offshore accounts and tax havens are still working and tax evasion and tax avoidance have not stopped. There is a need that the impact of inequality must be realized for making pro-equality policies. The fact is that, inequality creates: corruption (when it comes to demonstration effect from the rich), crimes (when the economic justice system fails), suicides (when disappointed from life and trust is shattered in the system, institutions and society), poverty (which further creates sub-evils like migration, crimes, corruption, complexes, and other social evils). "*Income inequality remains a major driver of inequality in other dimensions of material well being, but other factors, such as the quality of governance, social spending and social norms, matter as well*" (UNDP, 2013, Overview).

1-Corruption: Money earned through illegal means is always kept in offshore tax havens accounts. Due to fear, such earnings are hidden and not declared in the tax returns. Therefore, if used within the country, taxing authorities may detect the illegality. So, most of this money is moved out in safe havens, used for leisure trips, and entertainment purposes. An IMF study concluded that corruption indirectly creates higher inequality of education and distribution of land whereas it directly increases inequality and poverty (Gupta, Davoodi, & Alonso-Terme, 1998). In fact, corruption is a multifaceted element because money with ill-origin causes many negative values created within the society. There are various other factors: bad governance, injustice, lack of accountability which in turn create an atmosphere of illegitimate sources of income and abnormal profits, which add to inequality.

2-Informal economy: In many developing countries, informality is a common phenomenon in the state institutions and informal practices have become a parallel system with time whereas laws and procedures remain in the books. In these economies informal business is almost equal to the formal business. The governments cannot afford to make restrictions as economies are undocumented, if they do so economic activity comes down and unemployment is created. Some studies believe that informal economies at least generate growth and provide employment. Studies also confirm a positive relation between informality and inequality (Mishraa & Ray, 2011; Binelli, 2016).

3-Governance: The social contract has been transformed into a constitution of the states which is passed by the elected representatives. The state becomes the guarantor for upholding the provisions of that sacred document. Whereby laws, policy making and enforcement thereof and providing rights and securities as envisaged under the constitution are the duties of the state.

2.3. Economic Inequality

Economic inequality is generally defined as the phenomenon when income and wealth is not fairly distributed. However, there is no standard criterion for limits on income or wealth for categorizing equality and inequality. Instead, economists and statisticians have adopted an open criterion to be calculated between the distances of incomes of two or more groupings. This is only done for the income and wealth data available if economies are properly documented. Regarding undocumented income and wealth—literature is silent. Variables like honesty, integrity, corruption, ease of doing business and governance should be used to arrive at a more factual level of inequality.

Despite the fact that after wars and revolutions or crises, inequality value goes down whereas during development time it is high (Piketty, 2015). Sometimes questions are raised that inequality and its consequences are determined and pointed out by economists regularly. Then, why have states failed all the time to remove or minimize the inequality gap? Perhaps this area should be defined as no man's land. Optimistically, one day we would be able to include inequality value along with inflation, national income, interest rate, etc., while adjusting fiscal and monetary policies.

In fact, when we go through the factors that determine inequality, we find that there are certain common factors which increase inequality and at the same time they contribute to growth. In fact, the ultimate determinants of inequality are the factors within the economic system. A centrally controlled system has inequalities which are created by the system as it has its own standards of economic justice and justifications thereof.

Every system provides checks and balances to keep the different sectors of the economy, institutions and businesses within the binding limits under regulatory framework. This is to ensure that fair competition, open opportunities for all and independent market forces, may operate on principles of justice and equality.

2.4. Karl Marx

Marx believed that men are not equal in their strength and working capacity therefore a flat rate of wages is based on inequality as different people work in different quantities and quality. In fact, the word "exploitation" which Marx frequently used and is a core point in his theory also covers the definition of inequality. According to Marx, exploitation of the poor in a system is based on the structure of unequal distribution of capital and labor (Marx, 1964). Naming inequality as exploitation, which the entrepreneur commits with a clear mindset for maximizing profit and does everything in the line of business to do so.

2.5. Keynes

John Maynard Keynes found that economic inequality and the concentration of wealth damaged capitalism as a result of independent forces. The theory of "saving is equal to investment" supports the idea that when there is income inequality, those who earn more, save extra money and that money is used for investment. Keynes considers income inequality as a driver of the capital accumulation, which in turn drives economic growth. Pre-war, it was driven by the growth in capital. What accounted for this capital growth? Keynes wrote that on the contrary if the income inequality gap is lesser, people buy more products. More spending contributes to economic growth (Keynes, 1920). The only difference is in the first case the beneficiaries are the businessmen and in the latter case, that title is shared by the low-income group.

2.6. Rousseau

Rousseau believed in two types of inequalities. One is of physical nature that depends on the physical capacities and mental capabilities of humans as nature is not equal in its origin and traces its progress in the successive developments of the human mind. The other is political, which has been legitimized through convention, adoption, and practices (Rousseau, 1755). If we follow the progress of inequality in these various revolutions, we shall find that the establishment of laws and of the right of property was its first term, the institution of magistracy.

2.7. Amartya Sen

The work of Amartya Sen (1980, 1985, 1992, 1999), stresses on equality of opportunity rather than equality of income. He indirectly agrees with Marx regarding low wages and does not rule out some level of governmental interference in controlling inequality, however mainly relies on social and political affairs for tackling inequality.

3. Causes of Inequality

There are two types of inequalities in terms of economic development perspectives. Sometimes informality of developing countries is exploited by developed countries and in this process both work side by side cushioning each other. This happens particularly in a situation when, rich people shift their undocumented money earned through informal means from offshore accounts to those countries. There, they can then buy property, show bank balances for fulfilling conditions and qualify for citizenship or green card applications on a monetary basis. Thus, rich people from developing countries migrate to developed countries through these attractive policies. After the Panama leaks, a lot of cases of such type were highlighted in the media through the ICIJ.

In developed countries, inequality is due to their global policies. The USA and other western countries allow foreign labor to work there which pulls wages downward which is a cause of inequality. Secondly, investment is also made by multinational corporations (MNCs) in developing countries where cheap labor and raw material is available. MNCs open their corporate headquarters in those countries where the corporate tax rate is low.

In former British colonies in Asia, there are obvious factors which determine the inequality created by British rule. For getting political support to rule the sub-continent, the British Government bestowed upon many people the title of "Sir," allocated big lands to these people and many other perks and privileges to buy their support to rule the sub-continent without any crisis. So much so that for crushing the 1857 movement, local people were used to kill the protesting people. As a result, a rich elite was developed, controlling the big lands and that class still exists and their children are still ruling politically nowadays in India and Pakistan.

In South Asian countries, and other regions, income inequality has been rising due to development in big cities, but in rural areas such development has no role. The prices of agricultural produce are not competitive as most of the crops are perishable.

In certain areas of Asia and Africa, old cultures still prevail; early marriages are carried out when boys and girls attain the age of puberty. In these communities, a need-based approach is established for socio-economic security of family's vis-a-vis relying on old myths that early marriage and early sons are considered as the real supporters of poor cultivators. The elder male child is always thought to be the future head and financial supporter for the whole family in joint family systems. Even today in India, Pakistan and many other Asian countries, girls and boys are not equally treated in lower middle class families, the person who is to be the head of the family—the male child—and works in the fields and manages agriculture is given priority and importance. There is inequality built into some old practices themselves.

3.1. The State, Policies, and Inequality

In many EU countries where government spending is more, the inequality index is lesser as compared to the USA and UK. It is clearer that developed countries if they spend more on public expenses, they reduce inequality and those who do not spend more rather give tax benefits to rich people remain in high GINI index range. Moreover, those countries where tax to GDP is high have less inequality as compared to those having low tax to GDP ratio, as per OECD data. It means in developed countries high inequality is a policy phenomenon with an overwhelming shift in economic policy.

The market in developed countries was saturated from many aspects. After World War II, demand for goods increased therefore between the period after the war and the late sixties, huge investments were made in the industrial sector. However, with the passage of time, demand for goods was not increasing due to the saturation of markets, so new markets were needed to sell the goods. Moreover, availability of cheap raw material and labor pushed manufacturers in developed countries to make investments and develop a system of supply chain to further maximize profitability.

The fact is that many states remained negligent and rigid towards accepting the due evolutionary challenges. They neither preempt systematic informalities developed with time and did not formulate policies nor make relevant laws in time to limit inequality. As a result, those who are pro-inequality become powerful and the state gradually withdraws from its duties and the private sector fills the vacuum in their own style to create socio-economic justice. Low income groups are then left at the mercy of the free market system or alternatively the charity or social security system to cover the loophole in regulating the system. And the role of trade and business associations—composed of businessmen is increased as they are powerful stakeholders. Even the state cannot make or implement laws without taking them onboard.

However, in the fourth industrial revolution it appears that the importance of the regulatory framework has been becoming crucial. This is due to the way development is taking place, if it is not regulated efficiently through regulatory agencies, with the active cooperation of the private sector, it may take us to an unknown destination. As governments have not been so quick in coping with the speed of technological developments through the regulatory laws and taxing the new businesses in a fair way and "to do so, governments and regulatory agencies will need to collaborate closely with business and civil society" (Schwab, 2016, para 23).

3.1.1. Economists and political scientists are aware that inequality is not inflicted by any individual rather it is the prevailing economic governing system or paralleled political system which plays this part. It uses the free market ideology to accumulate capital in a smooth way so that nobody can do anything against this smooth transfer, done in the name of the system. The prices are fixed, predetermined and labeled, customer has no role in influencing the prices in the case of some items customers have lost their role of which they earlier acquired as being representing demand side, at the time of purchase, rather sometimes customer can only opt from the given discount packages if schemes are offered in certain situations in the market. This is the digital age; these factors are to be determined by the seller, an airline company or grocery stores owners. It appears that both the market forces which determine prices are in the control of the seller only. As market behavior is planned and not on the basis of real time situations rather advance surveys and market strategies for knowing demand and making supply are actually pre-planned and dictating the consumers.

3.2. Informality and Inequality

Economic crises have their own causes and reasons that gradually grow within the society and when conflicts reach the peak, public resistance starts. Sometimes it takes the shape of popular mass movements against the status quo which in one way or another reshape the economic and political structure of the society accordingly. It is then updated, so that it can deliver in line with the predetermined objectives, wishes and requirements of the citizens. If we observe the whole evolutionary process while sitting outside the sphere, the pen-picture thereof will show that it is in fact autonomous in nature and everyone therein is working in a natural way but cannot feel the gradual process of decay. It transpires that every efficient economic system constantly eliminates informalities developed with time as change is a permanent feature of this world. Today's informalities will no longer be the issues of tomorrow and this is how the development process remains in the continuum. If basic issues are not addressed in time, it penetrates the whole system which makes the institutions unworkable and serious crises start emerging. The length of the cycle depends on how the system works. If formalization is not taken place in due time, it will make the life cycle of the system comparatively short.

The slogan "no taxation without representation" was in fact due to the disappointment and mistrust of citizens on the state institutions for not being served as guaranteed in the constitution. The principle of universal rights does not support development of inequalities. Alberto Chong and Mark Gradstein find a positive link between income inequality and the size of the informal sector (Chong & Gradstein, 2007). States are interconnected economically and now most of the time important decisions are made by compromising sovereign powers for state's benefits.

3.3. Credit Buying and Inequality

When the paying capacity of the consumers is not enough to buy household goods, they have to rely on credit buying schemes which were introduced for those who could not afford. This policy generates economic activity, increasing production demand for goods but it does not decrease inequality as the policy itself is based on a credit buying strategy wherein income is not increasing but expenditure is increasing with interest. These schemes attract the consumer in a trap and most of them are forced to live under debt throughout their life. In fact, it is a marketing plan prepared by banks and other beneficiaries to fill the gap created through inequality of income by introducing credit buying policies.

The banks, manufacturers and suppliers or sellers will benefit from these schemes but not the consumers. One argument in favor of such schemes could be that after credit buying, consumers try to earn more money due to loan repayment pressure but it happens in limited cases as not all the consumers have the capacity, opportunities and efficiencies and neither is the market so favorable all the time for those consumers.

It has been observed that some consumers, even without having the capacity to buy things, manage to buy those products on credit or loans. They do this by working extra hours, saving more money by spending less on essential items so that eventually they may purchase household goods, and buy a house which otherwise is not possible under normal circumstances.

New trends in lifestyle, new fashion and new gadgets are a requirement of today's consumer and they want to buy these at all costs, even if they have to wait for the products to go on a discounted sale—which is also decided by the seller and this way they end up ruling the minds of the people. Consumers are thus almost under the control of the seller through channels of the media. On the one side they psychologically induce credit buying and on the other hand, banks and manufacturers/sellers are the beneficiaries through artificial ways of creating demand for their goods which is the main reason for rising inequality which ultimately turns into economic crises.

Whether it was the Great Depression or the Economic Recession, both times, the banks and market panicked due to credit expansion beyond limits. Financial innovations do not prove to be smart moves all the time—when there is a higher ratio of debt to equity and financial agents expecting profit greater than the interest. "*Thus, the Recent Crisis (the Great Recession) and the Great Depression can be seen as siblings that were born under different circumstances*" (Aiginger, (2010), p.34). Regarding the 2008 financial crisis, Mr. Alan Greenspan, Chairman of the Federal Reserve admitted before the congressional committee: "*I made a mistake in presuming that the self-interest of organizations, specifically banks, is such that they were best capable of protecting shareholders and equity in the firms*" *and* "*that he regretted his opposition to regulatory curbs on certain types of financial derivatives which have left banks on Wall Street and in the Square Mile facing billions of dollars' worth of liabilities*" from the hearing, The Financial Crisis and The Role of Regulators (2008).

4. After-Effects of Inequality

Inequality is in fact a silent peril of society. Its carriers are the economic systems as well as rich people and policy makers. Since it is not a direct phenomenon therefore it uses many negative values of the society to move in a way so that it should not be visible. It either moves through the system, bypassing the system or ignoring the moral, social, political and economic principles which create a lot of problems in the society.

4.1. Suicides

Justice is hope for the wellbeing of humans. Human nature is not so rigid rather it adopts whatever is required for living. Disappointment is the last factor to instigate the person towards suicide, when all other doors are closed. An Indian study suggests that both rural inequality and rural poverty have been significantly contributing to increasing farmers' suicides in Indian states over the twenty-year period (Banerjee, 2016). Many studies on the subject have been published and they have found a positive link with inequality and high suicide rates. A South Korean study showed the relation between inequalities and mental health of the people including depression, suicidal ideation and suicide attempts (Hong, Knapp, & Mcguire, 2011). It also finds that inequalities in each instance have doubled over the past 10 years, accompanied by widening income inequality following the nation's economic crisis in the late 1990s. Suicide has been associated with economic inequalities and economic shocks, both rapid booms and recessions (McDaid & Kennelly, 2009).

5. Economic Crisis and Inequality

The world has learned a lot from the crisis and wars. Economic crises in the past were seen as a consequence of not realizing the necessities of making timely changes in economic policies. These were required in the economic governance system for uplifting socio-economic conditions of people and reducing disparities and informalities in the system with time. Kingdoms vanished, dictatorship died, communism became irrelevant and now capitalism is under threat. It can be argued that the intellectual movement against growing inequality, also points towards the apprehensions regarding the future of capitalism and the new shape of regional or global capitalism in the 21st century. A fair approach to assess these questions will be based on the ground realities; which doctrine prevails and who will be leading the world.

The result will be according to the policy, philosophy and the creative spark of leadership. If we want to further polarize the world with a global vision that benefits the rich, the situation will not be better. According to UN Chief António Guterres the world is confronting "*the harsh realities of a deeply unequal global landscape,*" where unstable job markets and inequalities have resulted in protests across the world (United Nations, 2020).

Crisis followed by revolution has its own unique causations, which depend on many political, economic, and social factors. However, one thing is common in all the revolutions—the causations were created within the governing system by the inefficient governments and bad governance. People were denied the basic necessities of life, leading to protests breaking out on the streets. During the Russian revolution, low wages was one of the causes of mass protests where the state failed to provide basic necessities of life to the citizens.

During the French revolution, the main slogan was—Liberty, Equality, Fraternity—this need resurges when rulers become ignorant of the economic situation on ground, when policy makers lack wisdom. When there is a communication gap between the legislature and people, it contributes to arrogant behavior as people do not have the forum to communicate with authorities and make them realize their issues. The inner feelings of those with low wages are stormy—due to inequality and economic injustice. An emerging problem before civil society is how to rebuild trust in government institutions, private sector and public representatives, policy makers, as the capitalism–democracy model is in practice for more than a century. People have trust in the system but those who run the system manipulate it to their personal interest. In many countries the tolerance level of the lower middle class has been at the peak. Due to financial problems cab drivers burn their vehicles; farmers burn their agricultural produce as they receive prices less than the cost of producing the commodities. In Pakistan, farmers took out a procession to protest the low market price of agricultural products. In Bangladesh a similar situation also took place (The Financial Express, 2019).

There are three stakeholders in the scenario, one who is creating inequality; must be responsible for maintaining beneficiary. Another, who is facing it and is a victim; always suffers but cannot do anything as it has been inflicted upon them through a system, in an organized way that is not challengeable. A laborer cannot file a case that his wages are less, as it is settled on market rates, where all the laborers are getting the same amount. The third stakeholder is the state, which is gradually withdrawing from its role.

The capitalism–democracy model of today supports and guarantees to protect the rights and benefits of the rich class. Today's poor are not only blaming capitalism rather are more vocally criticizing democracy. Practically speaking, the common man is betrayed all the time, and it doesn't look possible all the time that citizens will start a protest on the street or before the Parliament house to let the government know what they want and what their problems are.

Inequality is not a direct phenomenon, it is created by various indirect means and it is the inefficiency on the part of state institutions that they do not make pro-equality legislation, improve governance, justified income distribution system through effective tax system:

1. Low income
2. Improper taxation
3. Informal economy
4. Corruption
5. Bad governance
6. Lack of accountability
7. Powerful private sector
8. Role of representative legislature and policy makers
9. Weak law enforcement
10. Discrimination

These factors in turn create an atmosphere of illegitimate sources of income, abnormal profits, and unequal distribution of wealth which indirectly create inequality. However, there is a difference between the inequality in the developed and developing countries, the issues and causations thereof are different, so the solutions are also different.

According to OECD "higher inequality drags down economic growth." It is obvious that when income of lower income groups increases, they spend more to improve their standard of living, quality education, buy those goods which otherwise were not affordable previously. Whereas when income of higher income groups increase, they save more which is ultimately used for investment and that money contributes to growth. Whereas many economists believe that during crisis and wars inequality is high and so growth remained high as in the case of the USA during the World Wars in the 20th century (Lindert & Williamson, 2016).

During the French revolution in the 17th century, King Louis was openly criticized, and a movement took place that he was protecting the interests of the elite, ignoring the poor masses. "*In a way, the French Revolution of 1789 was more ambitious. It abolished all legal privileges and sought to create a political and social order based entirely on equality of rights and opportunities*" (Piketty, 2015).

6. Globalization and Inequality

Globalization is the next phase of capitalism through direct integration on economic terms among the states to bring national economies under the canopy of globalization through global regulations. The creation of the World Trade Organization (WTO), a global institution that has been entrusted the role for smooth free flow of global trade as envisaged in the global rules of trade between nations.

The economic aspects of globalization include movement of goods, services, capital and now technology and data as well. As markets expand outside their borders, exchange of goods and funds are easier due to reduced restrictions of cross-border trade allowing for globalization of markets.

In developing countries local industries were not in a position to compete the global influx of goods moreover there were a lot of FTAs and PTAs in the field which actually harmed many local industries. In Pakistan the textile industry was badly affected due to free trade agreements and many small units were closed which created unemployment in the textile sector. In the same way the leather industry was also impacted, due import of cheap shoes from China and Thailand. Mixed trends out of globalization have been witnessed. Trade causes a decrease in inequality. Whereas, due to investments and financial globalization, inequality increases (Jaumotte, Lall & Papageorgiou, 2013).

According to Stigliz, the US and EU benefited from globalization through trade agreements with developing countries. "*The problem is that trade agreements advanced corporate interests at the expense of workers in both developed and developing countries*" (Stiglitz, 2019, para 1). Though, a smooth start has been taken to sail into the next phase of the evolution process, from capitalism to global capitalism without experiencing a major crisis with the realization that after effects of globalization on the US economy was not correctly estimated and inequality in the USA increased but developing countries benefited from that loss (Milanovic, 2016).

7. Taxation and Inequality

In the digital age, taxation has developed multifaceted features. It is the main tool to generate revenue but through tax policies, state expand economy, facilitate trade and business, prioritize industrial manufacturing and sectoral developments, incentivize export industries and protect local industry. However, the crucial role of taxation should be to discourage the accumulation of wealth and capital by few in the same way monopolies and oligarchies are not allowed worldwide. The basic philosophy behind these restrictions is to level the playing field for all (Stiglitz, 1989). Through balanced fiscal policies, the state should make progress in such a way that on the one hand, more tax revenue is collected and at the same time growth targets are achieved—"*the impact of economic growth on the lives of people is partly a matter of income distribution, but it also depends greatly on the use that is made of the public revenue generated by economic expansion*" (Drèze & Sen, 2013).

The state is required to impose duty/taxes in any manner to shift the money from the rich to the poor in a legal way to abide by the socio-economic responsibilities. There are three main policies by the governments to keep the economy in balance and on the development track—monetary policy, fiscal policy, and trade and investment policy. Here we will only discuss how fiscal policy is an effective tool for the government to check inequality. Many economists make their research on rising inequality being the cause of the economic crisis, whereas some believe that using only monetary policy as a tool to avoid financial crises is not the answer rather fiscal policy needs to be suitably tailored accordingly. Economists from the Keynesian school of thought believe that fiscal and monetary policies are the main tools required to run a better economy. Ricardo suggested tax on land rents so that landlords could not increase rent as in that case more tax will be payable by the landlord as it is the direct tax (Ricardo, 1817).

It is important to understand that direct taxation is the main balancing tool in the hands of the government to tax rich people and big companies. It can help check the accumulation of wealth in a few hands that ultimately tends to limit inequality and makes the income distribution fair as it is to be done through comprehensive fiscal measures. A major portion of tax revenue is collected by developing and developed countries from indirect taxes which is mainly paid by low income groups who are already living at subsistence level. Since they are in the majority, more tax is collected from them. In Pakistan, from 2009 to 2019, the share of direct tax remains around 40%, whereas indirect tax remains around 60% during the whole decade; government policies could not enhance the share of direct tax even in a decade (Government of Pakistan, 2019). The share of income tax to GDP is also decreasing because the government is bound to give tax relief under the trade and investment agreements to bring in foreign investment in view of globalization.

Because the number of taxpayers in the developed countries is higher the leading tax head is income tax, whereas in developing countries, a major portion of tax revenue is generated through indirect taxes as due income tax is not fairly paid by the wealthy individuals and companies.

> "*Direct taxes make the system more progressive through balancing income distribution and narrowing down the inequality gap. The use of such redistributive tools has remained under-utilized and the tax structure remained tilted towards the regressive indirect taxes, contributing around 60 percent of total taxes*" (State Bank of Pakistan, 2018).

Efficiency of tax administrations remained many steps behind to deal with new business models, and particularly digitally oriented business models and digital services as we have seen that it is becoming difficult to tackle tax issues in digital service tax. It is interesting to read that nowadays public representatives are vocal against inequality and they have the suggestion that the rich should be taxed to reduce inequality. US Senator Elizabeth Warren proposed an "ultra-millionaire tax" at 2% annual wealth tax (Rogoff, 2019). In developing countries, sometimes governments give relief to the commoners by slashing customs duty or sales tax rates on certain goods but incidence of tax relief does not reach the grassroots level as prices are not lowered in the market. Distributors or wholesalers are the main groups who play the game and do not transfer the benefits to the people as there are no checks by the government in this regard.

7.1. Tax Evasion and Tax Avoidance

In the high moral standard societies or states, it is believed that the rate of social crimes is very low as they are well educated and civilized. For example, in Nordic countries, the rate of crime is low. However, even in those states where crime is high, law firms do not advertise that they will give you guarantee for doing crime, but a strange thing is seen in the case of consulting firms, openly advertising their expertise in transfer pricing, providing services to avoid tax or minimize tax liability by way of transfer pricing.

Tax evasion is a worldwide phenomenon being orchestrated due to weak law enforcement, inefficiency of the tax administrations, lack of proper training to the tax staff and in some cases, connivance make it possible. However in the case of tax avoidance, companies make their accounting and tax calculation in such a way that the figures and amounts are declared somewhere but in those heads and in those jurisdictions where tax rate is either zero or minimum as compared to that where the tax was payable .The amount involved in tax evasion or avoidance if added with the household income it will change the whole scenario including change in Inequality index value, income figures and their ratios etc. Moreover in absence of undocumented and concealed amounts of income, wealth and tax thereon, researchers are helpless to calculate the actual figures of inequality and other variables, therefore our research will remain incomplete and vague and policy makers will not be able to make correct policies using incomplete data as informally those undocumented money remain in circulation in one way or another also generate economic activities. The link between inequality and tax evasion has been established by many economists through various studies. "*On average about 3% of personal taxes are evaded in Scandinavia, but this figure rises to close to 30% in the top 0.01% of the wealth*" (Alstadsæter, Johannesen, & Zucman, 2017). There are two studies on the subject which reveal that an EU study on offshore accounts and tax evasion found that global offshore wealth was estimated at $7.8 trillion in 2016 (€7.5 trillion) or 10.4% of the global GDP, which is a considerable amount. Other valuations are in line with this estimate. The EU share is valued at $1.6 trillion (€1.5 trillion), or 9.7% of the GDP. The related estimated revenue lost due to international tax evasion was €46 billion in 2016 (0.32% of GDP) (Vellutini, Casamatta, Bousquet, & Poniatowski, 2019).

Tax evasion or tax avoidance may not be the motive behind undeclared wealth in tax havens. There may be some other reasons to hide receipts of illegal or criminal activities including crime and corruption so that tax agencies and business partners may not know; so, indirectly tax is evaded. International investments are facilitated through offshore accounts, which are otherwise legal in the country of investment (Forstater, 2018). According to an EU study, global financial wealth held in offshore accounts constitutes about 8 or 10% of the world's GDP (Vellutini, Casamatta, Bousquet, & Poniatowski, 2019).

For more than two decades OECD was clustering the approach on how to limit inequality as it is evolving through the loopholes of the system. The problem with the economists is, how to evaluate certain variables in digits to use those values for determining the actual value of inequality.

Especially those variables which do not relate to economics but have a certain role in adding inequality and need to be treated as important variables including governance, institutional integrity, social environment, political system, informal economy, etc. "*The Gini coefficient, for instance, does not measure the relational equality of treatment afforded to individuals*" (Saarinen, 2019, para 5). The objectives of the research on inequality cannot be achieved unless we have a proper inclusive approach for calculations. This is because every country has a different socio-economic situation, corruption level, standard of governance, volume of informal economy—which are the main negative factors which impact the economy. The real income of a person is unknown in many cases, moreover, perks and privileges have not been added with the income of rich people when calculating the inequality. Without knowing all sources of income, we cannot actually calculate the inequality.

8. Life Cycle of Economic System and Inequality

Rules for the rulers are written in the rule books. The splendor of rulers lies at the cost of wellbeing, prosperity and happiness of common man. They build their political empire on the petty theories of contradictions and socio-economic inequalities. To bring down inequality, there needs to be amendment in the rule books. Historically, it is evident that every economic system ends when it fails to deliver to the people. Lord Macaulay drew certain rules for the rulers in the 17th century. We know the reasons why the big dynasties ruined. When a system completes its natural cycle its life is over, and it is to be obsoleted along with its structures. This cycle is like a manufacturing factory—can run at least, when cost of production is recovered from the sale of goods, beyond that it will not remain feasible to operate. In the same way the system runs its cycle up to that limit where inequality of opportunities does not surpass inequality of outcome.

System and Cycle
Every economic system reaches a point of full capacity utilization. Once a system completes its natural economic cycle, the working forces of the system require a new track to run on as previous tracks no longer

synchronize with it. Regional integration and globalization are two such tracks that are followed by economic systems to make use of free market forces. These operate on a broader scale which require a new form of economic governance through regional or global institutions so that trade, investment and supply chain systems through the countries may run smoothly, by removing trade barriers between the countries, agreeing on mutual cooperation for investment, reducing tariff and promoting free trade or preferential trade.

Regional integration proved successful so far however the impact of globalization in the short run may not be ideal but, in the long run it will be beneficial globally, as it is a natural way of moving into the next stage of capitalism. Sensing the need for change, many countries, even China, have adjusted their economic systems without changing the centrally controlled structure and have still been able to transform their productive capacity, whereas the USSR faced a crisis due to its rigidity towards change and development during hard core communist rule. The system jolted once when Gorbachev tried to introduce Perestroika and Glasnost in the 1980s. The rulers are aware that they cannot adopt a capitalist leaning model with the same ease as China.

Nevertheless, change is inevitable as the economic system is a complete structure with built-in pillars and posts that are backed by a certain ideology. However, this depends on how we see the importance of the timing for this change. There came a point when the communist ideology started to become irrelevant. People were looking for economic change in the form of a better model. In order to prevent a mass revolution that would break the system, the rulers began reforming and artificially jacking up the old communist structure. This was to please the masses who were struggling and looking for reform. The communist ideology introduced by Lenin and Stalin had completed its cycle and required structural reform, but dictators never learn from history rather they make history.

9. What is Next

Development never stays in one sector or in one area, it has to penetrate through all sectors of life spontaneously, but we can see today, we lack this in our health sector. We reached Mars and the Moon, but we could not predict the Coronavirus which has disrupted life globally. This is a very crucial age, where technology and human compatibility has attained a permanent direction, vectoring with high speed, leaving behind many ideas and objects which cannot compete. We have to pre-empt the future and design the course of action as the way things are advancing with digital accuracy and scientific perfection leaves no chance for errors in modeling for the next decade. And they will not allow for time to go back to recheck the mistakes. We have to design our model of governance perfections with digital accuracy for all sectors of life. Based on these ideas, intellectuals have to be honest in their thinking for a *human dignity-inclusive framework* for development. We have seen that regulating agencies are very far behind in tracking tax evasion or while leaving digital services tax, administrations were not even aware of the digital services taxability issue.

We have ignored the individual, it is the time to re-think about them - how they will rely on technology and its social effectiveness and dependence on others (Sen, 1999). They will be independent in decision-making and that stage will be very alarming when an individual will cast their vote online and will ask too many questions. They will initiate movements without any fear and will be dealing with their cases individually and will not be able to stop. People will not look at the causes, rather will ask for the outcome because they are not the part of causation. The errors of the policy makers will also come under the scrutiny by the individual who is educated and able to convey their message, feedback and criticism through technology. The trend of whatever is said loudly must be true, seems to be taking a back seat where there are now many loud voices and opinions. The history of economic development transpires that many economists were not accurate in their assessment of future developments and partially some theories and assessments were not proved correct. However, some living economists including Alan Greenspan have admitted to incorrect outcomes out of their assessment which caused a big crisis.

What has been said is that certain aspects are within human control; providing opportunity like social justice, equal opportunity for jobs, free education and health care so that inequality of outcome may be decreased. These efforts indirectly contribute to better results as the state controls all the institutions. Development in the economic fronts without reforming social and political institutions is just like putting the cart before the horse.

At the same time the state has to play an essential role to limit inequality, a fair taxation system is the effective tool to limit income inequality and accumulation of wealth. Government may generate more direct tax and public spending may be increased on health, education and infrastructure may be increased to cushion the low-income groups (IMF, 2017). It requires

Implementation and enforcement of laws for an effective governance system to eliminate informalities so that unlimited accumulation and unaccounted earnings may be recorded and brought into tax net. Aranchey Gonzalez is optimistic about positive change. According to her, wellbeing budgets are being presented, some countries are tackling wealth inequality and working on "relational equality" (Gonzalez, 2020). There is hope yet.

References

Aiginger, K. (2010). The Great Recession Versus the Great Depression: Stylized Facts on Siblings that Were Given Different Foster Parents. Economics. 4. 10.5018/economics-ejournal.ja.2010-25.

Alstadsæter, A., Johannesen, N., & Zucman, G. (2017). Tax Evasion and Inequality. doi:10.3386/w23772

Banerjee, D. (2016). Inequality and Farmers' Suicides in India (NIAS Working Paper WP5.

Bhandari, Medani P. (2020) Second Edition- Green Web-II: Standards and Perspectives from the IUCN, Policy Development in Environment Conservation Domain with reference to India, Pakistan, Nepal, and Bangladesh, River Publishers, Denmark / the Netherlands. ISBN: 9788770221924 e-ISBN: 9788770221917

Bhandari, Medani P. (2020), Getting the Climate Science Facts Right: The Role of the IPCC (Forthcoming), River Publishers, Denmark / the Netherlands- ISBN: 9788770221863 e-ISBN: 9788770221856

Bhandari, Medani P. and Shvindina Hanna (2019) Reducing Inequalities Towards Sustainable Development Goals: Multilevel Approach, River Publishers, Denmark / the Netherlands- ISBN: Print: 978-87-7022-126-9 E-book: 978-87-7022-125-2

Binelli, C. (2016) Wage inequality and informality: evidence from Mexico. *IZA J Labor Develop* 5, 5. https://doi.org/10.1186/s40175-016-0050-1

Butcher, M. (2019, January 23). World Economic Forum warns of AI's potential to worsen global inequality.Retrieved from https://techcrunch.com/2019/01/23/world-economic-forum-warns-of-ais-potential-to-worsen-global-inequality/

Chong, A., & Gradstein, M. (2007). Inequality and informality. *Journal of Public Economics, 91*(1-2), 159-179. doi:10.1016/j.jpubeco.2006.08.001

Cozzens, S. E. (2019, May). Inequalities and the Fourth Industrial Revolution. Retrieved from https://unctad.org/en/pages/newsdetails.aspx?OriginalVersionID=2068

Drèze, J., & Sen, A. (2013). *An uncertain glory: India and its contradictions.*

Filauro, S. (2018, May). The EU-wide income distribution: inequality levels and decompositions. *European Commission.*

Forstater, M. (2018). Illicit Financial Flows, Trade Misinvoicing, and Multinational Tax Avoidance: The Same or Different? *CGD policy paper, 123*, 29.

Gonzalez, A. (2020). Reinventing capitalism in the 21st century – for people, planet and prosperity. Retrieved from https://www.linkedin.com/pulse/reinventing-capitalism-21st-century-people-planet-arancha-gonzalez/?trackingId=s-fZhXl+cllgbdJQSFRBapA==

Government of Pakistan. (2019). Pakistan Economic Survey 2018-19 Retrieved from http://www.finance.gov.pk/survey/chapters_19/Economic_Survey_2018_19.pdf

Gupta, S., Davoodi, H., & Alonso-Terme, R. (1998, May). Does corruption affect income inequality and poverty?.IMF Working Papers, WP/98/76.

Hong, J., Knapp, M., & Mcguire, A. (2011). Income-related inequalities in the prevalence of depression and suicidal behaviour: A 10-year trend following economic crisis. *World Psychiatry, 10*(1), 40-44. doi:10.1002/j.2051-5545. 2011. tb00012.x

IMF (2017, October 01). Fiscal Monitor: Tackling Inequality. Retrieved from https://www.imf.org/en/Publications/FM/Issues/2017/10/05/fiscal-monitor-october-2017

Jaumotte, F., Lall, S. & Papageorgiou, C. (2013). Rising Income Inequality: Technology, or Trade and Financial Globalization? *IMF Econ Rev* 61, 271–309 https://doi.org/10.1057/imfer.2013.7

Keynes, J.M. (1920). The Economic Consequences of the Peace, (New York: Harcourt, Brace, and Howe: 1920). Retrieved from https://oll.libertyfund.org/titles/303

Kraus, M. W. (2018, December 1). The Roots of Economic Inequality. Retrived from https://insights.som.yale.edu/insights/the-roots-of-economic-inequalit

Levitz, E. (2020, April 07). Coronavirus Is Forcing the GOP to Admit Its Ideology Is Delusional. Retrieved from https://nymag.com/intelligencer/2020/04/coronavirus-trump-dhs-undocumented-workers-essential.html

Lindert, P. H., & Williamson, J. G. (2016). Unequal Gains. doi:10.1515/9781400880348

6- Oishi, S., Kushlev, K., & Schimmack, U. (2018). Progressive Taxation, Income Inequality, and Happiness. American Psychologist. 73. 10.1037/amp0000166.

Marx, K. (1964) . Das Kapital. Marx Engels Werke. Vols. 23-25.

McDaid D. & Kennelly, B. (2009). An economic perspective on suicide across the five continents. Oxford Textbook of Suicidology and Suicide Prevention. (ed. Wasserman D. and Wasserman C. p. 359–368).

Milanovic, B. (2016). *Global inequality: A new approach for the age of globalization*. Harvard University Press. 10-29.

Mishra, A., & Ray, R. (2010). Informality, corruption, and inequality. *Bath economics research paper, 13*(10).

Molander, P. (2016). *The anatomy of inequality: Its social and economic origins-and solutions*. Melville House. 53.

OECD (2020), General government spending (indicator). doi: 10.1787/a31cbf4d-en (Accessed on 01 May 2020)

Piketty, T. (2015). Capital in the 21st Century. P.25

Ricardo, D. (1871). On the Principles of Political Economy & Taxation.

Rogoff, K. (2019). Could a progressive consumption tax reduce wealth inequality? Retrieved from https://www.weforum.org/agenda/2019/09/the-benefits-of-a-progressive-consumption-tax/

Rousseau, J. J. (1755). Discoures on Inequality. Translated by G. D. H. Cole. p.9- 39. Retrived from https://aub.edu.lb/fas/cvsp/Documents/DiscourseonInequality.pdf879500092.pdf

Saarinen, S. (2019, August 28). The study of inequality and politics are intertwined – but how? Retrieved from https://www.helsinki.fi/en/news/nordic-welfare-news/the-study-of-inequality-and-politics-are-intertwined-but-how

Sen, A. (1980): "Equality of what?". In S. McMurrin (ed.), Tanner Lectures on Human Values, Cambridge University Press, Cambridge.

Sen A. (1985) Commodities and Capabilities. Amsterdam: North-Holland. Sen, A. (1992): Inequality Reexamined, Clarendon Press, Oxford

Sen, A. (1999). Development as Freedom. 19.

Schwab, K. (2016, January 14). The Fourth Industrial Revolution: What it means and how to respond. Retrieved from https://www.weforum.org/agenda/2016/01/the-fourth-industrial-revolution-what-it-means-and-how-to-respond/

State Bank of Pakistan. (2018). Annual Report 2017-18 - The State of Pakistan's Economy. Retrieved from http://www.sbp.org.pk/AnnualRepo/index-4.2.asp

Stiglitz, J. E, (1989). Using Tax Policy to Curb Speculative Short-Term Trading. 3:2-3 Journal of Financial Services Research 101-15.

Stigliz, J. E. (2019). US trade deals were designed to serve corporations at the expense of workers. Retrieved from https://www.cnbc.com/2019/04/22/joseph-stiglitz-us-trade-deals-helped-corporations-and-hurt-workers.html

The Financial Crisis and The Role of Regulators (2008), 110th Cong. (testimony of Alan Greenspan). Retreived from https://www.govinfo.gov/content/pkg/CHRG-110hhrg55764/html/CHRG-110hhrg55764.htm

The Financial Express. (2019). Farmers in a price-trap: A year-long phenomenon. Retrieved from https://thefinancialexpress.com.bd/views/farmers-in-a-price-trap-a-year-long-phenomenon-1568645656

UNDP. (2013, November). Humanity Divided: Confronting Inequality in Developing Countries. Poverty Reduction.

United Nations. (2015, October 21). Concepts of Inequality. Development Issues No.1. *Development Policy and Analysis Division Department of Economic and Social Affairs.*

United Nations. (2020). World Social Report 2020 - Inequality in a rapidly changing world. Department of Economic and Social Affairs. https://news.un.org/en/story/2020/01/1055681

Vellutini C., Casamatta, G., Bousquet, L., & Poniatowski G. (2019, September). Estimating International Tax Evasion by Individuals. *European Commission Taxation Papers.* Working Paper No. 76 –2019

Chapter 3

Religion and Gender Inequality in India

Salvin Paul and Maheema Rai***

*Peace and Conflict Studies and Management in Sikkim University

**Independent Research scholar.

ABSTRACT

Gender inequality that is structured in the religious and societal customs appears to exist all over the world. These inequalities that are sanctioned through the norms and values of cultures and religions have become a universal phenomenon where women themselves support such practices to make them officially invisible. In this backdrop, how women are unable to grasp and comprehend in their daily life the understanding on equality and its of late discourses in a religiously driven society like in India where secular understanding of equality has become a major issue of contemporary debate. The Indian society may have undergone considerable transformation; however, the position accorded to women in India is far from the rights enshrined in the constitution. This article would adopt an explorative approach in a broader qualitative framework to examine the experience of gender inequality among married women within the major religions in India. The objectives of this article would be to examine whether secular Indian state was successful in bringing change in the lives of women in general and why lives of women have become so deplorable and what role divergent personal laws of various religions play in perpetuating gender inequality among women in India.

1. Introduction

Gender inequality is one of the most pervading forms of institutionalized form of discrimination against women. In India, there are varied customs and communities who make religion as a first priority. The position of women is a major concern in the modern world. Intense social actions and intellectual exercises on gender equality have been noticed in last several decades. Religion, as a powerful source of collective identity, plays a central role in human civilization. As a particular system of faith and worship, it attributes sacredness and provides a normative basis to society. Every religion manifest in society through its system of beliefs and practices according to which stakeholders organize their lives, both individually and collectively. Indian society has undergone considerable changes in the modern times. However, the position accorded to women is far from the spirit of modern perception of equality. Pillars of equality and humanity are jeopardised in many ways and women face various problems through such corrosion of justice and equality. Personal laws, by being nurtured and validating patriarchal dominations, have reinforced the vulnerability of women's positions. So, this article highlights and examines the different personal laws present in all religions and analyzes their reinforcement of discriminatory and unequal laws that exploit women in the Indian society.

2. Religion and Inequality: Feminist Perspectives

Religion is one of the great sustaining forces that pervade major areas of people's lives. It is reflected in the values, norms, and cultural ethos of a society. Religion has always played an important role in shaping the society and its testament have influenced and been one of the most influential waves that have been carried through and acted in our society. Thus, the manifestations of religious behaviour are intricately related to various aspects of human life and are woven into the fabric of social life even in the modern world. The very nature of religion has gender biasness in its manifestation that subjugates the rights of women. Feminists argue that the philosophy of religion has ignored the gender ideology in the subject matter of religion. Religion is perceived as one of the patriarchal beasts that reinforces women's inferior position in society.

All the sacred literature of the world display an unvarying ambivalence on issues of women, as for every religious texts place woman on the high religious pedestal where another as the root of all evils. As per Elizabeth Cady Stanton, where she argues that there was moral degradation of women according to theological superstitions than other influences (Stanton, 1885).

Stanton's work mostly focuses on the Christian teachings; the power a status of the churches and their reinforcement to restrict the sphere of women's choice and action. Feminist religion has influenced the gender biasness which has been structuralized in our society. The patriarchal biasness in the society and within communities has perpetuated inequalities and subordination of women.

The feminists problematize the construction of gender which is imbued in the hierarchical relations of power, social norms, and religion. Placing women at the centre, the feminist theorists expose the pervasive androcentrism constructed through religious concepts. Religion was perceived by early feminist theorists as one of the patriarchal beasts that reinforced women's inferior position in society. Consequently, on the one hand, feminist theory tended to be staunchly anti-religious in its early articulations and engaged religious themes corrosively or only peripherally. According to Rita Gross, one of the factors that religion undertakes is the gender differentiation. One of such disparity would be seen in the women's position of religious leadership (Gross, 1997). Most of the feminists argue that the religious traditions are underpinned by the commitment to ontological equality which is been neglected in the male centric religious tradition Like, De Beauvoir who has argued that historically, men, who have traditionally controlled most institutions in society, also control religion. It is men who control religion beliefs, and they use God to justify their control of society (Beauvoir, The Second Sex). The discursive practices where the male is intimately connected to the production of ideologies have eliminated or devalued the status of women and initiated oppression of women (Frankenberry, 1993).

3. Defining Personal Laws

India is a land of diversity in terms of geography, religious faith, customs, and culture, hence there are diverse personal laws, governing people belonging to different religion and communities. These personal laws in definition are set of laws that administer and control the relations arising out of marriage, divorce, maintenance, succession, guardianship etc. as the foundation of these laws are based on the scriptural understanding that had been followed in the personal arenas. These laws also provide with norms, values, rules, and regulations in the areas of marriage, divorce, property inheritance, etc. (Wadje, 2013). The importance of personal laws can be seen in their very nature, composition, and their relations to which they are being applied. Personal laws are close to family values which have established the law pertaining to family affairs that an individual is entitled not just by virtue of being an individual but by being a member of certain religious or ethnic group or communities (Ghosh, 2007). Therefore, personal laws are close to family values and family affairs to which they are entitled by being a member of certain religious or ethnic communities.

3.1. Personal Laws and Gender Inequalities

The ideological and moral bases for the status delivered to women as well as the institutional roles of women in society are based on the religious ideas in India. The social restrictions on women, roles within the household and extra-domestic arenas are mainly derived from the religions. The Indian State adopted the principle of equidistance that conveyed an implicit rejection of the State's involvement in society that suggested the State's disposition toward religion. The colonizers did plan to sustain the compatibility of all the persisting religion in India. But the preservation of the religious personal laws into the post-colonial era may have been necessary but its continuity has caused many injuries mostly to the women. Along with the women the customary law started losing the position after the colonial power took the issues of personal law in their hand. They eliminated the indigenous people's rights, because they were not for the concern for the colonizers only, they took part where the legal issues were discussed (Ghosh, 2007). The preservation of the religious family law system in India had created biasness when the minorities are concerned. Failure to implement a uniform civil code has reinforced differences between Hindus and Muslims and left women more vulnerable as a result. Instead of moving toward a secular, equality-based legal system, the recognition of personal laws under the guise of protecting minorities from a dominant majority culture helped institutionalize patriarchal traditional practices that disadvantage Indian women (Sayeed, 2000).

India is a country of diversity. There are many arguments that project all the religious and customary laws as anti-women having the patriarchal norms, but women's rights are also constrained through several economic, social, and political underpinnings. Personal laws whether based on Hindu, Muslim, Christian, and Parsi traditions, custom or religion were drafted by male hierarchies and did not involve the participation of women. Religion is seen to control affairs of the home and family where the oppression of women remains the same. The gender stereotypes pervade all aspects of human existence making women at the tip of risk in terms of personal laws.

Hindu Personal Law: The Historic evolution of Hindu personal law is rooted in the Shastri law as it is not derived from a single scripture rather it is the amalgamation of different religious texts (Deshpande & Seth, 2009). The marriages in ancient Hindu tradition were regarded as a sacramental tie followed by sacred rituals between two people and their families that cannot be broken even after death. Before the enactment of Hindu Marriage Act, 1955 polygamy was quite frequent within the Hindu community but after the Hindu Marriage Act polygamy is considered as void. As for marriage, the early Hindu law and later Hindu personal law has one similarity i.e., inter caste and inter religious marriage is prohibited. A marriage involving a person converted to other religions i.e., Muslim, Christian, and Parsi is not considered as a marriage according to Hindu Marriage Act, 1955, which is stated in Section 2(3) (Hindu Marriage Act, 1955). The Hindu Marriage Act prohibits marriage between two persons within the degrees of relationship, for example, a Hindu cannot marry his own brother's or sister's daughter, but in South India there is a custom to marry one's sister's daughter while in North it is prohibited (Chaudhary, 2017). The Hindu Marriage Act, 1955 does not render a marriage void or voidable in the event that the boy has not completed the age of twenty-one years or the girl has not completed the age of eighteen years. These kinds of marriages that are not registered have created a large number of abandoned spouses in India. As marriage is considered as sacramental where there was no room for discussion of divorce according to Hindu law, but the Hindu marriage Act, 1955 recognized divorce. But these rights created new kind of complications in the society because of the husbands converting to Islam and getting married for second time. The case of Sarla Mudgal v. Union of India, where a Hindu husband solemnized for the second marriage by embracing Islam due to its polygamy system of marriage. The other issue that curtailed the dignity of woman is the restitution of conjugal rights under Section 9 of the Hindu Marriage Act that violates an individual's right to privacy.

The idea of inheritance rights was basically associated with Hindu laws, in ancient period. In this context there existed two schools of thought in Hindu tradition i.e., Mitakshara and Dayabhaga. The property of a Hindu male may devolve jointly through survivorship, upon four generations of male heirs by birth in Mitakshara law system and in Dayabhaga School neither accords a right by birth, nor by survivorship, through a joint male ownership of coparcenary and joint property. As Hindu law is gender biased in nature because it undermines the property rights of female. Hindu Succession Act (HSA), 1956 was enacted on the basis of equality of inheritance rights for both daughters and sons. However, it considerably does not guarantee full rights in the joint family property, in which sons have coparcenary rights by birth but daughters have no such rights. This is because they are not considered as a part of coparcenary (Parashar, 2005). But there were subsequent changes where Section 14 of the Hindu Succession Act, 1956 provided female with absolute property and in 2005 reformed Hindu Succession Act to establish egalitarianism, by making daughters as a coparcener in the joint family. It sets women out to be equal coparceners and the inherent discriminatory Mitakshara system of exclusive membership in joint family formally eliminated. However, the legislation, at the same time leaves women with lesser rights, than men in terms of inheritance rights (Patel, 2007; Saxena, 2008).

Muslim Personal Law: Islam provides guidance in every segment of human life and formulates laws for its followers. A part of these laws are those pertaining to family life and in which the rights and responsibilities of the members of a family have been fixed. The primary source of Islamic law is the Quran, which Muslims believe to be God's words. Under the Muslim law the term marriage is given definite meaning. Marriage (nikah) among Muslims is a solemn pact between a man and a woman soliciting each other's life companionship. Marriage is an institution ordained for the protection of the society and in order that human beings may guard themselves from foulness and unchastity (Mahmood, 1965). Under the Muslim law, a man is permitted to have four wives at the same time. If he marries a fifth wife when he has already four, the fifth marriage would be merely irregular. However, a marriage with a woman, who has her husband alive and who has not been divorced by him, is void. And when woman does the same as men then she is liable for bigamy under section 494 of the Indian Penal code and the offspring will be considered as the illegitimate (Varghese, 2015).

Muslim law permits an oral marriage where the registration of marriage is not obligatory or voluntary. In the case of divorce or talaq where the husband has the sole authority to get divorce by just pronouncing triple talaq whereas the woman is prohibited. The husband has more power than the women when it comes for dissolution of marriage in Muslim community. As just pronouncing triple talaq is enough to get divorce from their wives by the Muslim men which is not the case with the Muslim women. A husband who is convinced that his marriage has irretrievably broken down can quietly pronounce a divorce as there is no complexities or inequities in this process but the same is not allowed for female (Chaturvedi, 2004). The Allahabad High Court has held that the practices of the triple talaq is unlawful and void (Chawla, 2006). Though efforts are being made to curve social evils like the triple talaq, there are many more cases coming up which is unique and discriminatory.

Like the Hindu's tradition of Stridhan, in Muslim law too they deliver women with certain Stridhan to which they call as mehr. The Quran, sanctions mehr to married women, which is a sum of money or other property given to a woman by her husband at the time of marriage, and she has full right over it. It is meant to protect the wife against the arbitrary exercise of the husband's power to divorce, as well as it also stipulated at the time of marriage, future security (Agnes, 2011). The Muslim social structure is basically rooted on male dominance, which is considered as the reason for diminishing women's access to property rights. Under Sunnis, if the husband is dead and his wife and children are alive, then the wife is entitled to $1/8^{th}$ shares of the property; and if the wife dies and husband is alive, then the husband gets $1/4^{th}$ of the shares of the property. In case the son and the daughter inherit the property together, the son gets twice than the daughter. The husband gets $1/4^{th}$ share and the wife gets $1/8^{th}$ share when there is a child; and when there is no child, they get $1/2^{th}$ and $1/4^{th}$ respectively (Muslim Personal Shariat Law Application Act, 1937). The above-mentioned property sanction of women is very vague in nature. According to Islamic law there are three types of marriages: valid, void, and irregular marriage. In this context the valid or Sahih marriage is commonly called as the only marriage where they get rights of inheritance. In the other two forms of marriages they are not allowed the right to inheritance (Chawla, 2013).

Maintenance is seen as women's inheritance rights that are compensated during the time of dissolution of marriage. The social structure in Muslim community is based on the roots of male dominance, which is the reason for women's diminishing rights toward access to property (Moors, 1995). Although, Islamic laws offer property rights as compared to other religious communities, yet they still have many issues related to property and other matters. Though uncodified, the State did make certain changes in Muslim law like the Shariat law 1937, Dissolution of Muslim Marriages Act, 1939, and the Muslim Women (Protection of Rights on Divorce) Act, 1986 in India, that delivered absolute property rights to Muslim women (Malik, 2009). Though the government tried to codify the Muslim law and to bring justice to the Muslim women, but it is denied by the term laid down by their norms, customs to they are used to and the stern followers (Fazalbhoy, 2012). One historic incident related to maintenance rights of women is the case of Md. Ahmed Khan vs. Shah Bano Begum, whose maintenance rights was curtailed by Muslim personal law. But she was entitled to get maintenance beyond the period of Iddat under Section 125 of the code of criminal procedure (Parashar, 1992).

This decision gave Muslim women great relief and led to the enactment of Muslim women Act, 1986. But this Act was severely criticized and challenged the Islamic law. The Muslim fundamentalists criticized and argued that it unreasonably interfered and questioned Quran. This religious turmoil led to the amendment of Muslim Women's (Protection of rights on Divorce) Act, 1986, which denied right of maintenance to Muslim women under section 125 Cr.P.C, which was doubtless a negative step toward justice and rights of women (Calman, 1992).

Christian Personal laws: The Christians are governed by the English Law. "The law of was governed by distinct colonial influence [(Anglo-Saxon jurisprudence and the Continental system introduced by the French and the Portuguese within their respected territories" (Agnes, 2011)]. In Christianity, the marriage is very sacramental which is instituted and ordained by God as a life-long relationship between husband and wife. The Christian doctrine also states that "What God hath joined together, no man shall put as under." With this the authority to dissolve a marriage was taken out of human control. Like all the other religion the Christianity is also based on the patriarchal structure, where the rights of women are very limited. Inter religious marriage is not given preference according to Indian Christian Marriage Act, 1872. The Section 4 of the Indian Christian Marriage Act, 1872 states that marriage maybe solemnized by any person who are Christians or who have received Episcopal ordination. But dissolution of marriage is out of question in Christian law. The Section 10 and 10A discussed about the grounds for divorce and mutual consent divorce according to Indian Divorce Act, 1869. However, Section 10 of the Divorce Act curtailed the rights of women to sue for divorce as the law has given more rights to men than women.

As far as Christian women are concerned, the community and the Church with its strong patriarchal tradition have compelled women to remain subjugated. Socialization starts at the early life of a girl to become submissive and not to assert their rights. Hence, Christian women are in general reluctant to assert or demand their rights (Agnes, 2000). The Bombay High Court in 1907 in the case *Francis Ghosal verses Gabri Ghosal on property issue, prevented daughters to inherit parental or joint family inherit property*. Consequently, the cases that come to court asking for share of the family property are also very few. Most of the Indian Christians are governed by their own customary law where there is injustice in terms of the property is concerned. Having covered by the Indian Succession Act, 1925, discriminatory personal law still exists. One of the most important examples was the case of Mary Roy, *Mary Roy v. State of Kerala, AIR 1986 SC1011* who showed the courage to challenge the discriminatory personal law of the Christians namely The Travancore Christian Succession Act, 1916, almost forty-seven years after the commencement of the Constitution. Travancore

Christian Act, was biased which permitted to daughters to claim only one fourth shares of the son (Travancore Christian Succession Act, 1916). But this Act soon got repealed and Indian Succession Act 1925 was amended which provides for equal distribution of the properties among the children if the father dies intestate; however, women silently sign release deeds at the time of marriage thereby relinquishing their rights in the family property for ever without showing any resistance (Mridul, 2005).

Parsi Personal Law: In the modern period, the most extensive and well-documented body of law pertaining to Zoroastrians is the Parsi personal law of India. Zoroastrian personal law was the creation of elite Bombay Parsis lived in British India during the last century of colonial rule. The Parsi marriage is also considered very holy. Parsi husband or wife cannot remarry in the lifetime of his wife or husband until his or her marriage is dissolved by a competent court although he or she may have become a convert to any other faith. It shows the strict monogamy in Parsi marriages. Every marriage contracted contrary to the provisions of section 4(1) of the Parsi Marriage and Divorce Act, 1936 shall be void. If a Parsi, in violation of that section, marries again in the lifetime of his or her wife or husband before the dissolution of earlier marriage by the court, he/she shall be punished under section 494 and 495 of the IPC. The position of women in Parsi community about property rights is similar to other religious women communities. Parsi women are discriminated against by the law which has no basis in the community's religious belief.

In comparison, the Parsi women's position in terms of the inheritance is much worse than the above-mentioned religions. In Parsi law, when a Parsi woman dies, her son is entitled to the property, not her daughter as she is also debarred for the entitlement of her father's property when he dies (Shabbir, 1958). By the passing of the succession Act, 1925, the Parsi intestate succession was also incorporated and the amendment was made in the Indian Succession Act in 1991 (51), which gave equal rights to both son and daughter in the parental estate (Malik, 2009). Similar situation applies to Parsi women and other women due to the so-called religious law or the personal laws.

4. Uniform Civil Code toward Gender Justice

Personal laws incessantly affect women's lives and her rights in all communities, and this law also contradicts with the Constitution of India. Constitution of India grants equality to women but also empowers the State to adopt measures of positive discrimination in favor of women for neutralizing the cumulative socio-economic, educational, and political disadvantages faced by them. Fundamental rights, among others, ensure equality, equal protection, prohibit discrimination before law against any citizen on grounds of religion, race, caste, sex or place of birth, and guarantee equality of opportunity to all citizens in matters relating to employment. Articles 14, 15, 15(3), 16, 39(a), 39(b), 39(c) and 42 of the Constitution are of specific importance in this regard (Yunus & Varma, 2015). All these provisions directly and indirectly challenge the validity of Personal laws, which is one of the most important factors that hampers the life of women. Though various efforts are being made by the means of reforms of national laws and changing judicial trends, women are not treated equally and are discriminated against in the field of family law, particularly in the matters of marriage, divorce, maintenance, and inheritance. In these circumstances, a gender-just code is required to deliver justice and equality to women. So, the significance of Uniform Civil Code (UCC) is very important for the protection of oppressed women, to protect their rights, to eliminate prejudices against religion, community etc.

Uniform Civil Code is a set of laws that are homogeneous in nature in the arena of personal law. The Part IV, Article 44 of Indian Constitution, directs the State to provide a Uniform Civil Code throughout the territory of India. The Uniform Civil Code is a systematic code pertaining to private rights and remedies of a citizen. The common area where this code is generally covered is the personal status and rights of the citizens. It is an attempt to determine in advance, what legal exigencies will arise and to furnish the means for meeting them. The basic legal principles are worked out into very systematic details and great attention is given to its consistency. But the very nature of Uniform Civil Code always surrounds with debate and discussions.

The issue of UCC and its related debate has always been one of the most talked and discussed chapter since the colonial era till today. The spine of controversy revolving around Uniform Civil Code has been secularism and the freedom of religion enumerated in the Constitution of India. This issue raised doubts in the minds of the minority communities that resulted in rejection of formation of UCC. Their opposition to the UCC is articulated on the ground that it threatens the sanctity of their communities' religious law and therefore it directly threatens their religious identities. They perceive UCC as Hindu law's version that will be applicable to every religious community in India. In the other side of the picture, most of them argue that UCC will ultimately seize the privileges of the minority men and that will override with the version of the Hindu law hegemony (Shayeed, 2000). This idea of secularism has often been under attack from Hindu communalists.

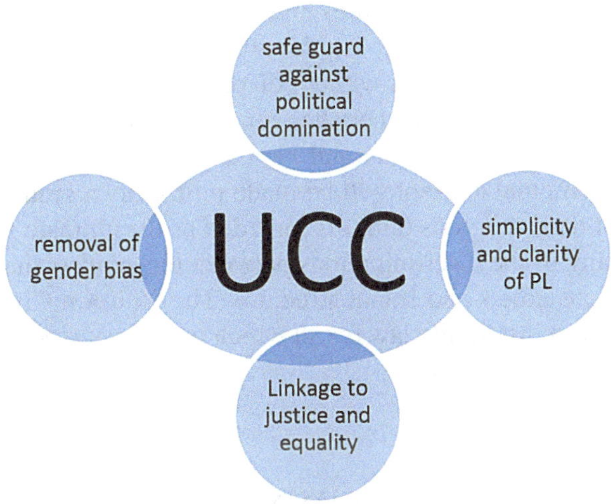

Figure 3.1. Nature of UCC

They have most severely attacked the State under the Congress government for appeasing minorities, especially Muslims, to generate a vote bank in the name of secularism. Through this debate the main agenda toward the formation of UCC was not discussed at all. The main issues that UCC works for is the uniformity of laws between different religious communities that ensures equalities between the rights of men and women.

The Uniform Civil Code has much relevance and required to obtain the true idea of equality and justice. The reason for its necessity is that UCC can serve as a strong instrument toward curbing the virus of communalism and gender injustice in our country. UCC is one of the finest manifestations of the civilized behavior. It will not give monopoly to a single religious community and rather it will monopolize ethical values. One of the most important features of the Uniform Civil Code is that it can restore the idea of women empowerment which India always talks about. Through UCC, the personal laws will take a new form which will be simple, clear, and uniform for all citizens of India (Gautam, 2017). In terms of succession and inheritance it will directly curb the system of Hindu Undivided Family property. It provides equal shares to sons and daughters from the property of the father whether self-acquired or joint family property. There will be no gender discrimination and gender inequality in terms of inheritance. Enactment of a Uniform Civil Code would impinge upon Muslim rights to polygamy. In almost all recent cases where the need for a Uniform Civil Code has been emphasized, women were at the receiving end of torture in the garb of religious immunity. Apart from the famous Shah Bano (1986) and Sarla Mudgal (1995) cases, there have been several other pleas by Hindu wives whose husbands converted to Islam only to get married again without divorcing the first wife.

The Uniform Civil Code cherishes constitutional goal by making single ground for divorce both for man and woman. The breakdown of marriage and divorce by mutual consent will be made uniformly a ground for divorce irrespective of their religious faithIt is regarded as the ultimate solution to do away with many of the discriminations between the genders that have swept through religious edicts and social structure. The Uniform Civil Code is an attempt to create a just family law for all which will protect the rights of every citizen.

5. International Initiatives for Gender Justice and Equality

Women have been trying for a very long time to gain emancipation through rights and opportunities in a male-dominated society. The concept of women's rights that has developed in a specific region reflects the socio-cultural, political, and religious dynamics of that region. The inclusion of the understanding of "gender equality" first referred to as "gender mainstreaming" at the 1985 Third World Conference on Women in Nairobi, Kenya and formally featured in 1995 at the Fourth World Conference on Women in Beijing in order to provide systematic solutions for the problems of women, has since been well established in international legislation (Ankara, 2011). Strategies and issues to curb gender injustice have become one of the main focal points in international agendas and are helping women fight for their right.

- **Convention on the Elimination of All Forms of Discrimination against Women:** As there are many articles that have been giving many contexts of violence against women but Articles 2, 4, 5 and 16 are most important when dealing with the status of women in society and how their rights are being submerged and their elaboration. These Articles condemn discrimination against women in all forms from socio-religious, cultural and economic, family affairs, and imposes obligation toward State to take suitable measures to stop discriminations against women. **Article 16** deals with the equal treatment toward women in almost every matter related to marriage and family relations. The most important provisions of this Article are the equal right of women to enter into marriage; the same right freely to choose a spouse and to enter into marriage only with their free and full consent; the same rights and responsibilities during marriage and at its dissolution; the same rights to decide freely and responsibly on the number and spacing of their children and all matters related to their children (Cusack & Pusey, 2013).

- **International Covenant on Economic, Social and Cultural Rights (ICESCR):** International Covenant on Economic, Social and Cultural Rights is a multilateral treaty that helps issues focusing on economic, social and cultural rights (ESCR) (Howard, 2010). Article 10 especially talks about family issues where the protection and assistance should be accorded to the family. In terms of marriage there has to be free consent between both the spouses. Here there is a special preference given to the mothers and children without any discrimination (Donald & Howard, 2015). Article 10 consists of the most important aspects of human rights i.e., the freedom of expression. Freedom of expression manifests one's religion or belief, but not an absolute right. Article 10(2) of ICESCR states that freedom of expression carries with its duties and responsibilities and allows formalities, conditions, restrictions, or penalties.

6. Conclusion

The evaluation of the status of women in India has been a continuous process of ups and downs throughout the history. The main reasons for the disbalance in the position and the status of women are the religious law or the personal laws that have guided our social and cultural customs. The Hindu (including Buddhist, Jains, and Sikhs), Islam, Christianity, and Parsi, they all have their own law that governs their life and their society at large. The areas to where the personal law operates are inheritance or property rights, divorce, marriage, adoption, and guardianship etc. The law is relating to marriage, divorce, maintenance, guardianship, and succession governing the Hindus, Muslims, and Christians etc., is different and varies from one religion to other. As a matter of fact, in all religions, personal laws are not favorable for the women rights. Until the codification of the Hindu law in 1955 and 1956, Hindu women did not enjoy equal rights. The women were deprived of the inherited property of the Hindu undivided family. Only men could turn out to be coparceners and property transfers to them by survivorship. Women are completely left out from inheritance. In many cases when women and man relation collapses women also lose the rights of property. The Indian Constitution has substantially given a framework to ensure equality toward its citizens as the Article 14 has delivered with the rights to equality and the protection. This view is strengthened by Article 15 of the Constitution, which goes on to specifically lay down prohibition of discrimination on any arbitrary ground, including the ground of sex, as also the parameters of affirmative action and positive discrimination. But there are violations of these laws as we can see according to different personal laws in India. To eradicate the inequalities associated with the personal laws is to have a unified Uniform Civil Code that envisages not just about secularism and communal politics, but more about securing justice and equality to women.

References

Agnes, F. (2000). Church, State and Secular Spaces. *Economic and Political Weekly, 35*(3), 2901-2902.

Agnes, F. (2011). *Family Law 1: Family law and Constitutional Law.* New Delhi: Oxford University Press.

Agnes, F. (2011). *Marriage, Divorce and Matrimonial Litigation Vol 2.* New Delhi: Oxford University Press.

Ankara (2011). *The Inclusion Of Gender Equality in The Main Plans and Policies in The European Union: The Analysis of Holland, Romania and Turkey.* Directorate General on the Status of Women.

Bare Acts (2010). *The Christina Marriage Act, 1872.* New Delhi: Universal Law Publishing.

Bare Acts (2010). *The Hindu Marriage Act, 1955.* New Delhi: Universal Law Publishing.

Bare Acts (2010). *The Divorce Act, 1869.* New Delhi: Universal Law Publishing.

Bare Acts (2010). *The Indian Succession Act, 1925.* New Delhi: Universal Law Publishing.

Bare Acts (2010). *The Indian Succession Act, 2005.* New Delhi: Universal Law Publishing.

Bare Acts (2010). *The Muslim Personal Law (Shariat) Application Act, 1937.* New Delhi: Universal Law Publishing.

Bare Acts (2010). *The Muslim Women (Protection of Rights on Divorce) Act, 1986.* New Delhi: Universal Law Publishing.

Bare Acts (2010). The *Special Marriage Act, 1954.* New Delhi: Universal Law Publishing.

Convention on the Elimination of all forms of Discrimination against Women (1979). Retrieved from www.un.org>daw>cedaw. Accessed on 19th August 2017.

Beauvor, S. (1983). *The Second Sex.* New York: Alfred Knopf.

Calman, L. (1992). *Towards Empowerment:Women and Movement Politics in India.* Oxford: Westview Press.

Chaturvedi, A. (2004). *Muslim Women and Law.* New Delhi: Commonwaealth Publishers.

Chawla, M. (2013). *Gender Justice;Women and Law In India.* New Delhi: Deep and Deep Publication.

Chaudhary, Renuka (2017). Hindu personal law and conflict in legitimacy: A social perspective, International Journal of advanced research and development, 2(6), 862–69.

Cusack, S., & Pusey, L. (2013). CEDAW and the Rights of Equality and Non discrimination. *Melbourne Journal of international Law, 14* .

Deshpande, S. & Seth, S. (2009). Role And Position Of Women Empowerment In India Society. *Journal Of International Referred Research, 1*(17), 24–27.

Donald, A. & Howard, E. (2015). *The Right to Freedom of Religion or Belief and its intersection with other Rights.* London: Middlesex University.

Fazalbhoy, N. (2012). Negotiating Rights and Relationship: Muslim Women and Inheritance. In F. A. Ghosh (Ed.), *Negotiating Sapces:Legal Domains, Gender Concerns and Community Constructs* . New Delhi: Oxford University Press.

Frankenberry, N. (1992). Classical Theism, Panentheism and Pantheism: The Relation of God Construction and Gender Construction, *Zygon Journal of Religion and Science, 28(*1), 29–46.

Gautam, V. (2017). the politics of muslim identity and the personal laws: safeguarding the rights of the iternal minorities. *IOSR Journal of Humanities and Social Science 22*(5), 11–19.

Ghosh, P. S. (2007). *The Politics of Personal Law in South Asia:identity,nationalism and the Unifrom Civil Code.* New Delhi: Routledge.

Gross, R. (1992). *Feminism and Religion,* Boston: Beacon Press.

Howard, E. (2010). *The EU race directive developing the protection against racial discrimination within EU.* New York: Routledge.

Malik, K. (2009). *Women and the law.* Haryana: Allahabad law Agency.

Mahmood, T. (1965). Custom as a source of Law in Islam. *Journal of Indian Law Institute*, Vol 7, 102.

Moors, A. (1995). *Women, Property and Islam: Palestinian Expriences, 1920-1990.* Cambridge: Cambridge University Press.

Mridul, E. (2005). Looking beyond gender parity, Gender inequalities of some dimensions of well being in Kerala. *Economic and Political Weekly, 40*(30) .

Parashar, A. (1992). *Women And Family Law Reform In India: Unifrom Civil Code And Gender Equality* . New Delhi: Sage.

Parashar, A. (2005). Just Family Law: Basic to all Indian Women. In Jaising, Indira (Eds.), Men's Laws Women's Lives: A Constitutional Perspective on Religion, Common Law and Culture in South Asia (pp. 286–322). New Delhi: Women Unlimited as associate of Kali for Women.

Patel, R. (2007). *Hindu Women's Property rights in Rural India: Law, Labour and Culture in Action.* England: Ashgate Publishing House.

Saxena, P.P. (2008). Succession Laws and Gender Justice. In Parashar,A. & Dhanda, A. (Eds.), *Redefining Family Law in India*, (282-305). New Delhi: Routledge.

Sayeed, B. (2000, July). Need for Codification. *The Hindu* .Retrieved from https://www.thehindu.com/thehindu/2000/07/25/stories/13250642.htm on 28/09/2017

Shabbir, R. M. (1958). *Parsi Law in India (5th Edition).* Allahabad: Law Book Co.

Varghese, A. E. (2015). Personal Laws In India: The Activisms Of Musslim Women's Organization. London, London, UK: Trinity College Of Arts And Sciences.

Wadje, A. (2013). Judicial Review Of Personal Laws Vis A Vis Constitutional Validity Of Peresonal Laws. *South Asian Journal Multidisciplinary Studies*, 99-111.

Yunus, S. & Varma, S. (2015). Legal Provision for Women Empowerment in India. *International Journal of Humanities and Management Sciences 3*(5), 367–370.

Chapter 4

Behavioral View on Inequality Issue: Bibliometric Analysis

Tetyana Vasilyeva, Anna Buriak, Yana Kryvych, Anna Lasukova
Sumy State University, Ukraine

ABSTRACT
Last decades' progress in behavioral economics and finance underpin new approaches to the problem-solving of global issues like poverty and reducing inequality. This study uses the concept of behavioral economics as basis for future policy initiatives to promote equal opportunities in society. It is agreed that nowadays policy of reducing inequality should extend beyond standard manifestations of inequality like income, gender, social status, and focus on processes and determinants that generate inequality in society. As the most widespread research view is about the consequences and causes of inequality on macroeconomic levels, this chapter encompasses conceptual framework for understanding behavior and decision-making of economic units under economic and social inequality in the country. Authors of the study analyze the latest streams in behavioral economics research and its applying opportunities for inequality issue using bibliographic mapping via VOSviewer. This paper presents results of thorough bibliometric and network analysis of almost 900 papers on the topic of inequality and behavior. As a result, the main areas and directions of the research were identified, which include publishing trends, influential authors, and affiliations.
Keywords: behavior, inequality, decision-making, bibliometric analysis, network analysis. JEL Classification: D14, D63, D91, E71.

1. Introduction

The issue of rising inequality has become one of the most challenging concern among 21st century development goals (Cingano, 2014, Dave, 2017, Kuzmenko and Roienko, 2017; Bhandari 2020; Bhandari and Shvindina 2019). Negative effects and outcomes of inequality underpin policymakers and analysts to provide and implement effective policy tools and measures to overcome this issue. Search of new policy tool kit and nonstandard frameworks for approaching inequality gaps is highly needed nowadays (Singh, 2018, Kolomiiets and Petrushenko, 2017, Balas and Kaya, 2019, Bilan et al., 2019, Vasilyeva et al., 2019, 2020). In the paper on global inequality, research authors point out the importance of multicausal and multidimensional approach, usage of qualitative case studies, incorporation of structural explanations, and actor-based studies (Christiansen and Jensen, 2019, Kyrychenko et al., 2018). Traditionally addressing the issues of poverty and inequality is based on conventional theories and models with assumption of rational economic behavior. At the same time, modern research and econometric models try to catch behavioral biases of economic agents, their preferences, incentives, and decision-making processes in general (Greco, 2018, Nur-Al-Ahad, 2019). As a result, last decades' advances and insights of behavioral economics started contributing in policy making on reducing poverty and inequality, trying to catch irrational features of human mind, and decision-making (Datta and Mullainathan, 2013, Hadbaa and Boutti, 2019; Njegovanovic, 2018). Understanding poverty issues through behavioral economics is given by the study of scarce mental resources—attention, self-control—and their influence on income generation and asset allocation problem (Bernheim et al., 2012; Banerjee and Mullainathan, 2008), behavioral patterns and economic conditions of poor etc. (Bertrand et al., 2004).

Conceptually, application of behavioral economics to the study of inequality can be regarded in the context of:

1. influence of decision-making errors and biases like emotions, social influence, incentives, and motivations on financial behavior and inequality ("behavior biases/errors and inequality"). Traditional drivers of global and cross-country inequality usually include technological change and existence of skill premium on the labor market, trade globalization and openness, financial globalization government policy, and education (Dabla-Norris et al., 2015, Kouassi, 2018). Modern trends show that increasing gaps between rich and poor have been exacerbated by a divergence in the behavior of the two groups. Inequality issues are

the byproduct of behavior and social-cultural norms (including policies, laws, and institutions). For example, according to Lusardi and others financial knowledge is found to be one of the determinants of wealth inequality pointing that 30–40 percent of retirement wealth inequality is accounted for by financial knowledge (Lusardi et al., 2017). Center for Economic Behavior and Inequality (CEBI) focuses on the role of behavior in generating inequality exploiting Danish data research infrastructure and conducting research, surveys, and experiments.

2. inequality impact on individual's outcomes and making decisions ("inequality and individual decision-making"). Effects of rising inequality on individual decision making could be a valuable mechanism of inequality understanding today (Moss, 2013). The paper on inequality and decision making by David Moss and others summarizes numerous empirical studies on influence of inequality on economic growth, population health, political outcomes, and financial situations (like financial fragility and crisis). Besides, authors focus on mechanisms and causal linkages by which income inequality effects human behavior and decision making—like household borrowing, risk taking, self-perception, consumption and saving, social capital and trust (Buriak, 2019). Robert Frank argues that concentrations of income and wealth provoke higher spending ("cascades"), and experimental study by Payene et al. proves that income inequality stimulates risky behavior driven by upward social comparisons (Frank, 2013; Payne, Brown-Iannuzzi, & Hannay, 2017).

3. household behavior and decision-making like savings, spending, borrowing under inequality conditions ("household finance and inequality"). Exploring financial vulnerabilities and behavior patterns of low-income households are intensively addressed in academic literature, like differences in saving behavior across various groups (Poliakh and Nuriddin, 2017, Morscher, C., Horsch, A., Stephan, J., (017). For example, study analyzes the debt profile of low-income households before and after the Great Recession (Kim, 2017).

Number of literatures addresses specific behavioral aspects of the topic inequality providing additional insights into the field, but thorough review of the papers on behavioral perspectives of inequality is lacking today. The current study provides bibliometric and network analysis for detecting existed and emerging topical areas, identifying the clusters of research and the most influential researchers. This paper presents a comprehensive literature review of almost 900 papers on behavioral view on the topic of inequality issue. Applying rigorous bibliometric tools—VOSviewer—to identify clusters of research and their publishing trends will be completed. The remainder of the paper deals with presentation of the research methodology and further implementation—results of bibliometric and network analysis.

2. Research Methodology

Identification of the search keywords

According to the thorough analysis of academic papers on application of behavioral economics to the study of inequality, data collection is based on the following key words and their combinations: "inequality," "behavior," "behavioral biases," "decision-making," and "household finance." The most common are "inequality" and "behavior," when other terms are used to catch specific behavioral areas of inequality research. This research approach for literature review is similar to the work by Fahimnia (2015), Guo (2019) and Shvindina (2019), and Pomianek (2018). The Scopus was chosen for search source as the most widely used data base.

Initial search results and their refinement

Collection of the studies was made using the search field of "title-abstract-keywords". Table 4.1 demonstrates main results of the search, pointing that inclusion-specific terms of inequality (like type of inequality) or behavior issue (like applied fields of behavioral economics) narrows search base for identification of established and emerging trends.

Table 4.1 Search outcomes (SCOPUS Database)

Search keywords	No. of Papers
Behavior AND inequality	15318
Behavior AND income inequality	579
Behavior AND economic inequality	452
Behavior AND wealth inequality	72
Decision making AND inequality	3087
Household finance AND inequality	48
Behavior economics AND inequality	48
Behavior finance AND inequality	5

A combination of the keywords "behavior AND inequality" was chosen as basic for further rigorous analysis to reveal all possible directions of behavioral application. A massive database of 15,318 papers requires detailed refinements as it consists of papers from almost all fields of research and dates back to 1928—paper in the Journal of the Textile Institute Transactions on textile fabrics, which is out of our research review.

The first refinement was made in fields of search results as papers from Medicine, Mathematics, Engineering, Computer Science, Physics and Astronomy, Arts and Humanities, Agricultural and Biological Sciences, Biochemistry, Genetics and Molecular Biology, Nursing, and Earth and Planetary Science were excluded. As a result, 2902 papers were concentrated in the fields of Social Sciences (1973), Economics, Econometrics and Finance (712), Psychology (474), Business, Management and Accounting (333), Multidisciplinary (112), and Decision Sciences (42). Second refinement was in type of documents (only articles and books were left excluding conference papers, reviews, etc.) and language (only English publications were reviewed). At this stage, search results comprised 2223 papers. After that period, analysis was limited to 2001–2019, as before 2001 papers were rare (only several publications per year)—1986 papers were retained. Third refinement was about excluding unrelated and insignificant keywords in papers—for example, names of the countries, exploited method or general words like "article," "interview," "statistics." Hence, the 899 publications were investigated further and saved in RIS format including essential data on abstract, title, author(s), keywords, and references.

3. Main results

3.1. Descriptive Data Statistics

As it can be seen from Figure 4.1, intensive expansion period has started from 2000s, though, 2001–2019 was chosen as the research period with increasing publishing trend. At this period, ideas of behavioral economics and inequality have been actively supported and developed—it can be proved by awards with the Nobel prize in economics: 1) Daniel Kahneman "for having integrated insights from psychological research into economic science, especially concerning human judgment and *decision-making* under uncertainty" (2002); 2) Angus Deaton "for his analysis of consumption, *poverty, and welfare*" (2015); 3) Richard H. Thaler "for his contributions

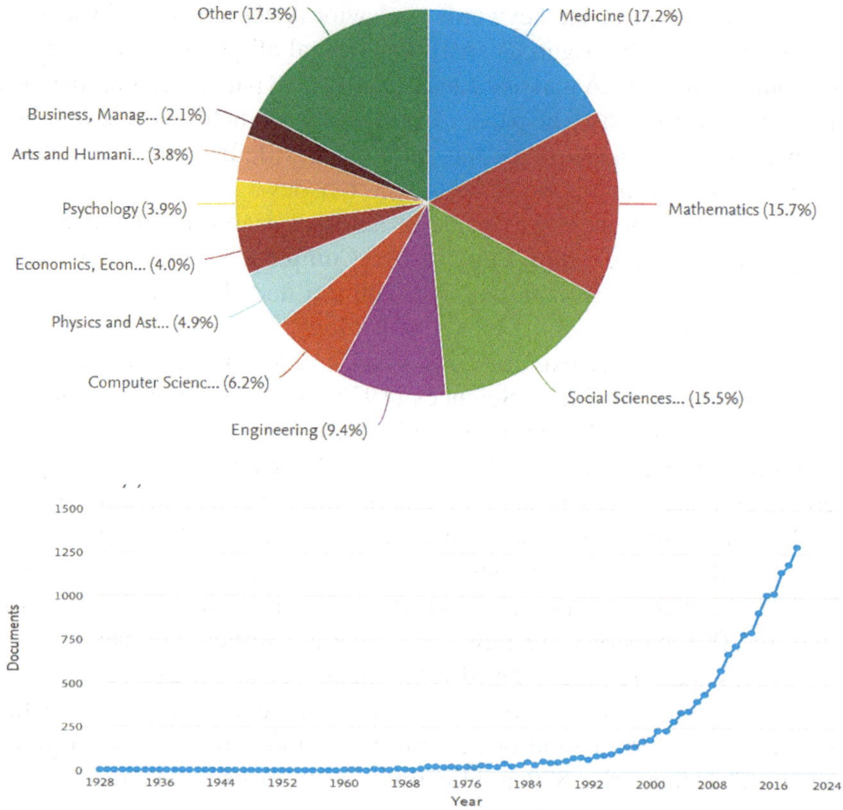

Figure 4.1. Search results in the field "behavior AND inequality" classified by the field and years

to *behavioral economics*" (2017); 4) Abhijit Banerjee, Esther Duflo, and Michael Kremer "for their experimental approach to alleviating global *poverty*" (2019).

Descriptive analysis points that 155 journals comprise research results of 899 articles. Top 10 journals contributed to the topic of behavior and inequality represents only 10% of all published papers, pointing that topic has multidisciplinary character—in the fields of Economic Behavior and Organization, Psychology, Social Science Research, Sociology, Public Economics, Demographic Research (Figure 4.2). The most prolific subject areas of published 899 papers are Social Science (49,6%), Economics, Econometrics and Finance (23,6%), Business, Management and Accounting (14,2%), Psychology (9,5%), and others (less than 2%).

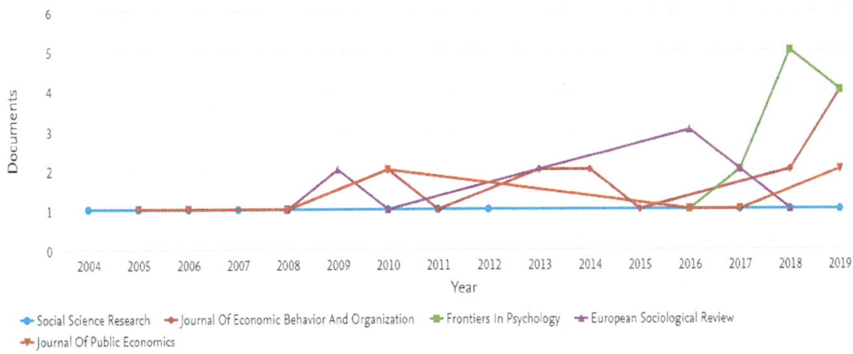

Figure 4.2. Publishing trend by the most popular journals with papers up to 10 sources per year

4. Bibliometric and Network Analysis

VOSviewer software package, developed in Leiden University, the Netherlands, (Van Eck and Waltman, 2010, 2014) was chosen to provide further bibliometric data analysis and visualization. As initial output data file (from Scopus dataset) contains all essential information for analysis—title, authors, journals, keywords, country, affiliation, references—several types of analyses were conducted.

Author and affiliations influence

Initial analysis of authors' contribution to the topic of behavior and inequality by the help of Scopus database and VOSviewer (bibliographic coupling analysis by authors) has revealed that maximum number of documents of an author was three. It proves interdisciplinary character of the issue and diversified contribution of authors from major fields of research. Among 1798 authors, 18 of them have 3 papers on the topic. Table 4.2 outlines the top five influential authors depend on the total link strength to show the author influence.

Among the most contributing (depend on the published papers) research institutions on the analyzed issue are: London School of Economics and Political Science (12 articles), University of Oxford (11), Northwestern University (10), Cornell University (10), Harvard University (9), KU Leuven (9), Universiteit van Amsterdam (8), CNRS Centre National de la Recherche Scientifique (8), University of Toronto (8), and Princeton University (7).

Table 4.2 The top five influential authors with three published articles (VOSviewer)

Author	Total link strength (TLS)	Citations
Edenhofer, O.	196	6
Rashotte, L.S.	160	60
Gintis, H.	40	105
Masclet, D.	39	27
Sefton, M.	26	71

If we look geographically on the most contributing countries and regions (bibliographic coupling analysis by countries), VOSviewer revealed 3 main centers (clusters):

1. United Kingdom (TLS 6997, 142 articles), Australia (TLS 2073, 32 articles), Germany (TLS 4865, 66 articles) and France (TLS 2234, 38 articles);
2. United States (TLS 10534, 345 articles), Canada (TLS 2859, 47 articles) and China (TLS 1264, 33 articles);
3. Spain (TLS 2961, 35 articles), Italy (TLS 2800, 35 articles) and Netherlands (TLS 1737, 28 articles).

Keyword statistics

Analysis of the most widespread keywords in the papers enables identification of the main research interests including current trends as well as perspective. Co-occurrence analysis of the 2675 keywords drawn from the 899 papers via VOSviewer program (frequency of co-occurrence more than 5) revealed the most popular terms using the abstracts of the papers. It has excluded phrases and words similar in meaning but in different forms—for example, gender gap and gender inequality, consumption and consumption behavior, social class and socioeconomic status, or general words like experiment, economics. Among the 10 most frequently used keywords are inequality (102 occurrence), gender (27), income inequality (24), education (21), poverty (21), consumption behavior (14), social inequality (13), voting behavior (10), social preferences (10), and culture (9). It shows from one side, the most popular types of inequality (income, gender, education etc.), and from the other side—types or aspects of behavior (consumption, political,

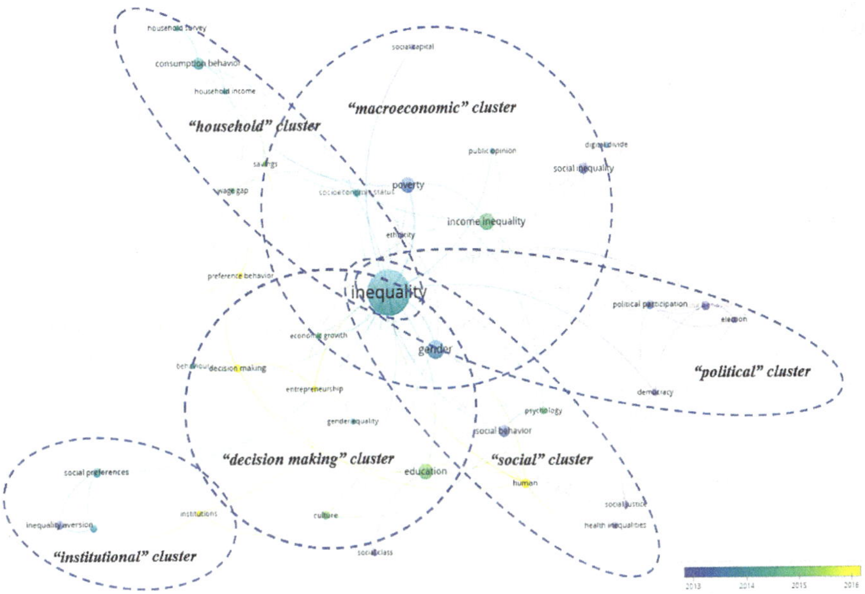

Figure 4.3. Overlay visualization of the most frequent keywords for the 889 papers with frequency more than 5 (score of the item is year of publication)

or social). Additional network analysis of the publications' keywords was realized with the unique technique of VOSviewer.

Figure 4.3 demonstrates visualization of the accomplished cluster analysis and shows interconnections between main terms of the research. Initially network consisted of 56 keywords, after refinements—37 words with further transformation in 6 clusters. Performed cluster analysis presents main directions of the research on behavior and inequality topic. The first cluster "household" (9 items—keywords: consumption behavior, household income, savings, socioeconomic status, wage, gender, and others) presents household finance and consumption network, where main research object is households and their behavior; among main inequality's determinants are socioeconomic status, wage, and gender of household. The second cluster "decision making" (7 items—keywords: culture, education, decision making, entrepreneurship, and others) is mainly about decision making research.

The third one (6 items like voting, election, and etc.) focuses on political behavior under inequality. The fourth cluster—"macroeconomic"—contains 6 items (economic growth, income inequality, poverty, public opinion ...), the fifth cluster—"social" characterized by 5 items: psychology, social justice, health inequality, and human and social behavior, the last cluster is "institutional," where 4 main items cover institutions, social preferences, experimental, and aversion. In the figure 4.3., the overlay visualization is shown through colored items of yearly of the publication. It allows to identify current trends in the academic literature. Analysis of the keyword's chronology (see please, figure 4.3) points that starting point in the research on inequality and behavior was about macroeconomic dimensions like poverty, democracy, political issues, then focus had shifted to the household finance and inequality influence on their financial behavior. Recent tendencies in the publication activity show interest in inequality impact on individual level—decision-making and main factors of this process.

5. Citation Analysis

One of the useful bibliometric analysis directions is citation, one to identify the most cited (as a result the most influential) authors or publications in the study of inequality and behavior. Citation analysis performed through VOSviewer, where minimum number of documents of an author is 3 and minimum number of citations of an author is 10, enables to identify the most cited authors in this field (Table 4.3).

Articles of the most influenced authors in the research of inequality and behavior are diversified, some focus on digital inequality depending on physical activity and disability (Dobransky & Hargittai, 2016) or socioeconomic background and available access points (Hargittai, 2019). Pampel Fred C. contributes substantially to the issue of disparities in health behaviors, and income inequality (Pampel et al., 2010), as well as the patterns of smoking influence on national income of 145 nations (Pampel, 2007). Gintis deals with political behavior and wealth dispersion impact on voter behavior (Gintis, 2016), Keister—with household well-being and consumption impact on behavior (Keister et al., 2016).

6. Co-citation Analysis

Additional type of literature review is co-citation analysis to fix and predict future directions of research. It can be made based on co-occurrence evaluation when two papers (authors) are both cited by the 3rd paper (author) and, as a result, point on the same research area. Using VOSviewer, 156

Table 4.3 The top five authors: citation measure (VOSviewer)

Author	Affiliation	Citations	Main papers
Hargittai, Eszter	University of Zurich, Zurich, Switzerland	428	• Unrealized potential: Exploring the digital disability divide (Dobransky & Hargittai, 2016) • From internet access to internet skills: digital inequality among older adults (Hargittai, 2019) • Connected and concerned: Variation in parents' online safety concerns (Boyd & Hargittai, 2013)
Pampel, Fred C.	University of Colorado Boulder, Boulder, United States	109	• Socioeconomic disparities in health behaviors (Pampel et al., 2010) • How Institutions and Attitudes Shape Tax Compliance: A Cross-National Experiment and Survey (Pampel et al., 2019) • National income, inequality and global patterns of cigarette use (Pampel, 2007)
Gintis, Herbert M.	Santa Fe Institute, Santa Fe, United States	105	• Homo Ludens: Social rationality and political behavior (Gintis, 2016) • Behavioral motives for income redistribution (Fong et al., 2005)
Keister, Lisa A.	Duke University, Durham, United States	79	• Lifestyles through expenditures: A case-based approach to saving (Keister et al., 2016) • Religion and wealth across generations (Keister, 2012) • Conservative protestants and wealth: How religion perpetuates asset poverty (Keister, 2008)
Van Laar, Colette	KU Leuven, 3000 Leuven, Belgium	73	• Perpetuating inequality: Junior women do not see queen bee behavior as negative but are nonetheless negatively affected by it (Sterk & Van Laar, 2018). • A social identity perspective on the social-class achievement gap: Academic and social adjustment in the transition to university (Veldman et al., 2019).

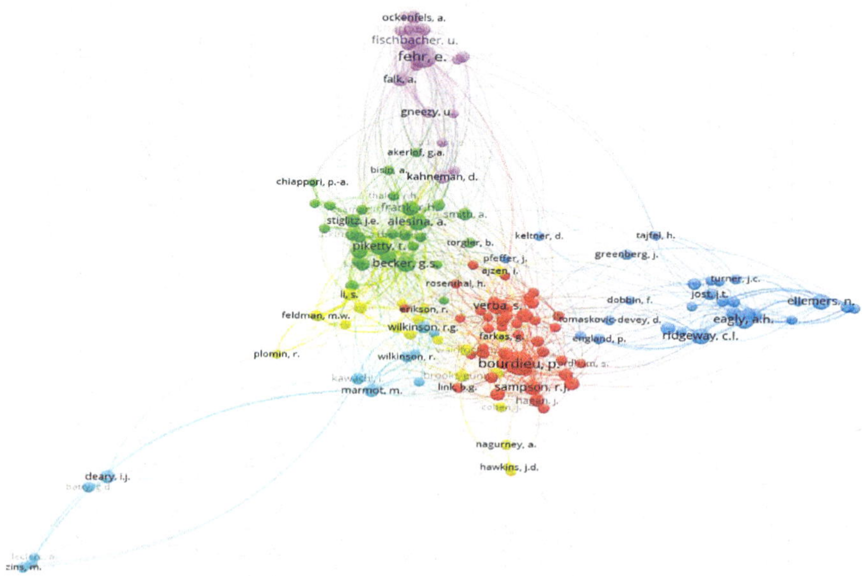

Figure 4.4. Co-citation network visualization of the most cited authors for the 889 papers with citations more than 25 (fractional counting)

nodes (fractional counting to be more accurate during analysis) of the papers produced co-citation network with 6 clusters of papers (authors) (Figure 4.4).

Cluster 1 (44 items) deals with political aspect of inequality and behavior, cluster 2 (29 items)—behavioral biases leading to inequality; 3rd cluster (28 items)—social side of inequality; 4th cluster (23 items)—determinants of inequality; 5th cluster (20 items)—neuroscience and decision-making; 6th cluster (12 items)—psychological side of inequality.

7. Conclusion

In this study, structured literature review of the academic papers is completed to fill the gap in the overview of behavioral aspects of the topic inequality. Since 2001, published papers started to grow substantially with increasing research interest and practical applications of behavioral economics. Applying VOSviewer as software package to provide bibliometric data analysis and visualization (using Scopus dataset) enables completion of the literature review by the authors, journals, keywords, country, and affiliation. Analysis based on the geographic criteria shows that United States of

America is the most contributing region in the research field as it accounts for 345 articles from 899 in general. Initially, preliminary content-review of the literature revealed only 3 main directions in the topic of behavior and inequality—“behavior biases/errors and inequality,” “inequality and individual decision-making,” and “household finance and inequality.” Later co-occurrence analysis of 2675 keywords provided with the help of cluster technique identified 6 clusters—research areas. It includes not only household finance, consumption and decision-making network, but also political behavior under inequality, macroeconomic, social, institutional effects of inequality and behavior. Recent tendencies in the publication activity show interest in inequality impact on individual level—decision-making and main factors of this process. Due to the citation and co-citation analysis we identified the most influenced authors in the field and found that they are rather diversified. It points to the lack of cross-sectional research and multidisciplinary character of the topic.

This work would not have been possible without the financial support of the Ministry of Education and Science of Ukraine. The paper was prepared as part of the Young Scientist Research on the topic “Economic-mathematical modeling of the mechanism for restoring public trust in the financial sector: a guarantee of economic security of Ukraine” (registration number 0117U003924; 0118U003569 and 0120U102001).

References

Balas, A. and Kaya, H. (2019). The Global Economic Crisis and Retailers’ Security Concerns: The Trends, SocioEconomic Challenges, 3 (2), 5-14. DOI: http://doi.org/10.21272/sec.3(1).5-14.2019

Banerjee, A. and S. Mullainathan (2008). Limited Attention and Income Distribution, American Economic Review, 98(2), 489–493.

Bernheim, B., Ray, D. and Yeltekin, S. (2012). Poverty and Self-Control, Mimeo, Department of Economics, New York University.

Bertrand, M., Mullainathan, S. and Shafir, E. (2004). A Behavioral Economics View of Poverty, American Economic Review, 94(2), 419–423.

Bhandari, Medani P. (2020) Second Edition- Green Web-II: Standards and Perspectives from the IUCN, Policy Development in Environment Conservation Domain with reference to India, Pakistan, Nepal, and Bangladesh, River Publishers, Denmark / the Netherlands. ISBN: 9788770221924 e-ISBN: 9788770221917

Bhandari, Medani P. (2020), Getting the Climate Science Facts Right: The Role of the IPCC (Forthcoming), River Publishers, Denmark / the Netherlands- ISBN: 9788770221863 e-ISBN: 9788770221856

Bhandari, Medani P. and Shvindina Hanna (2019) Reducing Inequalities Towards Sustainable Development Goals: Multilevel Approach, River Publishers, Denmark / the Netherlands- ISBN: Print: 978-87-7022-126-9 E-book: 978-87-7022-125-2

Bilan, Y., Brychko, M., Buriak, A., Vasilyeva, T. (2019). Financial, business and trust cycles: The issues of synchronization, Zbornik Radova Ekonomskog Fakultet au Rijeci, 37(1), 113-138 doi.org/10.18045/zbefri.2019.1.113

Bilan, Y., Raišienė, A.G., Vasilyeva, T., Lyulyov, O., Pimonenko, T. (2019). Public Governance efficiency and macroeconomic stability: Examining convergence of social and political determinants, Public Policy and Administration, 18(2), 241–255

Bilan, Y., Vasilyeva, T., Lyeonov, S., Bagmet, K. (2019). Institutional complementarity for social and economic development, Business: Theory and Practice, 20, 103-115 https://doi.org/10.3846/btp.2019.10

Boyd, D., & Hargittai, E. (2013). Connected and concerned: Variation in parents' online safety concerns, Policy and Internet, 5(3), DOI: 10.1002/1944-2866.POI332

Buriak, A., Vozňáková, I., Sułkowska, J., Kryvych, Y. (2019). Social trust and institutional (Bank) trust: Empirical evidence of interaction, Economics and Sociology, 12(4), 116-129 DOI: 10.14254/2071-789X.2019/12-4/7

Christiansen, Ch. O., Jensen, Steven L.B. (2019). Histories of global inequality: new perspectives, Palgrave Macmillan, 352 p.

Cingano, F. (2014). Trends in Income Inequality and its Impact on Economic Growth, OECD Social, Employment and Migration Working Papers, No. 163, OECD Publishing. http://dx.doi.org/10.1787/5jxrjncwxv6j-en

Dabla-Norris, M., Kochhar, M., Suphaphiphat, M., Ricka, M., Tsounta, E. (2015). Causes and Consequences of Income Inequality: A Global Perspective, International Monetary Fund.

Datta, S. and Mullainathan, S. (2014). Behavioral Design: A New Approach to Development Policy, Review of Income and Wealth, 60, 7–35.

Dave, H. (2017). An Inquiry on Social Issues – Part 1, Business Ethics and Leadership, 1 (2), 78–87. DOI: 10.21272/bel.1(2).78-88.2017

Dave, H. (2017). An Inquiry on Social Issues – Part 2, Business Ethics and Leadership, 1 (3), 45–63. DOI: 10.21272/bel.1(3).45-63.2017

Dobransky, K. & Hargittai, E. (2016). Unrealized Potential. Exploring the Digital Disability Divide, Poetics, 58:18-28, October. https://doi.org/10.1016/j.poetic.2016.08.003

Fahimnia, B., Sarkis, J., & Davarzani, H. (2015). Green supply chain management: A review and bibliometric analysis, International Journal of Production Economics, 162, 101–114.

Fong, Christina M. and Bowles, Samuel and Gintis, Herbert (2005). Behavioural Motives for Income Redistribution, Australian Economic Review, Vol. 38, No. 3, pp. 285–297.

Frank, R. (2013). Falling Behind How Rising Inequality Harms the Middle Class, With a New Preface, University of California Press; First Edition, Reissue, 176 p.

Fred C. Pampel, Patrick M. Krueger, Justin T. Denney (2010). Socioeconomic Disparities in Health Behaviors, Annual Review of Sociology, 36:1, 349–370.

Fred P., Giulia A., Sven S. (2019). How Institutions and Attitudes Shape Tax Compliance: a Cross-National Experiment and Survey, Social Forces, Volume 97, Issue 3, Pages 1337–1364, https://doi.org/10.1093/sf/soy083

Fu Guo, Guoquan Ye, Liselot Hudders, Wei Lv, Mingming Li & Vincent G. Duffy (2019). Product Placement in Mass Media: A Review and Bibliometric Analysis, Journal of Advertising, 48:2, 215-231, DOI: 10.1080/00913367.2019.1567409

Gintis, H. (2016). Homo Ludens: Social rationality and political behavior, Journal of Economic Behavior & Organization, Elsevier, vol. 126(PB), pages 95-109.

Greco, F. (2018). Resilience: Transform adverse events into an opportunity for growth and economic sustainability through the adjustment of emotions. Business Ethics and Leadership, 2 (1), 44-52. DOI: 10.21272/bel.2(1).44-52.2018

Hadbaa, H. and Boutti, R. (2019). Behavioral Biases Influencing the Decision Making of Portfolio Managers of Capital Securities and Traders in Morocco, Financial Markets, Institutions and Risks, 3 (1), 92-105. DOI: http://doi.org/10.21272/fmir.3(1).18–29.2019

Hargittai, E., Piper, A. & Morris, M. (2019). From internet access to internet skills: digital inequality among older adults, Univ Access Inf Soc 18, 881–890, https://doi.org/10.1007/s10209-018-0617-5

Keister, L. (2012). Religion and Wealth Across Generations, Keister, L., Mccarthy, J. and Finke, R. (Ed.) Religion, Work and Inequality (Research in the Sociology of Work, Vol. 23), Emerald Group Publishing Limited, Bingley, pp. 131–150. https://doi.org/10.1108/S0277

Keister, L. , Benton, R., & Moody, J. (2016). Lifestyles through Expenditures: A Case-Based Approach to Saving, Sociological science, 3, 650–684. https://doi.org/10.15195/v3.a28

Kim, Kyoung Tae and Wilmarth, Melissa and Henager, Robin (2017). Poverty Levels and Debt Indicators Among Low-Income Households Before and After the Great Recession, Journal of Financial Counseling and Planning, 28(2), 196–212.

Kolomiiets, U. and Petrushenko, Y. (2017). The human capital theory. Encouragement and criticism, SocioEconomic Challenges, 1 (1), 77-80. DOI: 10.21272/sec.2017.1-09

Kouassi, K. (2018). Public Spending and Economic Growth in Developing Countries: a Synthesis, Financial Markets, Institutions and Risks, 2 (2), 22–30. DOI: 10.21272/fmir.2(2).22-30.2018.

Kuzmenko, O. and Roienko, V. (2017). Nowcasting income inequality in the context of the Fourth Industrial Revolution, SocioEconomic Challenges, 1 (1), 5–12. DOI: 10.21272/sec.2017.1-01

Kyrychenko, K., Samusevych, Y., Liulova, L., Bagmet, K. (2018). Innovations in country's social development level estimation, Marketing and Management of Innovations, 2, 113-128. http://doi.org/10.21272/mmi.2018.2-10

Lisa A. Keister (2008). Conservative Protestants and Wealth: How Religion Perpetuates Asset Poverty, American Journal of Sociology, 113, no. 5, 1237–1271.

Lusardi, A., Michaud, P., & Mitchell, O. (2017). Optimal Financial Knowledge and Wealth Inequality, Journal of Political Economy, 125 (2), 431-477. http://dx.doi.org/10.1086/690950

Morscher, C., Horsch, A., Stephan, J. (2017). Credit Information Sharing and Its Link to Financial Inclusion and Financial Intermediation, Financial Markets, Institutions and Risks, 1 (3), 22-33. DOI: 10.21272/fmir.1(3).22–33.2017.

Moss, D., Anant, T., and Howard, R, (2013). Inequality and Decision Making: Imagining a New Line of Inquiry, Harvard Business School Working Paper, No. 13-099.

Njegovanovic, A. (2018). Artificial Intelligence: Financial Trading and Neurology of Decision, Financial Markets, Institutions and Risks, 2 (2), 58-68. DOI: 10.21272/fmir.2(2).58-68.2018.

Nur-Al-Ahad, Md. (2019). New Trends in Behavioral Economics: A Content Analysis of Social Communications of Youth, Business Ethics and Leadership, 3 (3), 107-115. DOI: http://doi.org/10.21272/bel.3(3).107–115.2019

Pampel, F. (2007). National Income, Inequality and Global Patterns of Cigarette Use. Soc. Forces; 86:445–466. doi: 10.1093/sf/86.2.445.

Payne, B., Brown-Iannuzzi, J., & Hannay, J. (2017). Economic inequality increases risk taking, PNAS Proceedings of the National Academy of Sciences of the United States of America, 114(18), 4643–4648.

Poliakh, S. and Nuriddin, A. (2017). Evaluation Quality of Consumer Protection by Financial Markets Services, Financial Markets, Institutions and Risks, 1(3), 75-81. DOI: 10.21272/fmir.1(3).75–81.2017.

Pomianek, I. (2018). Historical and Contemporary Approaches to Entrepreneurship, Review of Polish Literature, Business Ethics and Leadership, 2 (2), 74-83. DOI: 10.21272/bel.2(2).74–83.2018.

Shvindina, H. (2019). Coopetition as an emerging trend in research: perspectives for safety & security. Safety, 5(3), 61. https://doi.org/10.3390/safety5030061

Singh, S. (2018). Regional Disparity and Sustainable Development in North-Eastern States of India: A Policy Perspective, SocioEconomic Challenges, 2 (2), 41-48. DOI: 10.21272/sec.2(2).41–48.2018

Sterk, N., Meeussen, L., & Van Laar, C. (2018). Perpetuating Inequality: Junior Women Do Not See Queen Bee Behavior as Negative but Are Nonetheless Negatively Affected by It, Frontiers in psychology, 9, 1690. https://doi.org/10.3389/fpsyg.2018.01690

Van Eck, N., & Waltman, L. (2010). Software survey: VOSviewer, a computer program for bibliometric mapping, Scientometrics, 84(2), 523–538.

Van Eck, N., & Waltman, L. (2014). Visualizing bibliometric networks. In Y. Ding, R. Rousseau, & D. Wolfram (Eds.), Measuring scholarly impact: Methods and practice, Springer.

Vasilyeva, T., Bilan, S., Bagmet, K., Seliga, R. (2020). Institutional development gap in the social sector: Crosscountry analysis, Economics and Sociology, 13(1), 271-294 DOI: 10.14254/2071-789X.2020/13-1/17

Vasilyeva, T., Kuzmenko, O., Bozhenko, V., Kolotilina, O. (2019). Assessment of the dynamics of bifurcation transformations in the economy, CEUR Workshop Proceedings, 2422, 134–146

Veldman, J., Meeussen, L., & van Laar, C. (2019). A social identity perspective on the social-class achievement gap: Academic and social adjustment in the transition to university, Group Processes & Intergroup Relations, 22(3), 403–418. https://doi.org/10.1177/1368430218813442

Chapter 5

Public Health Policy Communication with Other Policies in the Context of Inequality

Oleksii Demikhov, PhD of Public Administration,
Assistant Prof., Department of Management, Sumy State University, Sumy, Ukraine, ORCID iD: 0000-0002-9715-9557

ABSTRACT
Public health is a new field of knowledge and human activity that is being developed in Ukraine today. In Ukraine and globally, public health is one of the highest priority areas of human development that falls into the category of systematic social inequality. Public health sector is socially important as it creates a health-preserving lifestyle for the population. The basic elements of this approach are population economic status, ecology, education, territorial settlement (urban or rural), and housing quality. At these basic levels, there is already a feasible scientific debate about the existence and growth of inequality. First of all, inequalities in access to:

- health care quality, prevention and treatment;
- healthy food quality;
- and furthermore, the opportunity to lead a healthy lifestyle, especially in urban areas being the most vulnerable.

Recognizing the effects of such inequalities in access to health-preservation, national and regional public authorities of the EU and Ukraine have begun to develop and implement public health concepts and programs at different levels.

The purpose of our research is to study public health sector of the EU and compare it with Ukraine in order to formulate proposals for mitigating health inequalities in access to health services, as well as developing new standards and to have an integrated approach to work out an effective public health policy.

A topical aspect is the processing and synthesis of information of public policy instruments in the context of preserving and promoting the health of the population, increasing the expectancy and quality of life, preventing diseases, and promoting a healthy lifestyle.

Methodologically, chapter uses multidisciplinary and systematic approach to analyze the data, mostly obtained from the secondary sources.

Data were sourced from the surveys of Ukrainian and foreign scientists, national statistical agencies of the EU and Ukraine, and associations of cities of the leading countries around the world. We are interested in the indicators such as the level of urbanization, the level of gross domestic product, area pollution, the level of mortality, and other economic, social and health characteristics.

1. Introduction

Socio-economic inequality around the world is continuously growing. This applies to the economic, social, and public health sectors of the population (Demikhov, Dehtyarova, Rud 2020; Barchan, Cherkashyna, Shklyar 2020). Socio-economic inequality leads to the spreading of diseases among urban population, as well as environmental, demographic, and social problems (Demikhov, Dehtyarova 2020). Uncontrolled processes of socio-economic inequality are already leading to humanitarian and business depression (Barchan, Demikhov, Cherkashyna 2020). Considering these processes, a new format for the development of the health sector is being introduced (Demikhov, Shipko, Heera Harpreet Singh 2020). This format is referred to as "the inclusion of public health issues into all of the public policies of the state." The study of the impact of socio-economic inequality on prevention, monitoring, and evaluation of the diseases among population, the quality of life, and the possibility of maintaining a healthy lifestyle will allow us to develop new approaches toward the formation and implementation of public policy for the public health sector in Ukraine (Loboda, Smiyan 2020). The study of social policies will help to understand the roles of the state, the local government institutions, the public and stakeholders in these processes (Demikhova, Smiianov, Prikhodko 2016; Prokopenko, Osadchenko, Braslavska 2020).

2. Methods

Methodologically, chapter uses multidisciplinary and systematic approach to analyze the data, mostly obtained from the secondary sources.

3. Results and Discussion

According "Health 2020: European policy and strategy for the 21st century" (Health 2020 … 2013), Europe has adopted the following standards in strategic planning:

1. **Strategic objectives of Health 2020**
   Health 2020 recognizes that successful governments can achieve real improvements in health if they work across government to fulfill two linked strategic objectives:
   - improving health for all and reducing health inequalities
   - improving leadership and participatory governance for health.
2. **Countries, regions, and cities setting common objectives and joint investment between health and other sectors can significantly improve health and well-being.**
   Priority areas include preschool education, educational performance, employment and working conditions, social protection, and reducing poverty. Approaches include addressing community resilience, social inclusion and cohesion; promoting assets for well-being; mainstreaming gender and building the individual and community strengths that protect and promote health, such as individual skills and a sense of belonging. Setting targets for reducing health inequalities can help drive action and is one of the principal ways of assessing health development at all levels.
3. **Addressing social inequalities contributes significantly to health and well-being.**
   The causes are complex and deeply rooted across the life course, reinforcing disadvantage and vulnerability. Health 2020 highlights the increasing concern about tackling poor health within countries and across the Region as a whole. The lowest and highest life expectancies at birth in the WHO European Region differ by 16 years, with differences

between the life expectancies of men and women; and maternal mortality rates are up to 43 times higher in some countries in the Region than in others. Such extreme health inequalities are also linked to health-related behavior, including tobacco and alcohol use, lack of diet and physical activity, and mental disorders, which in turn reflect the stress and disadvantage in people's lives.

4. **Taking action on the social and environmental determinants of health can address many inequalities effectively.**
   Research shows that effective interventions require a policy environment that overcomes sectoral boundaries and enables integrated programs. For example, evidence clearly indicates that integrated approaches to child well-being and early childhood development produce better and fairer outcomes in both health and education. Urban development that considers the determinants of health is crucial, and mayors and local authorities play an ever more important role in promoting health and well-being. Participation, accountability, and sustainable funding mechanisms reinforce the effects of such local programs.

There are important inequalities within countries across key lifestyle indicators, including: smoking rates, obesity, exercise, and limiting long-term illness. In addition, the 20% of the population with the lowest income is most likely to delay seeking care because of fear of financial catastrophe from out-of-pocket payments. Considerable evidence supports the claim that education and health are correlated. Data indicate that the number of years of formal schooling completed is the most important correlate of good health. According to the 2003 Human development report (United Nations Development Programme): "Education, health, nutrition and water and sanitation complement each other, with investments in anyone contributing to better outcomes in the others" (Health 2020 … 2013).

Thus, Figure 5.1 presents in more detail the relationship of various factors.

Thus, Figure 5.1 shows that the health of the population depends on various factors and environmental conditions. All of these factors are important for human health and have a corresponding impact.

1. According to leading British scientists (Official website of the Kingsfund 2019), research is being carried out in the following areas: the complex interaction between individual characteristics, lifestyle and the physical, social and economic environment changing, and future trends in health care.

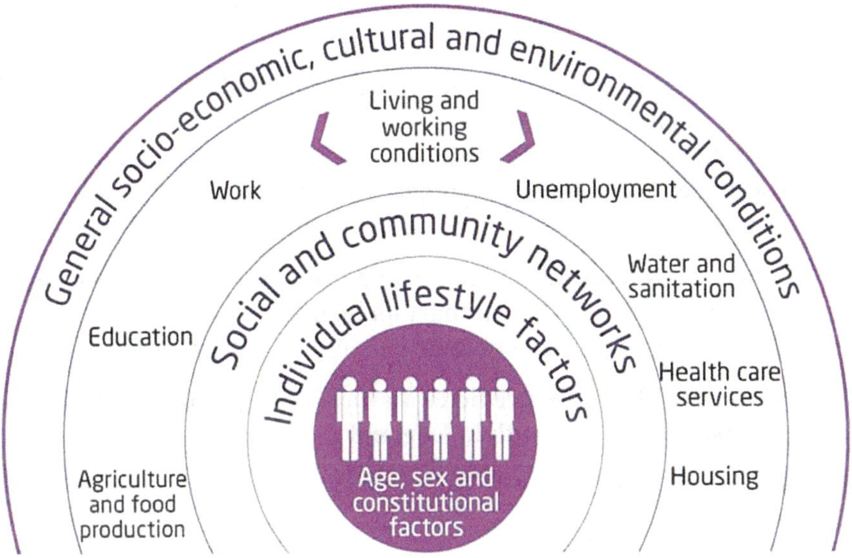

Source: (Dahlgren, Whitehead 1993).

Figure 5.1. Health is dependent on our genes, our lifestyles, environment, and health care. *Source:* (Dahlgren, Whitehead 1993).

These are the main aspects of scientific activity:

1. **Key messages:**

   - **Health is determined by a complex interaction between individual characteristics, lifestyle and the physical, social and economic environment**

   Most experts agree that these "broader determinants of health" are more important than health care in ensuring a healthy population.

   - **Economic hardship is highly correlated with poor health**

   The current downturn—the deepest since the Great Depression—threatens individual and family wellbeing, especially for the unemployed and those experiencing wage and benefit cuts.

   - **Increased levels of education are strongly and significantly related to improved health**

   Recent rise in the overall number of people in higher education and more people from poorer background in higher education should have long-term benefits for population health. However, there is uncertainty about whether improvements in access will continue.

- **Work-related illness is decreasing, particularly among people with manual occupations**

Employers are also showing a growing interest in the health of their workforce. While these trends may continue, the economic environment could exacerbate work-related stress and have a negative impact.

- **Improved housing conditions and greater access to green spaces should have a positive impact on health**

However, the future outlook is uncertain for the most disadvantaged.

- **Climate change is predicted to have both positive and negative implications for health in England**

2. **Key uncertainties:**

- **Wider economy**

It is difficult to predict how the UK and global economy will develop in the next 20 years, and the overall effect on employment and income.

- **Work environment**

Pay and working conditions could deteriorate markedly during the economic downturn. However, some large employers are recognizing the benefits of investing in their staff's welfare and could act positively to improve their employees' health.

- **Education**

Recent increase in the number of people going to universities may stall over time. Following the introduction of higher tuition fees, applications for English universities this year are down 10 per cent.

- **Environmental change**
  1. Carbon reduction targets are likely to drive considerable technological and social change, with significant health implications. There is, however, considerable uncertainty around the scale and timing of these effects (Official website of the Kingsfund 2019).

In addition, scientists are also interested in studying the broader causes that affect the health of nations (Figure 5.2).

Several studies are trying to evaluate how the broader determinants of health affect our health. Three pie charts show the main results of three scientific papers.

For example, the McGinnis et al. (McGinnis, Williams-Russo, Knickman 2002) think that mental health (40%), social conditions and

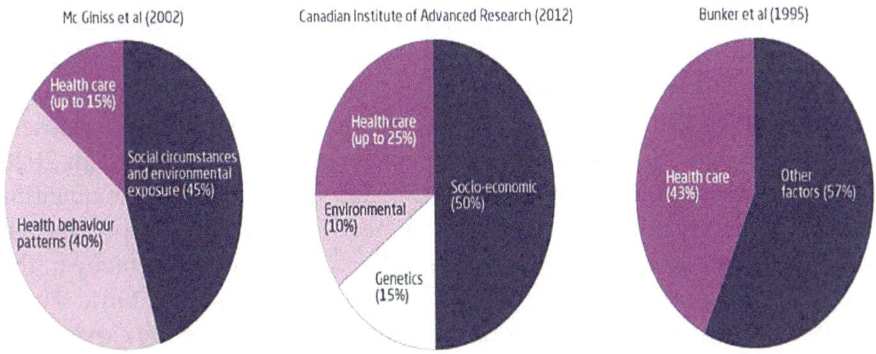

Figure 5.2. Estimates of the impact of the "broader determinants of health" on population health. *Sources:* (Agreement about the association … 2014; Bunker, Frazier, Mosteller 1995; Healthy places: Councils leading … 2012; Publication of two great reports ... 2019).

ecological location (40%) are influenced to a greater extent by the health status of nations, and to a lesser extent by health care (about 15%). Canadian Institute of Advanced Research (Healthy places: Councils leading on public health 2012) explored that the main factors are socio-economic living conditions (50%) and health (about 25%). Finally, Bunker et al. (Bunker, Frazier, Mosteller 1995) suggest two main factors: health care (43%) and all others (57%). Thus, along with the level of health care, all researchers note the leading role in influencing the health status of nations on the socio-economic factors of their lives (40 to 57%). That is, socio-economic conditions of life of people (and inequality of access to these qualitative conditions) are influential factors and lead to differences in health indicators of the population in different target groups.

Analysis of the statistics of the WHO Health Report in Europe in 2018 (WHO 2018 European Health Report 2018) shows that governments in Europe are strategically focused on achieving a number of health targets under the Health 2020 policy.

Therefore, the goal of "Increasing life expectancy in Europe" should be considered in greater detail.

The regional average life expectancy has steadily increased over recent years and the gaps in average life expectancy between the sexes and between countries are getting narrower. However, there are still considerable differences which require continued monitoring to ascertain consistent improvement. In 2015, the difference between countries with the highest life expectancy and the lowest was more than a decade. Women still live longer than men at all ages.

Despite increase in life expectancy at birth, the difference between countries with the highest and lowest life expectancy in the Region is still more than a decade. The aim for this Health 2020 target is to increase overall life expectancy while reducing differences in life expectancy between countries.

Target "Increase life expectancy in Europe" is linked to Health 2020 policy area 2 "Healthy people, well-being and determinants." The quantification for this Health 2020 target is a continued increase in life expectancy at the current rate, coupled with reducing differences in life expectancy in the Region. The core indicator for this target is life expectancy at birth. There are two additional indicators for this target: quantification of life expectancy at ages 1, 15, 45 and 65 years, and healthy life years at age 65. Now let's take a closer look at the constant increase in life expectancy at ages 1, 15, 45 and 65.

Table 5.1 shows the average regional life expectancy (years) at ages 1, 15, 45 and 65 for both sexes over time.

As we can see from Table 5.1, women are still living longer than men. The data in Table 5.1 indicate that the gender gap reduced slightly between 2010 and 2015 for regional average life expectancy at ages, 1, 15 and 45 (see Table 5.1). However, it increased slightly for regional average life expectancy at 65 years old. Moreover, there are still differences between countries in life expectancy at ages 1, 15, 45, and 65 years (WHO 2018 European Health Report 2018).

All citizens have the right to good health, well-being, education, and equal opportunities to thrive where they live.

Therefore, let us look at Task 3 "Reducing health inequalities in Europe (the goal of social determinants)". The aim is to reduce the health gaps associated with social determinants in Europe. This goal has five key metrics:

- Infant mortality per 1000 live births
- Percentage of primary school-age children not enrolled
- Unemployment rate
- National and/or sub-national health inequalities policies established and documented
- Gini coefficient (revenue sharing)

Health inequities are unfair distributions of health and well-being outcomes. Social determinants of health include all political, social, economic, institutional, and environmental factors which shape the conditions of daily

Table 5.1 Regional average life expectancy (years) at ages 1, 15, 45 and 65 years, by sex

Year	Life expectancy at age 1		Life expectancy at age 15		Life expectancy at age 45		Life expectancy at age 65	
	Female	Male	Female	Male	Female	Male	Female	Male
2000	77.5	69.9	63.9	56.3	35.1	29.1	17.9	14.3
2001	77.7	70.2	64	56.5	35.3	29.3	18.1	14.6
2002	77.7	70.2	64	56.5	35.2	29.3	18	14.5
2003	77.6	70.2	63.9	56.6	35.1	29.3	17.9	14.5
2004	78.1	70.7	64.4	57	35.6	29.7	18.4	14.9
2005	78.1	70.8	64.4	57.1	35.6	29.7	18.4	14.9
2006	78.6	71.4	64.9	57.7	36	30.3	18.8	15.3
2007	78.9	71.7	65.1	58	36.3	30.5	18.9	15.4
2008	79.1	72	65.3	58.3	36.4	30.7	19.1	15.6
2009	79.4	72.5	65.6	58.7	36.7	31	19.3	15.7
2010	79.6	72.8	65.8	59.1	36.9	31.3	19.4	15.9
2011	80	73.2	66.2	59.5	37.2	31.6	19.7	16.1
2012	80.2	73.6	66.4	59.9	37.4	31.9	19.8	16.3
2013	80.4	73.8	66.6	60.1	37.6	32.1	20	16.4
2014	80.7	74.1	66.8	60.3	37.8	32.3	20.1	16.6
2015	80.6	74.1	66.8	60.3	37.7	32.3	20.1	16.5

Source: (Health for All database 2018)

life, contribute to health and well-being, and the onset of illness throughout a person's life. National policies that address the reduction of health inequities by taking action on the social determinants of health can lead to improvements in the overall health and well-being of the entire population.

The number of countries in the European Region establishing and documenting national and subnational policies to address the reduction of health inequities has been increasing. In 2016, 42 countries out of 53 (79%) had a policy or strategy in place to address inequities or social determinants. This is an increase from 29 and 35 countries, respectively, in 2010 and 2013 (WHO 2018 European Health Report 2018).

Another important Target is "ensuring universal coverage and the right to health" (Target 5). Target 5 is focused on achieving universal health coverage by 2020 and envisages the "right to health" as a core policy construct and vision that fosters the idea that provision of equitable fair access to effective and needed services without financial burden is a fundamental right of all citizens. The WHO Director General, Dr Tedros, identified universal health coverage as one of the five key priorities of the World Health Organization. Ensuring universal health coverage without impoverishment is the foundation for achieving the health objectives of the SDGs. The World Health Assembly made universal health coverage one of WHO's three strategic priorities in the WHO's thirteenth general program of work 2019–2023, by setting the strategic priority (and goal) of "reducing persistent barriers to accessing health services and 1 billion more people benefitting from universal health coverage."

Task 5 "moves towards universal coverage (as defined by WHO: equal access to efficient and necessary services without financial burden) by 2020." To assess the performance of countries for this purpose, Member States collect and report data for a set of key indicators. One such data is the level of private household payments as a share of total health expenditure (WHO estimate).

The level of out-of-pocket payments or expenditure on health is expressed as a percentage of total expenditure on health. Private households' out-of-pocket payments on health are their direct expenses, including gratuities and payments in kind made to health practitioners and suppliers of pharmaceuticals, therapeutic appliances and other goods and services, whose primary purpose is to contribute to the restoration or enhancement of the health status of individuals or population groups. They also include household payments to public services, non-profit institutions or non-governmental organizations, non-reimbursable cost sharing, deductibles, co-payments, and fees for service. The evidence shows that there is a strong correlation between a country's public expenditure on health and private out-of-pocket payments. Countries with low levels of public expenditure on health usually experience high levels of out-of-pocket payments, which in turn may lead to financial hardship for households and adverse effects on health outcomes. International analysis suggests that once the share of out-of-pocket payments falls below 15% of total spending on health, very few households experience catastrophic or impoverishing levels of health expenditure. In 2014, 40 countries in the Region had proportions higher than this critical threshold, similar to the findings reported in the European health report for 2015. The regional average of private household out-of-pocket expenditure has slightly increased from 25.5% in 2010 to 26.6% in 2014 (Health for All database 2018).

This indicates an increase in inequitable access to health care along with an elevated level of financial risk, impoverishment and perpetuation of an economically vulnerable population. There were large differences in the proportions of private household out-of-pocket expenditure between the countries in the Region, which ranged from a low value of 5.2% to a very high maximum value of 72.1% in 2014-. There were also considerable differences between subregions in the proportions of private household out-of-pocket expenditure, which ranged from 14.7% in Nordic countries to 46.2% in CIS (Commonwealth of Independent States) countries in 2014 (WHO 2018 European Health Report 2018).

A detailed comparative statistical description of the EU and Ukraine key health indicators is provided in Annex 1 (Official website of the WHO 2019; addition).

It is useful to compare the main health indicators in countries such as Bulgaria, Ukraine, and Finland. Bulgaria was selected by the author of the article for comparison as a country of Eastern Europe, the former Soviet country of the Warsaw Pact, which is close to Ukraine in terms of economic potential and mentality but is now accepted by the EU and NATO. Finland was chosen as a developed capitalist country, however not as developed as Germany, allowing us to speak of a more relevant comparison to Ukraine. The comparison table below is drawn according to the annual statistical publication of the WHO Regional Office for Europe—2016 (Table 5.2) (the main indicators of health in the European region 2019).

It should be noted that these indicators pertain to the "Health 2020" policy makers and the Sustainable Development Goals.

The data presented in Table 5.2 was systematized, verified, and processed in accordance with the standard methodology used by Eurostat, WHO, and other UN agencies (The main indicators of health in the European region 2019).

Closer analysis of Table 5.2 indicates that the unemployment rate does not directly correlate with the other indicators of the nation's health due to the modern way of life, particularly in the urban areas, whereby the self-employed or the sole-traders have not been registered, allowing them to live with the "unemployed" status. The average unemployment rate in Ukraine, Bulgaria, and Finland is approximately the same. Life expectancy in Ukraine is shorter than in Bulgaria or Finland due to the high mortality rate in Ukraine (due to all the reasons indicated in Table 5.2) compared to the two other European nations. The total number of hospitals, as well as the government and general health expenses per 100 (one hundred) thousand people are much lower than European, whereas any related private hospital and health funding is higher in Bulgaria and Finland. There is some curious statistics as to the extent of bad habits such as smoking among the Ukrainian male population,

Table 5.2 Comparative table of key health indicators in Ukraine, Bulgaria, and Finland*

Health Indicator		**Ukraine**	**Bulgaria**	**Finland**
Unemployment rate (averaged 2008–2013)		8,0	12,0	8,0
Life expectancy in years (as of 2016)	Male	67,5	71,1	78,4
	Female	77,3	78,1	84,2
Number of healthy years of life from birth (as of 2016)		64,0	66,4	71,7
Mortality rate as a result of murder and assault (per 100,000 population), (as of 2016)		4,6	1,2	1,5
Suicide rate and intentional self-harm (per 100,000 population), (as of 2016)		15,8	7,8	13,6
Baby mortality per 1 thousand live births, (as of 2016)		8,1	7,7	1,7
Maternal death per 100 thousand live births, (as of 2016)		14,8	12,0	8,0
Health services expenses using own (private) funds as a percent of total health care costs, (as of 2016)		54,3	48,0	20,4
Total health care expenditure as percentage of GDP (as of 2014)		7,1	8,4	9,7
State expenditures on health care as a percentage to the total healthcare costs, (as of 2014)		50,8	54,6	75,3
Number of hospitals per 100 thousand population, (as of 2014)		4,0	4,9	4,7
The standardized extent (by age) of tobacco use (%) among people at the age 15 and over (as of 2016)	Male	47,0	44,0	23,0
	Female	14,0	30,0	18,0
Alcohol use among people at the age of 15 and over (liters) (as of 2016)		6,0	11,0	8,0
Standardized by age assessment level of the prevalence (%) of obesity (BMI ≥30) among people at the age of 20 and over, (as of 2016)		24,0	25,0	22,0
Children vaccinated against polio (%), (as of 2017)		48,0	92,0	89,0
Children vaccinated against measles (%), (as of 2017)		86,0	94,0	94,0

**prepared by the author based on the following source (The main indicators of health in the European region 2019)* which is much higher than that of the European population. However, it is completely the opposite for alcohol consumption. The European female population smokes more than the Ukrainian females. It can be assumed that this is caused by a high standard of living in Europe, which makes it possible to buy alcohol and tobacco products more frequently. Vaccination levels in Ukraine lag significantly behind those of Europe, especially for polio. Such trends affect the healthy life expectancy in Ukraine.

At the same time, the issues of unequal use of these targets between countries and regions of the country, as well as issues of gender and age inequality remain unresolved. In order to improve health policy instruments, it is proposed to create unified and functionally integrated information systems to integrate and coordinate efforts between countries to bridge the gap between research and public policy in this field (WHO 2018 European Health Report 2018).

Considering the abovementioned challenges of the modern world, the World Health Organization (WHO) and the UN-Habitat developed the "Urban HEART"—The Urban Health Equity Assessment and Response Tool. This tool facilitates the creation of an evidence database and is used by the city officials to assess the level of urban health inequity and to plan specific action to tackle it. This methodology for assessment and elimination of urban health inequity encourages the use of available data, which is broken down by criteria such as socio-economic group and area of residence. The UN-Habitat UrbanInfo program helps users store, present, and analyze urban metrics through a range of mechanisms (Official web-portal of the European Regional Bureau … 2010).

Thus by analyzing the global world trends the modern Ukraine is forming and implementing one of its main community goals—a public policy to preserve health and enrich people's lives, as well as to eliminate inequalities in these aspects. European integration in Ukraine conforms by the "Health 2020: the foundations of the European policy in support of the actions of the entire state and society in the interests of health and well-being" policy [6], and also guided by the requirements of the Ukraine–EU Association Agreement (Agreement about the association of Ukraine … 2014).

The measures to eliminate public health inequalities being developed in Ukraine deserve a separate mention:

1. Improving communication—this refers to informing and motivating the population to keep healthy;
2. Advocacy programs—these are the activities to represent and protect human health interests, which are directly related to the elimination of inequality.

4. Conclusion

It is evident that the world and Ukraine in particular, is creating and developing interdisciplinary and interconnected instruments to influence various public policies regarding the quality of life, health and the elimination of prerequisites for unequal access to these benefits.

The experience of developing public health policy instruments under the conditions of increased inequality that we discussed here can be applied in Ukraine. The comprehensive use of the aforementioned practices by the national officials to combat inequality through medical reform and setting a direction for public health will yield positive results. We consider it relevant to continue scientific research in the field of inequality of access to quality living conditions in the context of a new integral term, which is now being introduced into the mainstream European advanced research—"the health culture."

References

Barchan G., Demikhov O., Cherkashyna L. *et al.* (2020). A complex of regional ecological and medico-social factors: evaluation of dysplastic dependent pathology of the bronchopulmonary system // Polish Medical Journal - Polski merkuriusz lekarski : organ Polskiego Towarzystwa Lekarskiego. Vol. 48 (283). - P. 49-54.

Barchan G.S., Cherkashyna L.V., Shklyar A.S. *et al.* (2020). Forecasting of recurrent respiratory infections on the complex of constitutional-biological factors // Azerbaijan Medical Journal. № 2. – P. 80-87.

Bunker, J.P., Frazier, H.S. and Mosteller, F. (1995). The role of medical care in determining health: Creating an inventory of benefits. In, Society and Health ed Amick III et al. New York: Oxford University Press. pp. 305-341.

Dahlgren G, and Whitehead M (1993). Tackling inequalities in health: what can we learn from what has been tried? Working paper prepared for the King's Fund International Seminar on Tackling Inequalities in Health, September 1993, Ditchley Park, Oxfordshire. London, King's Fund, accessible in: Dahlgren G, Whitehead M. (2007) European strategies for tackling social inequities in health: Levelling up Part 2. Copenhagen: WHO Regional office for Europe: http://www.euro.who.int/__data/assets/pdf_file/0018/103824/E89384.pdf. (date of review: 15/12/2019).

Demikhov O., Dehtyarova I., Demikhova N. (2020). Actual aspects of public health policy formation on the example of Ukraine Bangladesh Journal of Medical Science 19 (3): 358-364, DOI: https://doi.org/10.3329/bjms.v19i3.45850

Demikhov O., Dehtyarova I., Rud O. et al. (2020). Arterial hypertension prevention as an actual medical and social problem Bangladesh Journal of Medical Science 19 (4): 722-729, DOI: https://doi.org/10.3329/bjms.v19i4.46632

Demikhov O., Shipko A., Heera Harpreet Singh *et al.* (2020). Intersectoral component of the healthcare management system: regional programs and assessment of the effectiveness of prevention of bronchopulmonary dysplasia // Azerbaijan Medical Journal. – №2. – P.88-96

Demikhova, N., Smiianov, V., Prikhodko, O. *et al.* (2016). Information and telecommunication technologies and problem-based learning in the formation of competitive competence in medical masters of Sumy state university. Azerbaijan Medical Journal, 2: 95-101

European Regional Bureau (2002). Oficial web-portal of the European Regional Bureau of the All-Union Health Protection Organization: Publication of two great reports. Retrieved from http://www.euro.who.int/en/media-centre/sections/press-releases/2010/11/who-releases-two-major-reports (date of review: 14/12/19).

Kingsfund (2019). Official website of the Kingsfund. Retrieved from https: https://www.kingsfund.org.uk/projects/time-think-differently/trends-broader-determinants-health (date of review: 12/12/2019).

Kuznetsova, D. (2012). Canadian Institute of Advanced Research, Health Canada, Population and Public Health Branch. AB/NWT 2002, as in Healthy places: Councils leading on public health. London: New Local Government Network. Available from New Local Government Network website.

Legislation of Ukraine (2014). Agreement about the association of Ukraine, from one side, that of the European Union, the European Union of goods from the atomic energy and the other Member States, from the other side: dated June 27, 2014, ratified of the declared law of Ukraine September 16, 2014, number 1678-VII. Database "Legislation of Ukraine". Retrieved from https://zakon.rada.gov.ua/laws/show/984_011 [in Ukrainian] (date of review: 25/12/2019).

Loboda A., Smiyan O., Popov S. *et al.* (2020). Child health care system in Ukraine // Turk Pediatri Arsivi, 55, p. S98-S104

McGinnis, J.M., Williams-Russo, P. and Knickman, J.R. (2002). The case for more active policy attention to health promotion. Health Affairs 21 (2) pp.78-93.

Prokopenko O., Osadchenko I., Braslavska O. *et al.* (2020). Competence approach in future specialist skills development // International journal of management, 11(4), pp. 645-656. https://doi.org/10.34218/ijm.11.4.2020.062

WHO (2018). European Health Report. More than numbers - evidence for all. Copenhagen, WHO Regional Office for Europe, 2018, pp.164

WHO (2018). Health for All database. European Health Information Gateway [website]. Copenhagen: WHO Regional Office for Europe. (https://gateway.euro.who.int/en/datasets/european-health-for-all-database/, accessed 19 August 2018) (date of review: 02/12/2019).

WHO (2019). The main indicators of health in the European region (2019). Annually statistical publication of the European Regional Office of the WHO, Copenhagen, 2016, 8 pp.

WHO (2020). Health 2020: European policy frameworks and strategies for the 21st century / World. org health care. Copenhagen: World. org Health, Europe. region. Bureau, 2013. VI, 16 pp. // World Health Organization, Regional Office for Europe: official website. Retrieved from http://www.euro.who.int/__data/assets/pdf_file/0018/215433/Health2020-Short-Rus.pdf (date of review: 01/01/2020).

WHO (2020). Health 2020: the foundations of European policy in support of the actions of the entire state and society in the interests of health and well-being / World. org health care. Copenhagen: World. org Health, Europe. region. Bureau, 2013. VI, 16 pp. // World Health Organization, Regional Office for Europe: official website. Retrieved from http://www.euro.who.int/__data/assets/pdf_file/0018/215433/Health2020-Short-Rus.pdf (accessed 12/07/2019) (date of review: 04/12/19).

Who (2020). Official website of the WHO [https://gateway.euro.who.int/en/country-profiles/ukraine/#h2020 ((date of review: 05/12/2019).

Chapter 6

Investigating Employment Discrimination and Social Exclusion: Case of Serbia

Mirjana Radovic-Markovic[1] *Milos Vucekovic*[2], *Aidin Salamzadeh*[3]
[1]Institute of Economic Sciences, Serbia and South Ural State University, Russia
[2]Singidunum University, Belgrade, Serbia
[3]Faculty of Management, University of Tehran, Tehran, Iran

ABSTRACT
This article considers the issues of employment discrimination and social inclusion and their relevance for the well-being of individuals and groups in Serbia. The article analyses the situation of particular groups in society—women, people over 55 years old, Roma people, and persons with disabilities. In some cases, comparisons are made with Slovenia in order to explain the prevalence of discriminatory experiences and their relevance for the study of social policy against discrimination. Research has shown that integrated approaches and strategies on all levels of society are needed to improve the social inclusion of marginalized groups. At the first place, the involvement of marginalized groups in education programs should be seen as an essential element of solving problems in this sector.
Keywords: discrimination, social inclusion, education, employment, marginalized groups

Povzetek: Prispevek obravnava vprašanja diskriminacije pri zaposlovanju in socialne vključenosti ter njihov pomen za dobro počutje posameznikov in skupin v Srbiji. Članek analizira položaj posameznih skupin v družbi - ženske, osebe, starejše od 55 let, Rome in invalide. V nekaterih primerih so primerjane s Slovenijo, da bi pojasnili razširjenost diskriminatornih izkušenj in njihov

[1] Corresponding author: Full Professor, contact email: mradovic@gmail.com

pomen za preučevanje socialne politike proti diskriminaciji. Raziskave so pokazale, da so za izboljšanje socialne vključenosti marginaliziranih skupin potrebni integrirani pristopi in strategije na vseh ravneh družbe. Vključevanje marginaliziranih skupin v izobraževalne programe je treba najprej obravnavati kot bistveni element reševanja problemov v tem sektorju.

Ključne besede: diskriminacija, socialna vključenost, izobraževanje, zaposlovanje, marginalizirane skupine JEL classification: I24; L26; M13.

1. Introduction

Realizing equality in a society is associated with several moral issues and imperatives. For this reason, the "Europe 2020 Strategy" includes and recommends numerous social programs which target the most vulnerable social groups and attempt to consider innovative educational programs (Radovic-Markovic, et al., 2014). Such programs must enable people to become employable by different employees, depending on their abilities and competencies. Moreover, some innovative programs are designed to fight different types of discriminations—including age, gender, racial and other possible discriminations. Addressing these issues is of paramount importance for the Serbian government as it is going to join the European Union. Despite the enactment of the anti-discrimination law in 2009, still, Serbia is far from being a tolerant society where differences are accepted, and all the citizens exercise their rights undisturbedly (EurActiv, 2013).

A new educational programs should be designed based on the capacity building logic in order to let individuals grow and become able to be themselves and flexible; also personality enables them to develop and show a clear path toward the following (Radovic Markovic, 2012): (i) promoting their achievements; and (ii) tackling barriers to social inclusion. Moreover, higher education institutions need to become more responsible in terms of providing support to disabled individuals and those who are not in an equal position with others (Furlong, Ferguson & Tilleczek, 2011). Our main goal is then to discuss the opportunities for marginalized groups to realize their right to education and to identify obstacles in their employment recruiting in Serbia.

2. Literature Review

Socio-economic inclusion (integration) of marginalized communities has the ultimate goal to provide for such communities an equal (or at least improved) access to jobs, education, and health services (Economic Commission, 2011).

The literature review shows that in all countries, the correlation between marginalized groups and discrimination is not recognized. Recent literature suggests that the age discrimination occurs in taking decisions in employment, (Gutek, Guillemard & Walker, 1994; Ngo, Tang & Au, 2002). Riger and Galligan (1980) have highlighted the visible socio-psychological and physiological differences that are taken into account in age discrimination (Radović-Marković, 2012).

For these reasons, there are no adequate strategies to support discriminated persons. Employment is essential not only for achieving individual economic security but also for his physical and mental health, personal well-being, and sense of identity. Numerous studies have shown that relevant education can lead to improved self-confidence (Carlton & Soulsby, 1999; Dench & Regan, 1999), communication skills (Emler & Fraser, 1999; Radović-Marković, 2011a), sense of belonging to the social group (Emler & Fraser, 1999; Jarvis & Walker, 1997; Bhandari 2020; Bhandari & Shvindina 2019), as well as the realization of personal identity (Radović-Marković et al., 2012b). Also, the education that follows the needs of the individual urges creative ideas and critical thinking (Radović-Marković, 2012c).

Accordingly, it can be concluded that appropriate education leads to improvements in social, economic, and personal life. Namely, the degree of qualification and possession of business skills are directly related to the degree of employability (UNDP, 2006). For marginalized and the young who abandoned schooling, lack of literacy and basic life and professional skills reduce their chances of getting hired and better socialized. In order to achieve this, it is necessary to support a society that should facilitate education in this segment of the population through various funding mechanisms. These groups are often more accessible to the informal form of education than the formal, organized by non-governmental organizations. In advance, it is necessary to define quality standards for informal training programs and to supervise them in terms of satisfying these standards. This will facilitate the integration of these groups in the labor market and lifelong learning (Bessette, 2011).

Accordingly, a commitment to promoting social inclusion makes an impact on the inclusion of all marginalized groups in lifelong learning and paid employment (IFSW, 2012). It will, in the long run, contribute to reducing inequality and achieving inclusive growth (UNDP, 2010). A study by the World Bank (2006) has shown that progress toward greater inclusion of marginalized groups consequently improves the economic and social development of a country.

3. Social Exclusion and Discrimination in Employment in Serbia

In the last decade, more attention has been given to improve the social inclusion of disadvantaged groups world-wide. The debate on social exclusion in the European context has led to the development of the concept in two ways: (a) focus on the factors that lead to poverty and (b) the multidimensional concept that cannot be directly associated with income (Berghman, 1995). The effects of social exclusion (NGEC, 2011), are as follows (Radovic-Markovic, 2016):

- Poverty;
- Political exclusion and discrimination;
- Limited approach to educational possibilities;
- Identity issues; and
- Negative stereotypes.

According to the National Employment Service, out of a total of 468,700 unemployed people in 2018, 14.1% were men, and 15.8% were women. The unemployment rate in the Republic of Serbia was 22.4% in 2018. According to the statistical yearbook of Serbia (2018), the unemployment rate of women was 14.3% and was higher by 1.5% compared to the men's unemployment rate of 12.8% in 2017, while in 2014 the unemployment rate for men was 18.3%, and for women 20.3%.

Although the unemployment rate has decreased for the last 3 years, the gender employment gap has not diminished. Discrimination in employment is particularly evident in the employment of younger women in their reproductive age. They are most often offered short-term jobs and jobs for a specific time, in order to reduce the cost of maternity leave or leave due to childcare. Older people also have a high risk of marginalization, as well as women older than 55. Women older than 55 years of age are mostly inactive in the labor market. According to the conducted surveys, discrimination among persons between 55 and 64 years of age regarding employment still exists in Serbia, which is reflected in their insufficient demand in the labor market (Radović-Marković, 2011). There are multiple reasons for this discrimination. First of all, the older workers are thought to be less productive, lacking adequate knowledge, accepting change more slowly, and are not able to make progress in their skills (Radović-Marković, 2012a). This high level of participation of women in the structure of unemployed persons older than 55 years can be explained by the existence of gender stereotypes. According to them, women of this age are not ready enough to fulfill their work tasks. They are not physically attractive enough; they have no modern knowledge and are not ready to improve. Despite these stereotypes, recent research has shown that the workers of both sexes are ready to improve and learn, for which they need adequate support (Radović-Marković, 2012). This support is vital at the same time by the state and educational institutions. In countries where stereotypes exist, it is difficult for women to get a job in a formal economy; they are deprived of their means of living (Hill & Macan, 1996). It forces them to work in the informal sector, which is the case with discriminated persons in Serbia as well. For these reasons, the involvement of the elderly in the development of society contributes to their well-being, as well as to the welfare of society as a whole. Therefore, their inclusion is necessary through the promotion of their social, economic, and intellectual contribution to society and by providing them with opportunities to make decisions at all levels (Radović - Markovic, 2016; Achakpa & Radović-Marković, 2018). Also, in the age group of 15 to 24, the unemployment rate is still high, 36.3%, but certainly less than in 2014 when it was 50%.

Gender Equality Index shows that Serbia is not yet halfway to gender equality (Government of the Republic of Serbia, 2016). The gender gap in employment not only persists in Serbia (40%), but also it remains in the EU countries. Namely, progress is slow in this area over the last 10 years in many countries in the EU. However, there is no vital change in Slovenia, Finland, Slovakia, Greece, and Italy (European Institute for Gender Equality, 2015). So, according to the European Institute for Gender Equality (2015), the employment rate in Slovenia for a cohort (20–64) was 65 % for women versus 73% for men. Because the total employment rate in Slovenia is 69%, it has not yet reached its national Europe 2020 strategy (EU2020) target (75 %).

Few people with disabilities are able to find employment in Serbia. In Serbia, in 2017 and 2018, nine new enterprises for persons with disabilities were established (NES, 2018), which is still a very small number. Employment of people of Persons with Disabilities of the Republic of Serbia (Official Gazette of the RS, 36/2009), are recognized as a group who suffer with special disabilities, including intellectual disabilities. It is very important, to note that disable people have been excluded from all social flows for decades, which resulted in a lack of skills and a lack of opportunities for them to be in the labor market. In the Law on Professional Rehabilitation and Employment, attention should be paid in view of the extremely low employment rate. This Law rate was created as an attempt to respond to the high unemployment (Milanović et al., 2012). Despite the Law being enacted, it is not consistently implemented in practice, as well as the inability to hire people with disabilities in Serbia. The far-reaching obstacle to employment is the attitude of employers toward people with disabilities. The measure by which a person with disabilities exercises the right to education and employment is a measure to which he can develop his potentials and become an active member of society (Odović, Rapaić, & Nedović, 2008, p. 189).

The report conducted by Kogovšek Šalamon (2018) confirmed that discrimination against people with disabilities still remains in Slovenia as well. Therefore, the Slovenian government undertook different measures to solve the problem. So, it has adopted the Act on Equal Opportunities for People with Disabilities (2014), which provides measures for people with disabilities, such as technical equipment for people with visual and hearing impairments and co-financing for the adaptation of vehicles (Kogovšek Šalamon, 2018, p.5).

Roma people and especially women face anti-Gypsyism, extreme poverty, exclusion, and discrimination, which reinforce their disadvantages, according to the latest report from the European Union Agency for Fundamental Rights (2019).

The social status of Roma in Serbia is also very unfavorable since they are the most vulnerable and marginalized minority. This shows the employment rates of Roma and displaced persons, which are between 5 and 10 percent lower than the average rate of total employment (Radovic Markovic, 2016a). Workers belonging to these groups of the population are poor and are more likely to be engaged in the informal sector of business (Radović-Marković, 2016). In addition, Roma has a very low level of education: even 55% of the Roma people have not completed primary education, 33% have primary education, 11% completed secondary education, and less than 1% completed high school or university (Cvejić 2014).

At the beginning of 2016, the Government adopted the Strategy for Social Inclusion of Roma men and women in the Republic of Serbia (2016–2025), as well as the Action Plan for the Implementation of the Strategy (2017–2018). There are approximately 27,000 members of the Roma national minority in the National Employment Service, and 11,000 of them are women, while, on the other hand, there are 100,000 Roma people capable of being employed in Serbia who are not in the records of the National Employment Service, while 90 percent of Roma men and women do not have permanent employment. During the first 10 months of 2018, the active employment policy measures included 4907 unemployed Roma men and Roma women from the NES record. By the end of September, the NES supported 134 Roma people in the start-up of their own business through the grant of self-employment subsidies (NES, 2018).

Positive action measures also exist in Slovenia in relation to the Roma people with disabilities in the field of employment. So, in the region, the Roma people have the best position in Slovenia (Kogovšek Šalamon, 2018). Namely, "the Roma community in Slovenia is positively discriminated, which means that it enjoys some special rights unavailable to the rest of the population" (Stropnik, 2011, p.4). However, it should be noted that in Slovenia, they participate with only 0.2% of the total population. However, they are living in similar conditions as in Serbia and other countries in the region. According to Stropnik research (2011), the majority of them are living in Slovenia in inadequate housing conditions and have poor education.

4. Obstacles that Marginalized Populations Face in Employment

In order to verify why the employment of marginalized population groups in Serbia is difficult, we offered four possible answers to the respondents:

- Timeliness of the adoption of appropriate legal regulations;
- The negative impact of employers' attitudes toward this group of the population;
- Stereotypes concerning their difficult adaptation and integration into the workplace by other employees; and
- Unsatisfactory mechanisms of protection against their discrimination in employment.

Our research conducted in 2016 (Radović-Marković, 2016a) has shown that the largest number of respondents said that because of the existence of stereotypes, it is difficult to get a job in the formal economy (50%). Besides, research has shown that, although there are appropriate laws, they are not respected enough. Given this, they feel insufficiently protected concerning their discrimination in employment. (Figure 6.1)

Respondents in Serbia see the highest chance in employment in the public sector, and they believe that it should be better used for their inclusion

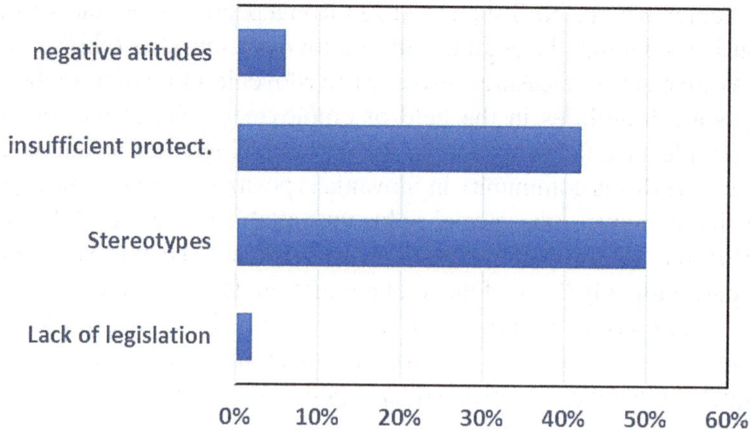

Figure 6.1. Different ways of discrimination in employment (*Source:* Authors)

in the work process (41%). Such a high level of commitment to employment in the public sector can be explained by the fact that the general unemployed in Serbia, including marginalized groups, prefer more permanent employment and choose the employer of a state-owned company rather than work in the informal sector or under a contract. Namely, the state is considered a desirable employer for the regularity of monthly income. Also, a large number of respondents believe that social entrepreneurship could absorb a significant number of unemployed, disabled people, elderly and refugees, as well as all those who are socially excluded or 35%.

Whereas, a smaller number of respondents shared the problem of having lack of entrepreneurial skills and knowledge that often hinders them in entrepreneurial intentions. Namely, the mentioned and similar research concludes that training in the field of entrepreneurship and acquisition of practical knowledge directly influences the entrepreneurial competencies (Radović-Marković, 2013). In the meantime, these findings can be very encouraging for educational institutions to invest in these programs and train at all levels of education, starting from the earliest age of children. In the longer term, this can have very positive effects on economic development, through the development of entrepreneurial culture among young people (Radović-Marković, 2012; Radović-Marković, & Salamzadeh, 2012).

5. Conclusion

Addressing inequalities will play a significant role in the future global development framework. In this context, the post-2015 development framework must be focused on a holistic and rights-based approach to development. A holistic approach to inequalities should include each type of inequality—social, political, economic, and material. Also, it must be based on equality and inclusive participation; ensures that the marginalized groups can benefit from development and growth.

A lack of suitable, well-designed, and well-tailored educational programs to meet the needs of the marginalized groups in Serbia is not the only major cause of social exclusion and the lack of employment opportunities. Besides the lack of appropriate educational programs, the reasons for a high unemployment rate of the marginalized groups might be sought in several factors that are intimately entangled. Factors such as a decline in GDP, decreased productivity, high unemployment rate, budget deficits, lack of social support mechanisms, as well as the lack of skill in tackling the problems among the marginalized groups are among the priority issues of society. In addition to this, one might mention the existence of stereotypes relating to their working and extra capabilities, the lack of measures and mechanisms to protect them, the lack of interest of companies to hire them, and the like are the other factors affecting this situation. According to all these factors, the economic development of Serbia would be guaranteed through investing more in the education of these groups and improving their knowledge base and qualifications. It might result in their higher rates of employment, as well as a decrease in their poverty and social inclusion levels. Thus, social exclusion issues should be dealt with mutually by the Serbian government. Also, the firms and educational sectors would be better to assist the marginalized groups on their way to their social inclusion.

The men and women with the disability, always face the discrimination, because they could not compete with the normal people.

That is crucial in their engagement. However, disabled persons can be useful both for themselves and their families and society as a whole through appropriate training and employment that matches their competencies and work skills. Accordingly, it is necessary to encourage employers to opt for the employment of persons with disabilities instead of paying the penalty fund, specifically to help large companies and systems that, within their organization, can have the whole workshop staffed by disabled persons. Partnerships with NGOs and local governments should be made in order to give employers better information about legal opportunities and working abilities of persons with disabilities. Also, Serbia as a candidate country for accession to EU should follow the experience of the Republic of Slovenia in the field of protection from discrimination and its different measures to improve the situation of people who are in a less favorable situation. The exchange of experiences and information between the two countries can be of great benefit for Serbia in the preparatory phase for joining the EU.

We expect that this research can trigger further research investigations and better inform policymakers toward reducing various forms of inequalities in the labor market.

References

Achakpa, P., & Radović-Marković, M. (2018). Employment Women Through Entrepreneurship Development and Education in Developing Countries. Journal of Women's Entrepreneurship and Education, (1/2), 17-30.

Action Plan for the Implementation of the Strategy (2017-2018). http://www.ljudskaprava.gov.rs/sites/default/files/dokument_file/ad_action_plan_eng_pdf.pdf

Bessette, G., (2011). 'Skills for the Marginalized Youth: Breaking the Marginalization Cycle with Skills Development', in NORRAG NEWS, Towards a New Global World of Skills Development? TVET's turn to Make its Mark, No.46, September 2011, pp. 66-68, available: http://www.norrag.org

Bhandari, Medani P. (2020) Second Edition- Green Web-II: Standards and Perspectives from the IUCN, Policy Development in Environment Conservation Domain with reference to India, Pakistan, Nepal, and Bangladesh, River Publishers, Denmark / the Netherlands. ISBN: 9788770221924 e-ISBN: 9788770221917

Bhandari, Medani P. (2020), Getting the Climate Science Facts Right: The Role of the IPCC (Forthcoming), River Publishers, Denmark / the Netherlands- ISBN: 9788770221863 e-ISBN: 9788770221856

Bhandari, Medani P. and Shvindina Hanna (2019) Reducing Inequalities Towards Sustainable Development Goals: Multilevel Approach, River Publishers, Denmark / the Netherlands- ISBN: Print: 978-87-7022-126-9 E-book: 978-87-7022-125-2

Carlton, S., & Soulsby, J. (1999). *Learning to Grow Older & Bolder: A Policy Paper on Learning in Later Life*. National Institute of Adult Continuing Education, 21 De Montfort Street, Leicester LE1 7GE, United Kingdom, Web site: http://www.niace.org.uk (13.95 British pounds).

Cvejić, S., Babović, M., Pudar, G. (2010). Studija o humanom razvoju – SRBIJA 2010-izvori i ishodi socijalnog isključivanja, UNDP, Srbija.

Cvejić, S. (2014). 'Platforma za sindikalno organizovanje osoba sa invaliditetom', Forum mladih sa invaliditetom, Beograd.

Dench, S., & Regan, J. (1998). Learning in later life: motivation and impact. Institute for Employment Studies.

Economic Commission (2011). Guidance note on the implementation of Integrated housing interventions in favour of marginalised communities under the erdf,28.01.2011. http://www.euromanet.eu/upload/86/94/Guidance_note_Housing_interventions_art__7_2_ERDF.pdf

EIGE(2015). Gender Equality Index 2017. https://eige.europa.eu/gender-equality-index/2015/domain/work

Emler, N., & Frazer, E. (1999). Politics: The education effect. *Oxford Review of Education*, *25*(1-2), 251-273.

EurActiv, (2013). Diskriminacija-problem srpskog društva. http://www.euractiv.rs/ljudska-prava/4514-diskriminacija-problem-srpskog-drutva-

European Union Agency for Fundamental Rights (2019). More targeted measures needed to support Roma women. https://fra.europa.eu/en/news/2019/more-targeted-measures-needed-support-roma-women

Furlong, A., Ferguson, B. and Tilleczek, K. (2011). Marginalized Youth in Contemporary Educational Contexts: A Tranquil Invitation to a Rebellious Celebration, CEA, Canada.http://www.cea-ace.ca/education-canada

Government of Republic of Serbia, (2016). Gender Equality Index for Serbia Measuring gender equality 2014

Gutek, B.A., Guillemard, A.-M. & Walker, A. (1994). Employers' Responses to Workforce Ageing -A Comparative Franco-British Exploration. Paris and Sheffield: University of Paris and Department of Sociological Studies, University of Sheffield

Hill, P.H.and Macan, S. (1996). Welfare Reform in the United States: Resulting Consumption Behaviors, Health and Nutrition Outcomes, and Public Policy Solutions, Human Rights Quarterly, 18(1), 142-159.

International Federation of Social Workers-IFSW, (2012). Ageing and older adults, Bern, Svajcarska.

Jarvis, P., & Walker, J. (1997). When the process becomes the product: Summer universities for seniors. *Education and Ageing*, 12, 60-68.

Kogovšek Šalamon (2018). Country report Non-discrimination Slovenia, European Commission, B-1049 Brussels.

Milanović, L. (2012). Preko zapošljavanja do inkluzije osoba sa invaliditetom u Srbiji, Asocijacija za promovisanje inkluzije – API SrbijeBeograd.

Nacionalna služba za zapošljavanje, Nezaposlenost i zapošljavanje u Republici Srbiji, No.16, Oktobar 2014. http://www.nsz.gov.rs/live/digitalAssets/2/2860_bilten_nsz_10_2014.pdf

Nacionalna služba za zapošljavanje,2014,Okrugli sto o socijalnom preduzetništvu-Model za smanjenje nezaposlenosti. http://www.nsz.gov.rs/live/info/najnovije-vesti/model_za_smanjenje_nezaposlenosti.cid17002?page=0

NES (2014). Okrugli sto o socijalnom preduzetništvu-Model za smanjenje nezaposlenosti. http://www.nsz.gov.rs/live/info/najnovije-vesti/model_za_smanjenje_nezaposlenosti.cid17002?page=0

NGEC 2011, Minority and Marginalised,Kenya.

Ngo, H.Y., Tang, C. & Au, W. (2002). Behavioural responses to employment discrimination: A study of Hong Kong workers. *International Journal of Human Resource Management*, 13, 1206-1223.

Odović, G., Rapaić, D., Nedović, G. (2008). Zapošljavanje osoba sa invaliditetom, Specijalna edukacija i rehabilitacija,2008, br. 1-2, str. 189-206.

Official Gazette of the RS, 36/2009

Radović –Marković, M. et.al. (2012b). Freedom, Individuality and Women's Entrepreneurship Education. У: CATALIN, Martin (ur.), DRUICA, Elena (ur.). Entrepreneurship education-a priority for the higher education institutions. Bologna: Medimond, 2012.

Radović –Marković, M. et.al. (2012c). Creative Education and New Learning as Means of Encouraging Creativity, Original Thinking and Entrepreneurship, World Academy of Art and Science. http://www.worldacademy.org/files/Montenegro_Conference/Creative_Education_and_New_Learning_as_Means_of_Encouraging_Creativity_by_Mirjana_Radovic.pdf

Radovic Markovic, M**.** (2016). The impact of education on entrepreneurs activity and employment among marginalized groups: an evidence of Serbia,9th Scientific Conference on Business and Management, Vilnius,Lithuania.

Radovic Markovic, M. (2016a). Empowering Employment of Women and Marginalized People Through Entrepreneurship Education in Serbia, Journal of Women's Entrepreneurship and Education, 2016, No. 1-2, 3-17.

Radovic Markovic, M., et. al. (2014). Virtual organisation and motivational business management, Alma Mater Europaea – Evropski center, Maribor Institute of Economic Sciences, Beograd.

Radović-Marković, M. (2011). Education System and Economic Needs in Serbia in Collected papers: Active measures on the labor market and employment issues (edited by Jovan Zubovic), Institute of Economic Sciences, Belgrade, pp. 27-42.

Radović-Marković, M. (2011a). Critical Approach towards the Employment Analysis in Theory Methodology and Research in Conference Proceedings: Influence of the Humanities and Social Sciences on Business and Societal Change, Naujosios Kartos Mokslo ir Verslo Klasteris", Vilnius, Lithuania.

Radovic-Markovic, M. (2012). Age discrimination in employment of women in Serbia, Institute of Economic Sciences, Belgrade

Radović-Marković, M. (2013). Female entrepreneurship: Theoretical approaches, *Journal of Women's Entrepreneurship and Education*, 1-2, 1-9.

Radović-Marković, M. (2018). Organisational resilience and business continuity: theoretical and conceptual framework, *Journal of Entrepreneurship and Business Resilience*, 1(1), 5-11.

Radović-Marković, M. et.al. (2009). The new alternative women's entrepreneurship education: e-learning and virtual universities, *Journal of Women's Entrepreneurship and Education*,1-2, 1-12.

Radović-Marković,M.(2012a).Diskriminacija–prvikoraknaputudoposla?,Internetportal „Šta žene žele ?„ http://stazenezele.rs/diskriminacija-prvi-korak-na-putu-do-posla/

Radović-Marković, M., & Salamzadeh, A. (2012). *The nature of entrepreneurship: Entrepreneurs and entrepreneurial activities*. Lap Lambert Academic Publishing: Germany.

Riger, S., & Galligan, P. (1980). An exploration of competing paradigms. *American Psychologist*, 35, 902-910.

Slovenia, Act amending the Act on Equal Opportunities for People with Disabilities (Zakon o spremembah in dopolnitvah Zakona o izena evanju možnosti invalidov), adopted on 30 June 2014, available at: http://www.uradni-list.si/1/objava.jsp?sop=2014-01-2080 (last accessed last accessed 10 July 2019).

Stropnik, N. (2011). Slovenia /Promoting Social Inclusion of Roma, Institute for Economic Research, Ljubljana

UNDP (2006). Poverty, unemployment and social exclusion, Zagreb, Hrvatska

UNDP (2010). Marginalised minorities in Development Programing, New York.

World Bank, World Development Report, (2006). Equity and Development, Washington D.C. World Bank,2006.

Chapter 7

Impact of Subsidies on Social Equality and Energy Efficiency Development: The Case of Ukraine

Sotnyk Iryna, Oleksandr Kubatko, Tetiana Kurbatova, Leonid Melnyk, Yevhen Kovalenko, Almashaqbeh Ismail Yousef Ali

ABSTRACT

Given the low incomes of the majority of population and high prices for utilities, the issue of reducing social inequality in Ukraine is extremely urgent. Today, it is being addressed by the government through the operation of a large-scale state program of utilities subsidies, a significant part of which is paid for energy. At the same time, the national economy has a high level of energy intensity and the housing and utilities sector is the main contributor to this outcome. The high energy intensity of GDP results in a vicious circle: high utilities prices, which are caused by low energy efficiency of both utilities suppliers and consumers, not only "wash out" low income of the population creating a non-payment threat but also make it impossible both to implement energy-saving measures by utilities suppliers and households and to improve the quality of services in the future. The existing subsidies system stimulates the further growth of utilities energy intensity, preventing both utilities consumers and suppliers from energy saving as well as creating a field for abuse, for instance, real income hiding. In this regard, the authors explore the possibilities of modernizing the current system of utilities subsidies in Ukraine in the context of its reorientation to support energy-efficient development of the national economy. Based on the data of the State Statistics Service of Ukraine and international organizations whose areas of interest include energy issues as well as on the results of the authors' research;

the chapter not only provides the analysis of prerequisites of origin and the essence of energy poverty of Ukrainian population but also substantiates its causes and consequences. Using the methods of factor, statistical, structural, and comparative analysis, the authors identify trends and estimate the factors for the development of the housing stock and communal infrastructure in Ukraine in 2010–2018. Besides, they evaluate the effectiveness of the state subsidy system for maintaining social equality and improving the energy efficiency of domestic utilities. Based on paired regression analysis, the relationships between utility prices, subsidies to the population, and the level of energy efficiency of the national economy and its housing and communal sector for 2014–2018 were established. To estimate the impact of utility costs and related factors on the poverty of the Ukrainian population and energy efficiency of the domestic economy in the long run, the corresponding econometric models were constructed, covering 1999–2018 data. Considering the empirical results, the problems of social equality observance in the country through subsidies and ensuring energy-efficient development of the state, as well as ways to overcome them have been identified. On this basis, the authors have proposed improvement for the state mechanism of utilities subsidies, which should (1) be combined with the existing state programs of energy-efficient development, in particular, the program of "warm" credits and feed-in tariff for renewable energy objects in households, (2) enhance the expansion of cooperation with international financial organizations to implement energy-efficient measures by Ukrainian households and utilities suppliers, and (3) allow comprehensive solutions to eradicate the energy poverty, ensure the growth of energy efficiency in the national economy and provide social equality among society members.

1. Introduction

Energy-efficient development of the countries is a priority for many national governments today, considering the increasing of states' energy independence, saving their energy resources, reducing greenhouse gas emissions from burning fossil fuels, meeting energy needs of the growing nations (Bilan et al., 2019; Cirman et al., 2012; Frondel et al., 2015; Lihtmaa at al., 2018; Shih et al., 2019; González-Eguino, 2015; Sineviciene et al., 2017; Sotnyk et al., 2015; Vasylieva et al., 2019; Bhandari 2018; 2020). Along with achieving economic, energy, and environmental development goals, remarkable accomplishments of energy efficiency improvement are social inequality reduction in the regions, energy poverty alleviation, and improvement of the quality of life of the population (Bouzarovski & Petrova, 2015; Bouzarovsky & Tirado Herrero, 2017; Braubach & Ferrand, 2013; Buzar, 2007; Hernández & Bird, 2010; Kaygusuz, 2011; Li, et. al., 2019, Melnyk et. al., 2019, Melnyk et. al., 2016, Petrova et al., 2013; Thomson et al., 2017; Urge-Vorsatz & Tirado Herrero, 2012). Simultaneously, social programs aimed to support socially vulnerable groups often create additional barriers to reducing the energy intensity of national and regional economies, stimulating the irrational consumption of energy resources. Among these programs are state utilities and energy resources subsidies for households intended to combat energy and fuel poverty of the population.

The problem of energy and fuel poverty of households has been widely researched in recent years by scientists given the relevance of this issue in many countries around the world (see, for example, (Boardman, 2012; Bouzarovski & Petrova, 2015; Bouzarovsky & Tirado Herrero, 2017; Fankhauser & Tepic, 2007; González-Eguino, 2015; Hernández & Bird, 2010; Hills, 2012; Kaygusuz, 2011; Petrova et al., 2013; Thomson & Snell, 2013; Thomson et al., 2017; Waddams & Deller, 2015)). Thus, Bouzarovsky & Tirado Herrero (2017) define energy poverty as "the inability to secure a socially and materially necessitated level of energy services in the home". Based on this definition, Lihtmaa et al. (2018) examine this phenomenon in the context of its relationship with efficiency-based renovation subsidies for apartment buildings in Estonia. Their research shows that many European Union (EU) countries are characterized by high levels of energy poverty: from 14.8% for the Czech Republic to 39.5% for Romania and 40.1% for Bulgaria in 2014. Estimates of energy poverty rates in European countries during different periods were made by many other researchers (see, for example, (Bouzarovski et al., 2016; Brunner et al., 2012; Buzar, 2007; Dubois & Meier, 2016; Frondel et al., 2015; Gerbery & Filčák, 2014; Legendre & Ricci, 2015; Papada & Kaliampakos, 2016; Phimister et al., 2015)) and have become the background for active political action. For instance, in 2017 the Republic of Ireland, which has high levels of energy poverty, announced the funding of €10 million for the Warmth and Wellbeing scheme that provides free energy efficiency upgrades for households who are classified as energy-poor and contain young children with chronic respiratory diseases (Health, 2017). In general, lower incomes are associated with worsened health indicators (Kubatko & Kubatko, 2017; Kubatko & Kubatko, 2019). Therefore, many states of the USA (DSIRE, 2019) propose different programs to their low-income households in order to increase energy efficiency of homes.

Estache et al. (2002) and Foster et al. (2000) used affordability indicators to measure energy poverty in Latin America. Considering affordability as the share of monthly household income that is spent on utility services, Fankhauser & Tepic (2007) study the affordability of electricity, district heating and water for low-income consumers in transition countries. Based on the consensual approach, proposed by Healy & Clinch (2002), Thomson & Snell (2013) research fuel poverty in EU countries. Healy & Clinch (2002) determine fuel poverty through "analysis of subjective and objective data on the presence of mold, the absence of central heating and ability to keep warm." With regard to this approach, Thomson & Snell (2013) estimate the fuel poverty levels of European countries, which in different scenarios reach the highest values for Bulgaria (31.1%), Cyprus (23.8%), and Romania (24.6%), while the lowest levels of fuel poverty are found in Denmark (2.7%), Finland (3.8%), and Sweden (4.1%).

By linking energy and fuel poverty directly with unsatisfactory levels of nations' energy efficiency, the UNECE identifies an energy inefficiency trap as "a situation in which countries with lower energy efficiency are unable to change their respective status due to lack of funds, experience, technology, motivation, and initiative" (UNECE, 2009, p. 7). To overcome this trap, Cirman et al. (2012) propose the proper subsidy targeting energy efficiency improvements for Central and Eastern European countries, where the problem of the inefficient housing stock is the most urgent. The authors argue that grants and subsidies for energy efficiency development serve as a powerful tool to encourage the co-operation of individual owners and improve the organizational performance of the management of privatized multi-dwelling buildings. However, Lihtmaa et al. (2018) and Mlaabdal et. al. (2018) point to the need to respect social equality both in providing access to such subsidies and in controlling its use, that includes the prevailing implementation of energy-saving projects in the residential sector of particular regions where the population is more socially active than in depressed territories. Violation of such a balance can lead to crisis deepening in the housing stock of certain territories against the formation of energy-efficient housing estates centers in other regions, increasing regional development disproportions. All in all, the above-mentioned subsidy type provides an efficient mechanism for solving the issue of energy and fuel poverty over the long term, but it is not always able to deliver the desired results in the short term.

Analyzing the energy poverty of American households, Hernández & Bird (2010) claim that energy burden for the low-income population is much higher than many policymakers would assume. Therefore, weatherization, utility and housing assistance policies provided by state and local authorities can significantly help in reducing energy poverty levels and require a coordinated, regional approach that integrates programs in each area of energy-efficient changes.

It is worth considering the paper by Frondel et al. (2015) in the context of studying the causes and effects of energy poverty. This study deals with the impact of green energy development on rising electricity prices in Germany and the energy poverty of households; thereby, it reveals new causes of poverty. Traditionally, energy poverty is considered as the inability or limited access of the population to energy sources and utilities due to the underdeveloped energy infrastructure, which is the case for developing southern countries. For developed northern states and many emerging economies that have high levels of energy infrastructure development, fuel poverty is more characteristic, when part of the population is unable to pay for consumed energy and utilities at current prices and accumulates debt or is forced to refuse or reduce energy consumption, resulting in the deterioration of the quality of life of families. In the case of countries that are actively developing green energy, such as Germany, Denmark, Spain and others, where preferential feed-in tariffs for energy producers are applied, one of the reasons for energy poverty may be the constant rise in energy prices for households due to the need to compensate feed-in tariff payments. Thus, in Germany, the use of a feed-in tariff, which has contributed to the significant development of renewable energy in the country, and its coverage by raising electricity prices for all energy consumers, has exacerbated the issue of the ability of many poor families to carry an increasing financial burden. To support social equality, restrain price rises and further development of renewable energy, the authors argue for the feasibility of changing energy pricing mechanisms by moving from feed-in tariffs to green auctions and directly subsidizing both energy-efficient measures and the construction of renewable energy facilities in poor households.

The direct negative consequences of energy and fuel poverty are the inability of the population to pay for consumed energy and other utilities due to low household incomes and high prices for such resources, as well as population stratification and intensification of social conflicts. Moreover, Urge-Vorsarz & Tirado Herrero (2012) point out that rising energy prices by national governments, for example, to stimulate greenhouse gas emissions reduction without adequately protecting vulnerable social groups can lead to social discontent and have the opposite effect in the form of changing into high-carbon economies with increasing levels of energy poverty. Therefore, to combat energy and fuel poverty outcomes, researchers pay particular attention to the impact of utility and energy subsidies on reducing energy poverty of the population (see, for example, Buzar, 2007; del Granado et al., 2012;

Fattouh & El-Katiri 2013; Hernández & Bird, 2010; Saunders & Schneider, 2000; Solaymani, 2016; Sovacool, 2017). It is the simplest tool used by countries around the world to quickly solve the problems of social inequality generated by high energy prices. Meanwhile, many scientists conclude that the mechanism of state subsidies for utilities and energy does not allow the households to be interested in energy-efficient development of the territories and needs significant reform (Fattouh & El-Katiri, 2013; Johnston at al., 2014; Kerimray et al., 2018; Poputoaia & Bouzarovski, 2010; Saunders & Schneider, 2000; Solaymani, 2016; Sovacool, 2017). For example, Saunders & Schneider (2000) believe that the advantage of implementing energy consumption subsidies is to ensure access to a minimum level of energy consumption for all consumers, especially poor households. Moreover, increasing energy consumption in such a way encourages industrial growth and reduce unemployment. On the other hand, payment of subsidies causes disruptions in the price mechanism and results in inefficient use of energy in the economy (del Granado et al., 2012; Fattouh & El-Katiri, 2013; Poputoaia & Bouzarovski, 2010). Despite these negative effects, many empirical studies prove that removal of energy subsidies would decrease the households' welfare and increases the level of poverty, especially in developing countries and emerging economies (BuShehri & Wohlgenant, 2012; Lin & Jiang, 2011; Siddig et al., 2014; Solaymani, 2016). Therefore, changes in utilities and energy subsidies mechanisms require balanced and visionary solutions.

As practice shows, the most vulnerable European countries concerning energy and fuel poverty and those in need of state utility subsidies for the population are the countries of Central and Eastern Europe, which have a socialist past, and many former Soviet Union republics. Inherited from those times, they received an outdated housing stock, which was often built without regard for the energy efficiency requirements because of the low cost of energy resources in the countries of the socialist bloc and the Soviet Union (Buzar, 2007; Fankhauser et al., 2008; Kerimray et al., 2018; Korppoo & Korobova, 2012; Lihtmaa et al. 2018; Petrova et al., 2013; Poputoaia & Bouzarovski, 2010; Thonipara et al., 2019). Renovation of this stock in modern conditions requires considerable investment to reduce energy costs for its maintenance. To encourage the population to upgrade their housing stock through the innovative energy-efficient technologies use, many national governments are resorting to increasing utility tariffs, which may also be dictated by rising world prices for imported energy. However, the vast majority of housing stock in the former socialist countries is now privately owned, and the funds of its owners are often not enough even to pay for utilities, not to mention investments in the weatherization of buildings. In this regard, there is a vicious circle of energy poverty: households cannot pay for high-value utilities that are often used inefficiently and are unable to reduce their consumption without significantly reducing their quality of life due to lack of funds. On the contrary, utility suppliers can't receive the funds for services provided on time and in full, which leads to quality deterioration of the latter, bankruptcy threat of public utilities and ultimately the collapse of the housing and utilities sector of the territories. As practice shows, deploying state subsidy programs by governments to solve the problem of energy poverty is not an acceptable way out, because subsidies do not have a remarkably positive effect on energy efficiency changes in households, dissuading the population from reducing energy costs and thereby resulting in energy poverty and getting the country into the energy inefficiency trap.

In this study, we explore the impact of increasing utility subsidies on the energy efficiency of housing and utilities sector development in Ukraine that stands as a stark illustration of a state in the energy inefficiency trap. To that end, we have examined the main trends, status and factors of development of housing stock and communal infrastructure in the country for the period 2010–2018. On this basis, we have identified the causes of energy poverty of domestic households. Further, we have evaluated the effectiveness of the state subsidy system in the context of overcoming social inequality and ensuring energy-efficient development of the country's housing and utilities sector. Our empirical contribution, which demonstrates how utilities subsidies and other related factors influence energy-efficient development and energy poverty during the periods of 1999–2018 and 2014–2018, fills a gap in scholarly research in explaining the impact of utilities expenditures and utilities subsidies on social and energy development of emerging economies. Based on the empirical results obtained, we offer conclusions and identify recommendations for improving the coherence between utilities subsidies, energy poverty and energy efficiency development mechanisms at the national level.

2. Methods

We have used the methods of factor, statistical, structural, and comparative analysis to identify trends, assess the status and determine factors for the development of housing stock and communal infrastructure in Ukraine in 2010–2018, as well as identify the causes of household energy poverty. The use of qualitative and quantitative analysis made it possible to assess the impact of the state subsidy system on social equality and energy efficiency in the country's housing and utilities sector. The method of paired regression analysis was used to establish the relationship between utility prices, subsidies to the population, and the level of energy efficiency of Ukrainian national economy and its housing and utilities sector in the 2014–2018 period, when the largest increase in prices and subsidies took place.

To find out the impact of utility expenditures and related factors on the poverty of Ukraine's population and energy efficiency of the domestic economy in the long run, we have constructed econometric models covering the 1999–2018 period. The chosen study period is explained, firstly, by the availability of comparative data that was taken from the database of State Statistics Service of Ukraine. Secondly, the 20-year period allows building econometric models characterized by acceptable reliability of the results. Thirdly, 1999 was the last year of the crisis decade, in which the subsidy system in Ukraine had been functioning steadily and the income of the population had started to increase gradually.

Considering repeated changes in the official methods of calculating statistics in Ukraine over the last 20 years as well as the introduction of new indicators and the removal of some indicators collected by the State Statistics Service of Ukraine and submitted to international databases, for modeling purposes we have taken the effective indicators of poverty headcount ratio below national poverty lines (% of population) as a proxy for the energy poverty indicator and GDP per unit of energy (constant 2011 PPP USD per kg of oil equivalent) as a proxy for the energy efficiency indicator.

For the same reason, model factors included GDP per capita (constant 2010 USD) as a proxy for household income, gross capital formation (current USD) as a proxy for investment in energy efficiency measures, expenditures for utilities (%) as a share of household expenditures on payment for utilities in their income, and crude oil average prices (USD) as a proxy for utilities consumer prices. So, we built the following models:

$$PR_t = \beta_0 + \beta_1 GDP\_pc_t + \beta_2 GFCFt + \beta_3 EXU_t + u_t, \quad (1)$$

PR_t – is the poverty headcount ratio below national poverty lines (% of population) in year *t*;
GDP_pc_t – is the GDP per capita (constant 2010 USD) in year *t*;
$GFCF_t$ – is the gross capital formation (current USD) in year *t*;
EXU_t – is the expenditures for utilities (%) in year *t*;
β_0 – is a constant term (cons);
$\beta_1, \ldots, \beta_3$ – are the regression coefficients of the model;
u_t – is an error term;

$$GDP\_E_t = \alpha_0 + \alpha_1 GDP\_pc_t + \alpha_2 GFCF_t + \alpha_3 EXU_t + \alpha_4 CRO_t + \upsilon_t, \tag{2}$$

GDP_E_t – is the GDP per unit of energy (constant 2011 PPP USD per kg of oil equivalent) in year *t*;
CRO_t – is the crude oil average prices (USD) in year *t*;
α_0 – is a constant term (cons);
$\alpha_1, \dots, \alpha_4$ – are the regression coefficients of the model;
υ_t – is an error term.

These models were analyzed using the ordinary least square technique for time series data within the software application Stata 14.0.

Based on the obtained empirical data and the results of qualitative analysis, the authors formulated recommendations for the state subsidy system transformation to ensure the growth of energy efficiency of the national economy, reduce the level of energy poverty and respect social equality.

3. Prerequisites of the Origin and the Essence of Energy Poverty of Ukrainian Population

Issues related to the energy efficiency improvement of the housing stock and the housing and communal sector enterprises in the context of reducing the utility cost and overcoming social inequalities such as energy and fuel poverty of the population are of great importance to Ukraine (Fankhauser et al., 2008; Komelina & Maksimenko, 2014; Petrova et al., 2013). However, in this country energy poverty should be considered precisely as the financial inability of households to pay for consumed energy and utilities, rather than as limited access to energy infrastructure, given its high development. Thus, in 2018, 37.4% of Ukrainian families had access to central heating, 45.2%—to individual heating, 80.9%—to water supply, 80.4%—to sewerage, 40.0%—to hot water supply, 77.8%—to centralized gas supply, 11.7%—to bottled gas, around 100%—to electricity (State, 2019).

After the collapse of the Soviet Union and given the scarcity of own energy resources, the state was forced to purchase energy at world prices that were significantly higher than the Soviet ones. Absence of sufficient financial resources for renewal of communal infrastructure and housing stock on energy-efficient basis for almost 30 years of independence, constant conflicts with the Russian Federation on purchase prices for imported natural gas, low income of the majority of households, and inefficient state policy of housing and communal sector development caused the accumulation of many structural problems of the sector that directly affected the population's life quality. In an effort to protect the socially vulnerable segments of the population from excessive financial burdens and avoid the crisis of non-payment for utilities, the Ukrainian government has been implementing the state utility subsidy program for several decades. However, while it was not massive till 2014, subsidies have become widespread due to the multi-stage increase in prices for basic energy resources for the population since 2014–2015.

It should be noted that today in Ukraine, a large part of the state utility subsidies for the population relates namely to the payment of energy: natural gas, heat and electricity (up to 80–85% in the heating season). Nonetheless, the domestic economy is characterized by a high level of energy intensity, which in 2018 exceeded the indicators of the developed countries by 2–4 times and the world average by 2.08 times (Figure 7.1). Even compared to the former Soviet Union republics (e.g., Russia and Kazakhstan), the energy intensity of Ukraine's GDP was 1.1–1.3 times higher. In the meantime, the housing and utilities sector, providing services to the population, is one of the most energy-intensive sectors, consuming more than 30% of the country's energy resources. For instance, in 2017, the share of final energy consumption in the household sector amounted to 32.8% of total final energy consumption by the national economy. This percentage share has increased by more than

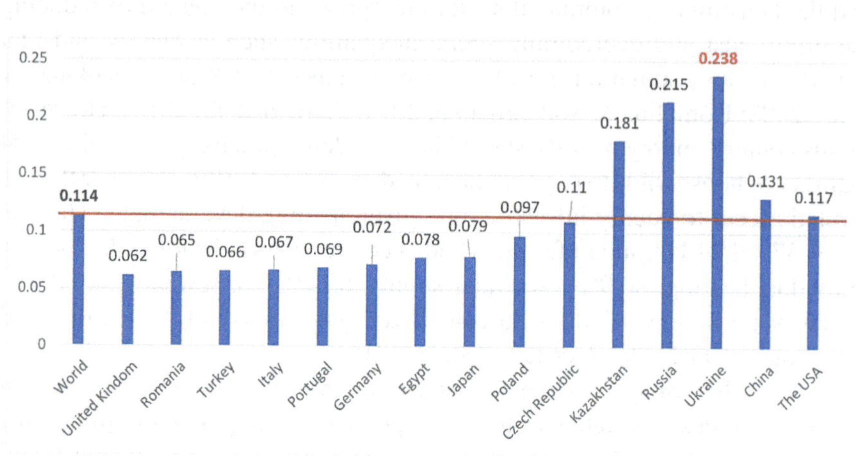

Figure 7.1. GDP energy intensity of some countries in 2018, koe/$2015P (Global, 2019)

22% over the last 10 years (from 26.8% in 2007) (State, 2019). The reasons for the growth are both outdated housing stock and housing and utility infrastructure, which do not meet current energy-efficient requirements and are not updated on time, causing a backlog of the housing and utilities sector in rational resource use from other areas of economic activity.

More than 25% of Ukrainian residential buildings are over the age of 60–70 years and over 85% of the housing stock was built by the 1990s. At least 70% of buildings need ongoing or major repairs and 90%—complete thermal modernization (MDL, 2018; Nova 2016; State, 2019). The annual heat losses in the heating of old homes are estimated at up to 400 kWh/yr compared to modern energy-efficient analogues with heat losses of 20–25 kWh/yr. Therefore, the potential for reducing heat losses reaches 16–20 times (Nova, 2016).

According to the State Statistics Service of Ukraine, as of January 1, 2019, the housing stock of the country amounted to 993.3 million m^2 of total area, with the number of dilapidated and emergency buildings amounting to 45.5 thousand units, in which 59.8 thousand people lived (State, 2019). On average in Ukraine at the beginning of 2019, 0.33% of the housing stock was considered dilapidated and 0.1% of the dwelling was in an emergency condition. The most problematic regions were Kharkiv, Zhytomyr, and Odesa regions with 0.64–0.77% of dilapidated buildings and Donetsk, Poltava, Cherkasy, and Odesa regions with 0.19–0.21% of emergency housing stock (i.e., almost twice as high as the national average) (State, 2019). Considering age and condition of the housing stock, even the most optimistic estimates anticipated that less than 9% of the dwellings today satisfies current energy efficiency requirements, which is explained by the chronic lack of funds of the population and the reluctance of owners, first of all, of apartment buildings to invest in its energy-efficient renewal based on shared responsibility. Comparing the state of the country's housing stock with previous years, it should be noted that the share of dilapidated and emergency housing stock remains almost at a stable level, undergoing slight annual fluctuations and witnessing the mismanagement of both local authorities and homeowners.

The issue of housing and utilities infrastructure is not much better: 2/3 of the fixed assets of the sector have exhausted their lifetime, heat losses and water leaks in external networks reach 60%, heat losses in housing stock exceed 30% while obtaining more accurate estimates is complicated by the lack of complete input heat metering in households, who are the largest consumers of heat (Green, 2015). Specific energy consumption is 2.5 times higher than in Europe, the number of breakdowns in the last decade has increased almost 5 times (Komelina & Maksimenko, 2014). On average in the country, more than 18% of heat and steam networks and about 35% of water supply networks are considered to be obsolete and emergency, while in some regions these figures reach 40–54% (in particular, in Dnipropetrovsk, Lugansk, and Lviv regions) (State, 2019).

The threatening condition of the housing and utilities infrastructure, which is constantly deteriorating, causes a decrease in the quality of public utilities along with increasing its cost. The consequence is the government's forced measures to raise energy prices in the face of growing consumer dissatisfaction with the quality of utilities provided and, therefore, a decrease in motivation to pay for consumed services of low quality. Thus, in 2014–2019, prices for such basic energy resources as natural gas and electricity increased almost 12 and 3 times respectively (Figures 7.2and 7.3). However, this increase in prices was not accompanied by an adequate increase in real incomes of the population, which led to the need to expand the state subsidy program during this period. Thus, in 2014–2015, the real incomes of the population showed a significant fall (to 88.5% and 79.6% respectively) with the increase of the share of utilities in household expenditures from 10.3% to 12.5% due to a rise in consumer price indices for energy by 134.3% and 203% respectively (Table 7.1). In 2016–2017, the share of utility expenditures of the population continued to increase to 17.7%, with prices rising to 147.2% and 110.6% in 2016 and 2017 respectively. Hence, during 2014–2016, the growth rates of energy and utilities prices were steadily higher than the dynamics of real household incomes (State, 2019), although its growth in 2017–2018 stagnated

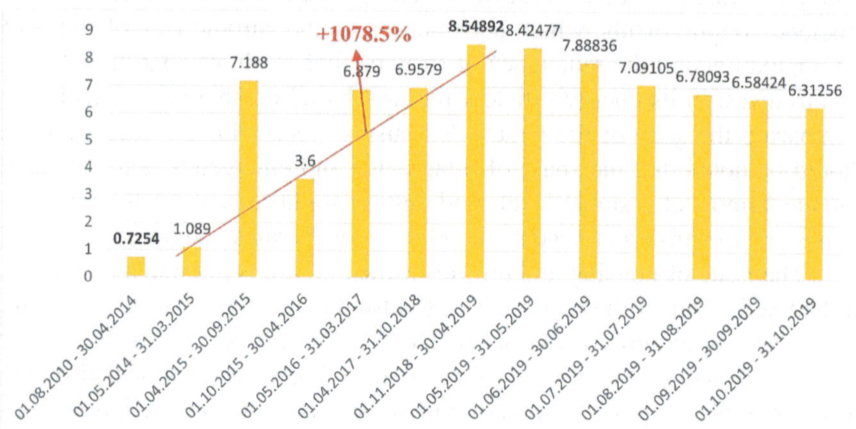

Figure 7.2. Growth in minimal natural gas prices for Ukrainian population in 2014–2019, UAH/m³ (Naftogas, 2019)

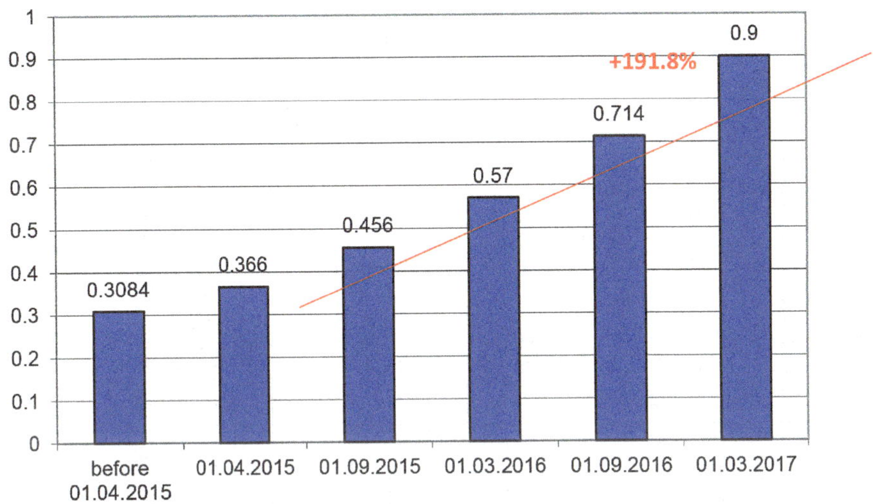

Figure 7.3. Growth in minimal electricity prices for Ukrainian population in 2015–2017, UAH/kWh (National, 2019)

somewhat (Table 7.1). However, considering the growth rates of real incomes in 2014–2018, the incomes have not even managed to catch up with rising energy and utility prices. Today, according to official estimates, about 40% of Ukraine's population is below the poverty line, receiving cash income that is below the actual subsistence level and could potentially be invested in energy-efficient projects, with the largest share of such population in 2016 (see Table 7.1). Considering the statutory subsistence level, which is practically half of the actual subsistence minimum, during the last 9 years the population of Ukraine was the poorest one in 2010, 2012, and 2014. Therefore, in a situation where 40 to 60% of country's households have cash incomes below the subsistence level, it is not advisable to expect large-scale investments in improving the energy efficiency of the residential sector. Besides, about one-third (23.7% in 2018) of Ukrainian households are made up of elderly women and men (59 years and older), who due to their low incomes are the main beneficiaries of subsidies (Expenditure, 2019). On the other hand, this category of the population cannot take credit for introducing energy-efficient measures, as banks do not risk lending to senior citizens.

An important obstacle to increasing the energy efficiency of housing and utilities sector and overcoming energy poverty in Ukraine is the mentality of the population, which is accustomed to expecting assistance from the state. Numerous attempts by the government and local authorities to shift

Table 7.1. Indicators of real incomes of Ukrainian households, its share of total energy expenditures and energy consumer price indices for the population in 2010–2018 (State, 2019)

Year	Real disposable households' incomes, % to the corresponding period of the previous year	Share of monthly total households' expenditures on housing, water, electricity, gas and other fuels per household, %	Consumer price indices for housing, water, electricity, gas and other fuels for households, % to December of the previous year	Percentage of households with an average per capita equivalent monthly cash income below subsistence minimum*, %:	
				statutory	actual
2010	117.1	10.0	113.8	24.3	-
2011	108.0	10.3	111.0	11.4	-
2012	113.9	10.7	100.7	12.1	-
2013	106.1	10.2	100.3	11.6	-
2014	88.5	10.3	134.3	12.0	-
2015	79.6	12.5	203.0	9.4	60.7
2016	102.0	16.6	147.2	9.3	63.1
2017	110.9	17.7	110.6	6.4	48.0
2018	109.9	16.0	110.6	4.1	38.8

*The average monthly amount of the statutory subsistence minimum in 2015 was 1227.33 UAH, in 2016—1388.08 UAH, in 2017—1603.67 UAH, in 2018—1744.83 UAH per capita per month; the average monthly actual subsistence minimum in 2015 was 2257.0 UAH, in 2016—2642.38 UAH, in 2017—2941.46 UAH, in 2018—3262.67 UAH per capita per month (State, 2019).

state's responsibility for multi-family housing stock, where for early 2019 49.7% of households lived, to its real owners are not always successful. It is evidenced by the slow dynamics of the share of condominiums formed (housing cooperatives), building cooperatives and other bodies of self-organization of the population in the regions of the country that voluntarily undertake the management and maintenance of own housing. Thus, as of October 2019, only 17.6% of all apartment buildings in Ukraine had housing cooperatives, 7% had own elected governors, 22.7% of the houses got a manager, appointed by local self-government bodies according to the results of competitions (Minregion, 2019). The passivity of the population to elect the managers of apartment buildings due to the necessity to incur additional financial costs for the maintenance of common property continues to generate mismanagement and worsens the condition of the housing stock.

In turn, the lack of housing stock's renewal on energy-efficient grounds leads to a vicious circle of energy poverty: high utility prices, caused by the low level of energy efficiency of both utility suppliers and consumers, "wash out" low income of the population, create a threat to non-payment for utilities as well as prevent the introduction of energy-saving measures by utility suppliers and households, provoking future price increases without improvement of the utility's quality. The existing system of subsidies only pushes further growth of utilities' energy intensity, depriving consumers and utility suppliers of energy saving, creating a field for abuse, in particular for concealing real income. In this regard, both the consumers' willingness to pay for provided utilities of unsatisfactory quality and the ability of providers to reliably supply qualitative utilities are significantly reduced. Therefore, there is a need to modernize the current utilities subsidy system of Ukraine in the context of the reorientation to support energy-efficient state development.

4. Analysis of the Effectiveness of the Modern Subsidy System in Ukraine

The current state subsidy system in the country is the largest social program of the government in terms of financing. This program was launched in 1995 to assist low-income families to pay for utilities. For more than two decades the program remains the main social protection mechanism in the context of increasing utility prices and tariffs. However, Ukrainian households have started to use subsidies most actively in the last 5 years (Figures 7.4 and 7.5), when the mechanism of its calculation was significantly changed, and

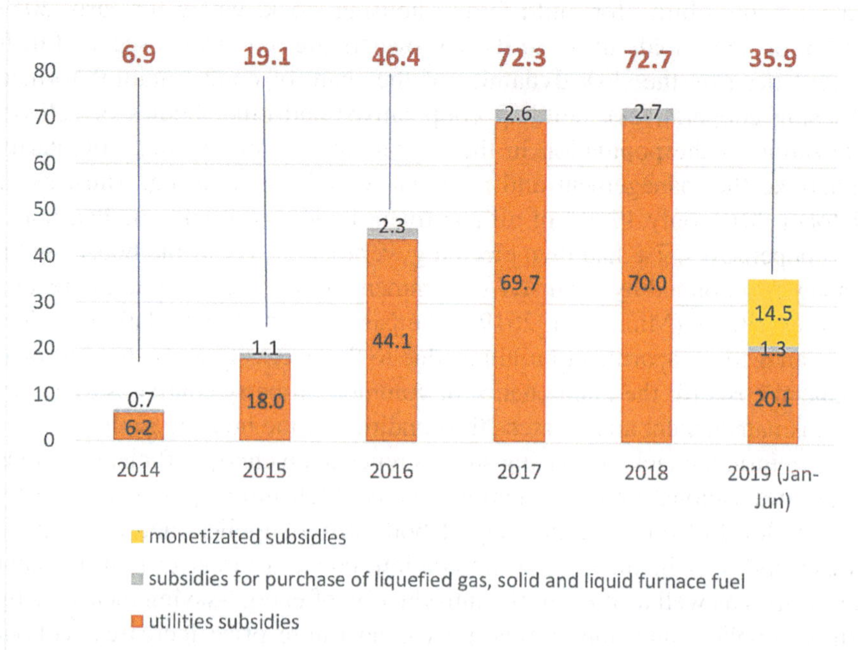

Figure 7.4. Utilities subsidies for Ukrainian population in 2014–2019, bl UAH (Subsidies, 2019; Zanuda, 2018)

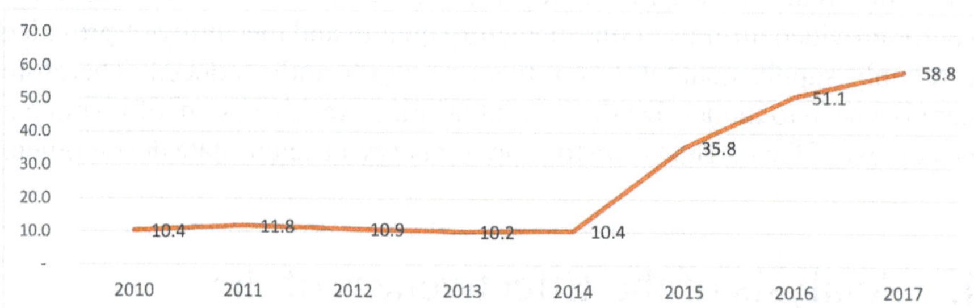

Figure 7.5. Percentage of Ukrainian households which received utilities subsidies in 2010–2017, % (calculated by the authors according to the data (State, 2019))

they became available to most of the country's population. While in 2010–2014 the share of subsidized households did not exceed 12%, in 2015 more than a third of Ukrainian families was subsidized by the state, and in 2016–2018 about 60% of households received utilities subsidies.

Along with changing the rules of the game, another and the main reason for the increase in subsidies was the significant increase in utility prices, especially for natural gas and electricity for households in 2014–2018 (see Figures 7.2 and 7.3). Consequently, while in June 2014 the average state aid per household amounted to 71.7 UAH, in June 2019 it was already 116.4 UAH. The number of families receiving subsidies has also increased: in June 2014, it was 116.3 thousand, whereas in June 2019, it counted 643.6 thousand (Subsidies, 2019). Depending on the region, the percentage of subsidized families reached up to 83% of the total number of households. Traditionally, subsidies are the most demanded by urban residents (66–78% of subsidies depending on the year). More than 65% of families receiving subsidies consist of one person (State, 2019).

The structure of subsidies provided during 2012–2018 has undergone some changes, which is explained by increasing prices for various utilities for the population. While in 2012, the shares of subsidies for natural gas and district heating were approximately equal to each other, since 2014, there has been an increase in the share of subsidies for natural gas payment due to the rapid raising in gas prices (in some years more than 40% in the structure of total subsidies, see Figure 7.2). As a result, subsidies for district heating and hot water supply were the second largest part (about 30%). During the above-mentioned period, due to the increase in the share of subsidies for natural gas and heating, the share of costs for the maintenance of houses significantly decreased from 21–22% in 2012–2013 to 15–16% in the following years (State, 2019). Mirroring the trends in changing shares of subsidies amounts for natural gas and heating, the households' payment levels for these utilities have changed too. While by 2014, the lowest payment level was typical for district heating and hot water supply (86–97%), after 2014, natural gas with the lowest payment rates (up to 73–74% in 2016 and 2018) ranked first among other types of utilities, which is explained by the largest amounts of payments for gas consumption. The increase in government subsidies for natural gas provided to the population did not have a positive effect on the payment discipline of Ukrainians: in 2012–2014 the payment level was consistently higher and reached 97–102%. Firstly, such dynamics can be explained by the low incomes of households, which, while surviving, prefer to spend more on food rather than pay for utilities. Secondly, debt collection requires a utility supplier to apply to a court and pay the initial legal fee. However, for small amounts of debt scattered over a large number of debtors, such costs are unprofitable. The duration of trials and the high probability of entering into debt restructuring agreements before delivery of a verdict hold utilities suppliers away from lawsuits and enable non-payer households to defer payment for a later period, violating financial discipline. Bearing these aspects in mind, the government steadily tightens its requirements for timely payment of utilities by grantees, making them liable for non-payment, in some cases even refuses to provide a subsidy.

Currently, government subsidies can be obtained for the following services (provided that the actual household utilities expenditure exceeds 15% of its total income):

- utilities—for natural gas supply and distribution, electricity supply and distribution, heat supply, hot water supply, centralized cold-water supply, sewage, household waste management (monthly);
- subscription fee for utility customers provided in apartment buildings under individual contracts (monthly);
- cost of managing an apartment building that is functioning as a housing cooperative or a building cooperative (monthly);
- cost for the purchase of liquefied gas, solid and liquid furnace fuel (once a year) (Ministry, 2018).

In contrast, those households are not eligible to receive state subsidies, any members of which have either made a one-off purchase of goods or services in the amount at least of 50,000 UAH (about $2,000) in the last 12 months or have owned a vehicle aged less than 5 years, or do not pay social security taxes, or have arrears of utility bills, or have heat supply area that is above the installed standard (over 200 m^2) (Kalachova, 2018; Ministry, 2018). These restrictions make it possible to obtain state aid for precisely those households that really need it, while at the same time disciplining the recipients of subsidies to pay their share of utility bills without delay and, therefore, to save the state budget funds for the payment of subsidies.

Comparing 2014 and 2019, it should be noted that the requirements for both the state and the population to pay and use subsidies have been strengthened. In particular, under the old system, the timing of delivering subsidies by the state had not been set, whereas under the new one a subsidy must be delivered to a household not later than 24th of every month. Besides, subsidies are nowadays calculated based on the actual consumption of utilities by the household, whereas previously they were calculated based on social standards, which led to the overpayment of utility bills of the most economical grantees. The principle of calculating the mandatory percentage of the household payment, which now depends solely on family income, has changed, as well as the social standard of housing has been increased, giving more households the right to get subsidies (Minregion, 2018).

Simultaneously, during 2018–2019 the number of households receiving subsidies decreased slightly. According to the estimates of the Ministry of Social Policy of Ukraine, the reasons were the more stringent government requirements for recipients of subsidies and the gradual increase in households' incomes over the last 2–3 years, which in 2017 slightly exceeded utility prices indices (see Table 7.1). On the one hand, the latter indicates a revival of economic activity in the country and improvement of macroeconomic indicators, which have a positive impact on household incomes. On the other hand, it attests to the reduction of the state budget expenditures on utility subsidies, while the released funds can be used for other programs, e.g., ensuring energy-efficient household development. However, the subsidy amounts provided by the state budget increased steadily in 2014–2018, which indicates that the energy poverty of the country's population during this period increased.

An important feature of the state subsidy system reform in 2014–2015 was its personification. Until 2014, there was a mechanism for paying utility bills for subsidized households, under which direct government compensation was transfered to utilities suppliers for reduced (social) utility bills for subsidized households. Thus, governmental payments did not rearch households and the latter could not appreciate its value and the importance of saving utility costs. Such a mechanism has led to over-spending of the state budget funds and problems of ensuring the profitability of state-owned energy companies. For example, in 2014, the deficit of the Ukrainian energy company Naftogaz, the main gas supplier to the population and enterprises in the country, was 120 billion UAH, which had arisen from the application of preferential gas prices and delays in budgetary payments (Zanuda, 2018). When the government introduced targeted subsidies for utilities in 2015, spending 18 billion UAH on it, the positive effect on the state budget from the implementation of the new mechanism was at least 120 – 18 = 102 billion UAH compared to 2014 (not including the increase in gas prices in 2014–2015). At the same time, "Naftogaz," a chronically unprofitable state-owned company, turned into one of the most profitable enterprises in the country in 2016–2018, receiving a record 39.3 billion UAH profit in 2017 (Naftogas, 2019a).

Considering low incomes of the population, further increase in utilities and energy prices in 2015–2018 led to an increase in government subsidies. The huge public funds, due to the lack of control mechanisms for funding, allocation and spending of subsidies, have opened up a wide scope for abuse. In response, the government began tightening subsidy requirements to eliminate the possibility of subsidizing high-income earners. Subsidies have become even more targeted, and nowadays they include the income of all family members, not just those living in an apartment or house. To stimulate grantees' interest in saving utility costs and implementing energy-saving measures, the government is constantly improving its subsidy mechanism making recipients responsible for timely and full payment for utilities. Thus, in 2018, a record year in terms of subsidies given to the population, there were significant changes in the way they were allocated (Ministry, 2019). In particular, the criteria for providing this type of state support were reviewed to improve efficiency and enhance targeting. At present, the housing subsidy program supports only those households that are unable to pay for services without government assistance.

Since July 1, 2018, significant innovations in subsidy accounting have been introduced as a result of the implementation of the electronic state register of subsidy recipients. It made the process of allocating housing subsidies more transparent and accessible to public scrutiny, and the Ministry of Social Policy was given the tool to analyze the characteristics of subsidized households and prepare appropriate changes to the program. Starting from January 1, 2019, the country launched another economic experiment to stimulate socially vulnerable groups of the population to save utility costs through the introduction of subsidy monetization. This mechanism involves the implementation of three stages, the characteristics of which are presented in Table 7.2. The implementation of the mechanism should result in the formation of economic interest of every Ukrainian in the economical consumption of public utilities and the reduction of government spending on subsidizing the population in the long term. However, at the end of April 2019, that is 2 months after the start of the second stage of cash subsidy monetization, only 75% of program participators paid in cash for utilities on time. The remaining 25% citizens who received funds but did not pay their bills on time run the risk of being left without a subsidy until they pay off the debt and will continue to be denied cash subsidies. In such a way the government plans to foster a responsible attitude of citizens to comply with payment discipline (Ministry, 2019a).

Along with the unconditional incentive for economical use of utilities by households, the proposed subsidy monetization mechanism has weak points because it does not interest population directly in implementing large-scale energy-saving measures. Considering the low incomes of most Ukrainians and comparatively high investment in energy-saving measures in the residential sector, which can have a significant economic impact on reducing utility consumption, subsidy monetization can only stimulate the introduction of low-cost energy-saving measures, e.g., replacement of room lighting system with energy-saving bulbs, draught proofing, etc. Given several waves

Table 7.2. Characteristics of the stages of subsidy monetization in Ukraine (compiled according to (UNIAN, 2019))

Stage	Start of period	Population segment	Stage description
First	from January 1, 2019	Households seeking social assistance for the first time	The state through PJSC "Oshchadbank" pays part of the bills within the allocated subsidy, and the families pay the other part of the bill according to the amount indicated by the bank in the text message.
Second	from May 1, 2019	Households receiving a subsidy for more than a year	All beneficiaries of subsidies (except for those who applied for a subsidy for the first time in 2019) submit a declaration of income and a claim for monetization to the territorial social protection department in May. Thereafter, a bank account is opened for each application and subsidies are charged to pay for heat, electricity, water and gas supply. Every month PJSC "Oschadbank" sends a text message about the receipt of the subsidy and its balance to the subsidiant. When the heating season is over, unused money will be credited to the beneficiary's account. Utility suppliers form payment orders, and the bank transfers the subsidy. The amount of subsidy, paid services and account balance can be checked in the personal account.
Third	from October 1, 2019	All households applying for or receiving a subsidy	Each beneficiary will receive in cash everything that has been saved during the heating season.

of rising energy prices, the majority of the population has already implemented small and partially medium-cost energy-saving measures in their homes. Large-scale and more effective measures, such as building weatherization, installation of individual heat stations, renewable energy mini-power plants, etc., require significant investments, which subsidiants do not have and most of them will not have in the future. Thus, the subsequent increase in energy prices and utilities will only lead to further increase the number of subsidies paid by the state, deepening of the energy and fuel poverty of the population. Therefore, in this case, simple cost savings will not have a long-term positive effect because, instead of investing in increasing the energy efficiency of households and utilities suppliers, costs will be consumed by families on other purposes. However, even such "drops in the ocean" can significantly help for energy saving in the housing sector, provided they are not simply consumed but properly targeted. Therefore, it is advisable to increase the effectiveness of the subsidy mechanism by combining it with other state levers to stimulate the country's energy-efficient development.

All in all, a comparison of the dynamics of prices for utilities, subsidies, and energy efficiency of Ukrainian national economy in general and its housing and utilities sector in particular (Figure 7.6) gives grounds for formulating the following conclusions. Firstly, there is a clear correlation between changes in subsidies and utility prices, which is quite logical given the low

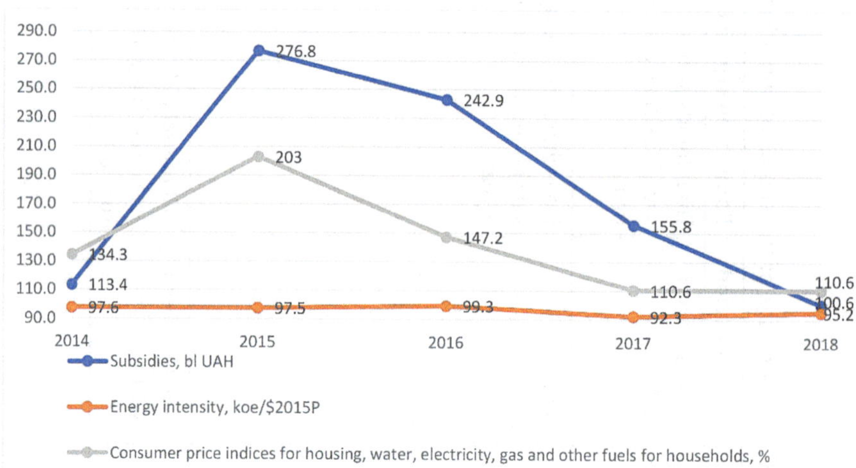

Figure 7.6. Dynamics of utilities subsidies, utilities consumer prices and energy intensity of Ukrainian economy in 2014–2018, % to the previous year (calculated by the authors according to the data (Global, 2019; State, 2019))

incomes of Ukraine's population. Secondly, the paired regression analysis for 2014–2018 revealed the lack of any close relationship between the dynamics of energy intensity of the national economy, the amount of final energy consumption by households, and utility prices ($R^2 \leq 0,4$). During this period, the energy intensity of the national economy declined somewhat, as did the absolute volume of final energy consumption by households. However, it was caused not by large-scale energy-saving measures, but primarily by the loss of control of part of the eastern industrial territories, where energy-intensive production is concentrated, and dilapidated communal infrastructure and housing stock are present. Therefore, in recent years, changing energy prices for the population has not been a strong incentive to increase the energy efficiency of the housing and utilities sector. In such a case, subsidies are impeding rather than supporting a more rational use of energy, playing a purely social equalization role. Thirdly, considering that the price factor is a catalyst for energy-efficient changes around the globe, it is advisable to revise the principles of subsidizing Ukraine's population and create a strong relationship among energy efficiency of households, utility prices, and subsidies.

To identify long-term trends in the impact of subsidies and other related factors on the energy poverty and energy efficiency of the national economy, we have evaluated the available empirical data on econometric models (1) and (2). The simulation results are presented in Tables 7.3 and 7.4.

The simulation results (Table 7.3) indicate that as households' expenditures for utilities grow by one percentage point the poverty headcount ratio below national poverty lines increases by 3 percentage points. In our assessment, this outcome shows that the majority of Ukraine's population has low incomes and therefore the increase in expenditures for utilities causes a marked disproportionate increase in poverty.

Table 7.3. Relationship between poverty headcount ratio, GDP per capita, gross capital formation and expenditures for utilities in Ukraine in 1999–2018 (authors' calculations)

Source	SS	df	MS			
				Number of obs	=	20
				F(3, 16)	=	58.80
Model	15808.9745	3	5269.65818	Prob > F	=	0.0000
Residual	1433.90748	16	89.6192174	R-squared	=	0.9168
				Adj R-squared	=	0.9012
Total	17242.882	19	907.520106	Root MSE	=	9.4667

PR	Coef.	Std. Err.	t	P>|t|	[95% Conf.	Interval]
GDP_pc	-.0596202	.0086136	-6.92	0.000	-.0778803	-.0413602
GFCF	2.42e-11	3.06e-10	0.08	0.938	-6.24e-10	6.73e-10
EXU	3.018121	.9587459	3.15	0.006	.9856702	5.050571
cons	174.2383	16.23436	10.73	0.000	139.823	208.6536

Table 7.4. Relationship between GDP per unit of energy, GDP per capita, gross capital formation, expenditures for utilities and crude oil average prices in Ukraine in 1999–2018 (authors' calculations)

Source	SS	df	MS
Model	7.74601213	4	1.93650303
Residual	.571956656	15	.038130444
Total	8.31796879	19	.437787831

Number of obs	=	20
F(4, 15)	=	50.79
Prob > F	=	0.0000
R-squared	=	0.9312
Adj R-squared	=	0.9129
Root MSE	=	.19527

GDP_E	Coef.	Std. Err.	t	P>|t|	[95% Conf.	Interval]
GDP_cp	.0010407	.0002478	4.20	0.001	.0005125	.0015689
GFCF	-1.57e-11	6.34e-12	-2.47	0.026	-2.92e-11	-2.15e-12
EXU	.1144372	.0225127	5.08	0.000	.0664524	.1624219
CRO	.0066722	.003486	1.91	0.075	-.000758	.0141023
_cons	-1.151111	.3811954	-3.02	0.009	-1.96361	-.3386122

With GDP per capita increasing by 100 USD, there is a decrease in headcount ratio below national poverty lines by 6 percentage points. It confirms our statement about the low incomes of most households that are bordering on poverty. That is, even a slight increase in household incomes has a significant effect on reducing poverty. Instead, gross capital formation is not statistically significant and does not affect the poverty indicator.

Thus, we have empirically confirmed the significant positive impact of rising household expenditures for utilities on raising the poverty level of Ukraine's population, accompanied by a corresponding increase in government subsidies. In turn, it is obvious that GDP per capita growth significantly contributes to the poverty reduction of Ukrainians.

The results obtained from Table 7.4 indicate that as GDP per capita increases by 1000 USD, energy efficiency improves by 1 USD per kg of oil equivalent. In 2018 the GDP per capita in Ukraine was 3110 USD and the GDP per unit of energy was 3.75 USD per kg of oil equivalent. That is, 1000/3110 = 32% increase in GDP per capita is associated on average with 1/3.75 = 26% improvements in the energy efficiency of GDP.

According to the outcomes from the second model, the increment in gross capital formation negatively affects GDP per unit of energy. It means that gross capital formation is not related to energy-saving technologies in Ukraine and even impedes energy-efficient progress. As households' expenditures for utilities grow by 10 percentage points, the energy efficiency indicator improves by 1 USD per kg of oil equivalent. That is each additional percentage of expenditures for utilities (in the structure of total household expenditures) leads to an increase in the energy efficiency of 0.11 USD per kg of oil equivalent. With rising crude oil average prices by 10 USD, GDP per unit of energy improves by 0.067 USD per kg of oil equivalent.

Therefore, the results of econometric modelling have confirmed the positive impact of an increase in GDP per capita, expenditures for utilities and crude oil average prices on improving energy efficiency in the long run. Nevertheless, for the lack of available data, it is not possible to trace the impact of subsidies on GDP per unit of energy change during the period under study. In contrast, the investment factor negatively affects energy efficiency, which indicates the gaps in the state's investment policy and the lack of energy efficiency goals in its priorities.

The outcome of the positive impact of crude oil average prices on energy efficiency increase slightly contradicts the results of the paired correlation analysis of utility prices and energy intensity of the national economy since the paired analysis shows a lack of close connection for the period 2014–2018. It may be explained by the error caused by the small number of years selected for paired analysis and the lack of comparable indicators in the medium and long term that most accurately represent the phenomena under study in the housing and utilities sector.

All in all, obtained empirical results should be taken into account when improving the system of state subsidies, national investment guidelines and other regulatory mechanisms to ensure energy-efficient development of Ukraine.

5. Improvement of the State Subsidy System in the Context of Ukraine's Energy-efficient Development

Providing social equality in society, eradicating energy and fuel poverty, and ensuring energy-efficient development of the state are the goals that should be achieved by the modern reform of the state subsidy system. Given the relevance of improving energy efficiency for utilities suppliers, utilities and the housing stock, it is appropriate to link government energy efficiency programs to the subsidy monetization mechanism. For this reason, it is necessary to improve this mechanism so that it would:

1. be combined with existing national and regional energy efficiency development programs, in particular, the weatherization assistance program and the feed-in tariff for renewable energy facilities in households,
2. envisage expanding cooperation with international financial organizations to implement energy-efficient measures for households and utilities suppliers.

3. allow a comprehensive solution to the problems of energy poverty, ensuring the growth of energy efficiency of the Ukrainian economy and maintaining social equality in society.

Low income of domestic households is one of the main reasons for slowing down the implementation of energy-efficient measures in residental sector. To overcome this problem in Ukraine, the State Targeted Economic Program for Energy Efficiency and Development of Renewable Energy Sources and Alternative Fuels for 2010–2020 has been working since October 2014 (better known to the public as the "warm" credits program (CMU, 2010)), which provides financial support to every citizen for implementing energy-efficient projects at home. The rise in utility prices in 2014 and subsequent years sparked a wave in population's demand for "warm" loans starting from 2015–2016. While in 2014–2015 the program funds allocated by the government on loans to individuals were sufficient by the end of the year, the funding constantly came to a halt earlier than scheduled during 2016–2019; already in late summer–early autumn in some regions of the country the program funding reserves had been exhausted, which led to several decisions being made to allocate additional funds from the state budget. In 2014–2019, more than 700,000 families became participants of the program, attracting 8.2 billion UAH for energy-efficient measures and receiving about 2.7 billion UAH of state compensation. So, for 1 UAH compensated by the program, Ukrainians have invested 3 UAH of their own funds. During the first program year, 67,000 families benefited from "warm" loans, in subsequent years this number was at least 140,000 families, and even before the end of 2019, 188,000 households participated in this program. Housing cooperative demand is even more illustrative. During the first three program years, 1200 housing cooperatives benefited from "warm" loans and received about 220 million UAH from banks. The figure, achieved by 2018, exceeded all previous years cumulatively; 1600 housing associations attracted 420 million UAH in loans. For the incomplete 2019, there is a record condominium demand for the program; more than 2100 condominiums have received 700 million UAH of "warm" loans. In total, over 5000 housing cooperatives across the country have benefited from the program, spending over 1.3 billion UAH in "warm" loans for energy efficiency activities in multi-apartment buildings. It is worth noting that the average amount of loan taken by a housing cooperative increases annually, that is, these organizations are gradually moving to the implementation of more complex large-scale projects for thermal modernization of buildings (SAEE, 2019; Sklyarov, 2019).

In 2017–2018, the government implemented annual monitoring and evaluation of the program's performance. Expert conclusions indicate that the limited amount of resources allocated to the program reduces the overall effectiveness of energy-saving measures and increases the level of social tension in society; it does not make it possible to reach all households and housing cooperatives that are the prime need in energy-efficient measures and could implement them. Besides, the availability of utilities subsidies for program participants weakens incentives to reduce utilities consumption (SAEE, 2019). Currently, there is another source for housing cooperatives to obtain loans for energy efficiency activities, such as the Energy Efficiency Fund, which recently started operations in Ukraine. Its budget is much larger than the "warm" loan program, but the requirements for projects are tighter, which not every housing cooperative can meet (Sklyarov, 2019).

Given the current popularity of the "warm" loan program, it would be advisable to extend it in the coming years and provide the families, which receive subsidies and ready to spend money on energy-efficient activities at home, with additional compensation under the program through monetized subsidies. Today, the state program provides 35% compensation for subsidiants that get "warm" loans for the purchase of non-gas or non-electric boilers and energy-efficient equipment or materials not exceeding 12,000 UAH. If a housing cooperative has subsidized families, such condominiums are reimbursed at a weighted average of between 40% and 70%, depending on the number of subsidized apartments. Also, in many regions of Ukraine, there are local programs to reduce the cost of "warm" loans, for which additional compensation (from municipal budgets) is provided for principal amounts or interest on such loans (SAEE, 2019).

To encourage energy efficiency measures in the poorest households, it would be advisable to increase the compensation to 50–70% for individual "warm" loans for subsidiants, which have been able to save money as a result of rational use of utilities and are willing to invest in improving the energy performance of their homes using "warm" loans. Moreover, funding for the program for population should be further expanded. To increase the interest of subsidies recipients in implementing energy-efficient measures and ensuring 100% transition of the population to commercial metering of resources and services consumption, maximum compensation (up to 90%) should be provided for loans for the acquisition and installation of metering devices in households along with gradual tightening of resource consumption standards. This step will create the preconditions for controlling the resource consumption by the population and potentially justify the indicators of its reduction based on the implementation of the energy-saving measures even by the poorest families.

It is advisable to differentiate the rates of compensation for "warm" loans to households, depending on such factors as (1) whether the borrower is the recipient of the subsidy (the level of the borrower's income, the availability of monetized subsidy and the willingness to spend it on energy saving); (2) what measures the loan is taken for (installation of metering devices; separate or complex energy-saving measures); (3) the loan amount; (4) the condition of the building where energy-saving measures are implemented (emergency, dilapidated, requires capital repairs, meets/does not meet energy efficiency standards, etc.); (5) expected economic and environmental impact of the measures. Thus, the amount of compensation should increase while implementing energy-efficient measures in old buildings, realizing complex high-cost projects with significant expected economic and environmental effects, attracting low-income households. Considering that the assessment of the energy-saving measure relevance for a particular house can be more reliably carried out at the local level, it should be entrusted to local authorities to determine the additional compensation premiums to the state rate of compensation with the allocation of appropriate funding.

The involvement of renewable energy sources in energy generation is one of the ways to increase energy efficiency in the residential sector (Cebula et al., 2018; Lyeonov et al., 2019; Yevdokimov et al., 2018). It is worth mentioning that the state incentive policy in the field of renewable energy, the main instruments of which are feed-in tariffs, tax, and customs privileges, was introduced in Ukraine in 2009, but it was applied only to legal entities (Sotnyk et al., 2019). In 2014, the Law of Ukraine "On electric power industry" (Verkhovna, 1997) was amended, whereby economic incentives were spread to private households.

Under (Verkhovna, 2017), owners of households have the right:

- to install generating facilities designed to generate electricity based on solar radiation and wind energy, the total installed capacity of which does not exceed 30 kW;
- to sell to an electricity supply company the electricity generated from the above-mentioned renewable energy sources at a feed-in tariff in the amount that exceeds the monthly electricity consumption by a private household.

The proliferation of economic mechanisms aimed at stimulating the electricity generation from renewable energy sources in the residential sector has contributed to the activation of the installation of renewable energy generation facilities. However, although motivational mechanisms extend to two types of renewable energy, solar and wind energy, nowadays private households prefer to install small solar power plants.

Since 2014, only three small wind farms have been put into operation in the residential sector, whose owners sell electricity at a feed-in tariff (SAEE, 2019a). The significant difference in feed-in tariffs is the biggest obstacle to promoting balanced solar and wind energy use by private households. When solar and wind energy generation is used simultaneously, the tariff difference requires a household to set up two electricity metering systems, which in turn leads to double costs, so private homeowners opt for solar power plants with a higher feed-in tariff (Verkhovna, 2017). As a result, only solar power plants have been actively launched in recent years, growing almost 355 times—from 21 plants in 2014 to 7450 plants in 2018 (SAEE, 2019a). However, as of the end of 2018, the number of solar power plants in households has more than doubled compared to 2017. One of the reasons for its dynamic growth in recent years is the planned reduction of feed-in tariff rates in 2020; therefore, to make greater profits in the future, the population with sufficient means is in a hurry to put generating facilities into operation by 2020.

Besides, the capacity of small solar power plants installed by private households was 157 MW by the end of 2018, demonstrating an increase of 3.1 times compared to 2017 and 1570 times compared to 2014, which also confirms the investment attractiveness of solar energy for the residential sector (SAEE, 2019a). The green electricity unconsumed by the households and sold at a feed-in tariff to energy supply companies has also had a positive upward trend over the last 5 years. In 2014–2018, it increased from 0.01 to 76.12 million kWh (SAEE, 2019a).

However, although the dynamic start-up of generating facilities is observed in the private sector, the share of these energy assets in the country's energy balance is negligible today. Thus, by the end of 2018, it constituted only 2.3%, the remaining 97.7% of green electricity was generated by privately owned high power renewable energy facilities (National, 2019a). One of the key reasons is the high initial investment in the construction of solar and wind power plants, which is overwhelming for the vast majority of the population with low incomes. Thus, only wealthy households and business entities have the opportunity to capitalize on feed-in tariffs, while the poorest citizens are forced to pay a rising price for green electricity since the feed-in tariff is offset by increasing the weighted average electricity price in the wholesale market.

A favourable way out of the current situation could be preferential lending to renewable energy projects for private households. It should be noted that some steps have already been taken in this direction. Several commercial banks in Ukraine have opened programs aimed at lending to renewable energy projects, but the lending conditions for such programs are not attractive to the public.

For example, under the program "Eco-Energy Simple" by JSB "Ukrgasbank," a loan is granted for the purchase and installation of solar power plants, heat pumps, and wind power plants. The interest rates on the loan under this program are directly dependent on advance payments and loan term. The minimum and maximum payments are 15% and 60% of the cost of the equipment/project respectively, the minimum and maximum crediting period are 1 and 5 years respectively. With the minimum advance payment and the minimum term of crediting, the loan rate is 9.29% per annum, with the maximum advance payment and the maximum term of crediting the loan rate is 14.79% (Ukrgasbank, 2019).

PJSC "Oschadbank" within its credit program "Green Energy" provides loans for the acquisition, installation, assembling and launching of solar heating facilities, solar panels for electricity production, heat pumps, solar batteries/accumulators, wind power plants and additional equipment and materials for the above-mentioned goods. The downpayment to this loan program is 15% of the cost of the equipment/project, the maximum loan term is 6 years, and the annual loan rate is 19.5% (Oschadbank, 2019).

Accordingly, given that the installation of green energy generation facilities is quite expensive today, such lending conditions are unacceptable to most private household owners. Therefore, at the state level, it is advisable to provide additional state compensation for such loans to the poorest households, which would allow subsidiants to become participants of the green energy market and increase their incomes. At the same time, considering the high investment in power plants and other renewable energy facilities, financial state support for each low-income private household can reach 90–95% of the facility cost, which will be an unbearable burden on the state budget. Consequently, it is necessary to extend the preferential terms of obtaining a feed-in tariff for the households at condominiums and building cooperatives, as well as to streamline the legal aspects of the activity of energy cooperatives created by the groups of individuals for the green energy generation (Kurbatova et al., 2018). In this context, it should be possible to use monetized subsidies as investments in the creation of new collective renewable energy facilities by the population, as well as to provide an increased amount of state compensation for construction loans in proportion to the share of project participants that are subsidies recipients.

The problem of involving the poor segment of the population in green energy market could be partially solved with the help of the European Bank for Reconstruction and Development credit lines opened in Ukraine, namely: Ukraine Sustainable Energy Lending Facility (USELF, 2019), Ukraine Energy Efficiency Programme (UKEEP, 2019), and IQ-energy (IQ-energy, 2019). However, nowadays none of these credit programs is applied in the private household sector. Since April 7, 2016, individuals could participate in the IQ-energy program, but on September 10, 2018, the validity period of the program was over. Furthermore, lending under the green energy program covered only the purchase of solid fuel biomass boilers, solar collectors, and heat pumps and had no significant impact on the deployment of renewable energy projects in the residential sector. Thereby, a promising way for Ukraine is to renew and expand cooperation with international financial institutions to open new credit lines, even for low-income households with collective participation in energy-efficient projects, as well as for utilities suppliers, housing and building cooperatives.

The financial support for energy-efficient measures in the residential sector of Ukraine to reduce the energy poverty of the population and increase the energy efficiency of the national economy should be supported by other organizational and economic measures that will holistically achieve the desired goals. Particularly, among them, there is the gradual abolition of government subsidies on natural gas, electricity and heat for the population, which will make it economically feasible to install renewable power plants by private householders and increase the cost-effectiveness of energy-saving measures. In the budget of Ukraine for 2019, 1.87 billion USD (Verkhovna, 2018) was earmarked for targeted subsidies, payment for utilities and the purchase of furnace fuel. In contrast, the government's focus should be on raising household incomes and reducing the attractiveness of traditional energy, as well as encouraging reduced energy consumption through price regulation and affordable energy-efficient measures.

In addition, currently, there is no economic leverage in Ukraine specifically aimed at stimulating green electricity consumption. Therefore, it is advisable to extrapolate foreign experience in applying support schemes based on increasing demand for such electricity, including the introduction of mandatory quotas for its consumption (Kurbatova et al., 2019). At the same time, as the share of green electricity in the country's overall energy balance increases, the problem of payments under the feed-in tariff will be exacerbated due to the increase in weighted average electricity prices. To prevent the increase of the energy poverty level of the population, it is advisable to consider the transition of green energy producing households to green auction mechanisms, which will be conducted separately from the industrial renewable energy sector, where the cost of producing green energy is lower through the scale effect.

Increasing public awareness on the advantages and practicalities of implementing energy-efficient and renewable energy projects as well as on the financial benefits that can be received from selling excess electricity at a feed-in tariff by households could greatly accelerate the enhancement of energy-efficient and renewable energy projects in the residential sector. While every Ukrainian is aware of subsidies through mass media, most citizens do not have any clue about loan programs for increasing the energy efficiency of their homes and detailed instructions for its implementation. This deficiency should be corrected as soon as possible.

Providing opportunities for every family, even the poorest one, to capitalize on energy efficiency and renewable energy in the face of rising utility prices and attracting monetized subsidies for these purposes will contribute to a comprehensive solution to the problems of energy poverty, ensuring the energy efficiency of Ukraine's economy and social sustainability.

The publication contains the results of research carried out within the framework of research works of the Ministry of Education and Science of Ukraine "System model of efficiency management and forecasting of electricity use" (No. 0118U003583) and "Innovation management of energy-efficient and resource saving technologies in Ukraine" (No. 0118U003571).

References

Bhandari, Medani P. (2020) Second Edition- Green Web-II: Standards and Perspectives from the IUCN, Policy Development in Environment Conservation Domain with reference to India, Pakistan, Nepal, and Bangladesh, River Publishers, Denmark / the Netherlands. ISBN: 9788770221924 e-ISBN: 9788770221917

Bhandari, Medani P. (2020), Getting the Climate Science Facts Right: The Role of the IPCC, River Publishers, Denmark / the Netherlands- ISBN: 9788770221863 e-ISBN: 9788770221856

Bhandari, Medani P. (2018) Green Web-II: Standards and Perspectives from the IUCN, Published, sold and distributed by: River Publishers, Denmark / the Netherlands ISBN: 978-87-70220-12-5 (Hardback) 978-87-70220-11-8 (eBook).

Bilan, Y., Streimikiene, D., Vasylieva, T., Lyulyov, O., Pimonenko, T., & Pavlyk, A. (2019). Linking between renewable energy, CO2 emissions, and economic growth: Challenges for candidates and potential candidates for the EU membership. Sustainability (Switzerland), 11(6) doi:10.3390/su11061528.

Boardman, B. (2012). Fuel poverty synthesis: lessons learnt; actions needed. Energy Policy, 49, 143–148.

Bouzarovski, S., & Petrova, S. (2015). A global perspective on domestic energy deprivation: Overcoming the energy poverty – fuel poverty binary. Energy Research & Social Science, 10, 31-40. https://doi.org/10.1016/j.erss.2015.06.007.

Bouzarovski, S., & Tirado Herrero, S. (2017). The energy divide: integrating energy transitions, regional inequalities and poverty trends in the European Union. Eur. Urban Reg. Stud., 24, 69–86. http://dx.doi.org/10.1177/0969776415596449.

Bouzarovski, S., Tirado Herrero, S., Petrova, S., & Ürge-Vorsatz, D. (2016). Unpacking the spaces and politics of energy poverty: Path-dependencies, deprivation and fuel switching in post-communist Hungary. Local Environ., 21, 1151–1170. doi: 10.1080/13549839.2015.1075480.

Braubach M., & Ferrand, A. (2013). Energy efficiency, housing, equality and health. Int. J. Public Health, 58, 331–332.

Brunner, K.-M., Spitzer, M., & Christanell, A. (2012). Experiencing fuel poverty. Coping strategies of low-income households in Vienna/Austria. Energy Policy, 49, 53-59. https://doi.org/10.1016/j.enpol.2011.11.076.

BuShehri, M. A. M., & Wohlgenant, M. K. (2012). Measuring the welfare effects of reducing a subsidy on a commodity using micro-models: an application to Kuwait's residential demand for electricity. Energy Econ. 34, 419–425.

Buzar, S. (2007). The 'hidden' geographies of energy poverty in post-socialism: Between institutions and households. Geoforum, 38 (2), 224–240. https://doi.org/10.1016/j.geoforum.2006.02.007.

Cebula, J., Chygryn, O., Chayen, S. V., & Pimonenko, T. (2018). Biogas as an alternative energy source in Ukraine and Israel: Current issues and benefits. International Journal of Environmental Technology and Management, 21(5-6), 421–438. doi:10.1504/IJETM.2018.100592.

Cirman, A., Mandic, S., & Zoric, J. (2012). Decisions to renovate: identifying key determinants in Central and Eastern European post-socialist Countries. Retrieved from https://www.researchgate.net/publication/258630062_Decisions_to_Renovate_Identifying_Key_Determinants_in_Central_and_Eastern_European_Post-socialist_Countries.

CMU (2010). State targeted economic program for energy efficiency and development of renewable energy sources and alternative fuels for 2010-2020: decree of the Cabinet of Ministers of Ukraine (CMU) No. 243, 01.03.2010. Retrieved from https://zakon.rada.gov.ua/laws/show/243-2010-п. (in Ukrainian).

del Granado, J. A., Coady, D., & Gillingham, R. (2012). The unequal benefits of fuel subsidies: a review of evidence for developing countries. World Dev., 40 (11), 2234–2248.

DSIRE (2019). Find policies & incentives by state. Retrieved from https://www.dsireusa.org/.

Dubois U., & Meier, H. (2016). Energy affordability and energy inequality in Europe: Implications for policymaking. Energy Res. Soc. Sci., 18, 21–35. doi: 10.1016/j.erss.2016.04.015.

Estache, A., Wodon, Q., & Foster, V. (2002). Accounting for poverty in infrastructure reform: Learning from Latin America's experience. World Bank, Washington DC.

Expenditure and resources of households of Ukraine in 2010-2018 (2019). Statistical Yearbooks. Retrieved from ukrstat.gov.ua (in Ukrainian).

Fankhauser, S., & Tepic, S. (2007). Can poor consumers pay for energy and water? An affordability analysis for transition countries. Energy Policy, 35, 1038–1049.

Fankhauser, S., Rodionova, Y., & Falcetti, E. (2008). Utility payments in Ukraine: Affordability, subsidies and arrears. Energy Policy, 6, 4168-4177.

Fattouh, B., & El-Katiri, L. (2013). Energy subsidies in the Middle East and North Africa. Energy Strateg. Rev., 2(1), 108–115.

Foster, V., Tre, J.-P., & Wodon, Q. (2000). Energy prices, energy efficiency and fuel poverty. World Bank, Washington DC.

Frondel, M., Sommer, S., & Vance, C. (2015). The burden of Germany's energy transition: An empirical analysis of distributional effects. Economic Analysis and Policy, 45, 89-99. https://doi.org/10.1016/j.eap.2015.01.004.

Gerbery, D., & Filčák, R. (2014). Exploring multi-dimensional nature of poverty in Slovakia: access to energy and concept of energy poverty. Ekonomický Časopis, 18, 579-597.

Global Energy Statistical Yearbook 2019 (2019). Retrieved from https://yearbook.enerdata.net/total-energy/world-energy-intensity-gdp-data.html.

González-Eguino, M. (2015). Energy poverty: An overview. Renewable and Sustainable Energy Reviews, 47, 377-385. https://doi.org/10.1016/j.rser.2015.03.013.

Green Paper on Municipal Energy Efficiency (2015). S. Pavlyuk (Ed.). Retrieved from https://cdn.regulation.gov.ua/3b/d3/cb/a5/regulation.gov.ua_File_185.pdf (in Ukrainian).

Health Service Executive Minister Naughten announces €10 m for Warmth & Wellbeing in 2017 (2017). Retrieved from https://www.hse.ie/eng/services/news/media/pressrel/minister-naughten-announces-€10m-for-warmth-wellbeing-in-2017.html.

Healy, J. D., & Clinch, J. P., 2004. Quantifying the severity of fuel poverty, its relationship with poor housing and reasons for non-investment in energy-saving measures in Ireland. Energy Policy, 32, 207–220. http://dx.doi.org/10.1016/S0301-4215(02)00265-3.

Hernández, D., & Bird, S. (2010). Energy burden and the need for integrated low-income housing and energy policy. Poverty Public Policy, 2(4), 5–25. doi: 10.2202/1944-2858.1095.

Hills, J. (2012). Getting the measure of fuel poverty: final report of the fuel poverty review. London School of Economics and Political Science; London, UK. CASE Report 72.

IQ-energy (2019). Retrieved from http://www.iqenergy.org.ua.

Johnston, A., Heffron, R. J., & McCauley, D. (2014). Rethinking the scope and necessity of energy subsidies in the United Kingdom. Energy Res. Soc. Sci., 3, 1-4. https://doi.org/10.1016/j.ecolecon.2016.12.009.

Kalachova, H. (2018). Subsidy-2019: All about monetization, new payout rules and income calculations. Retrieved from https://www.epravda.com.ua/rus/publications/2018/12/28/643944/ (in Ukrainian).

Kaygusuz K. (2011). Energy services and energy poverty for sustainable rural development Renewable and Sustainable Energy Reviews, 15 (2), 936-947. https://doi.org/10.1016/j.rser.2010.11.003.

Kerimray, A., de Miglio, R., Rojas-Solórzano, L., & Ó Gallachóir, B. P. (2018). Causes of energy poverty in a cold and resource-rich country: evidence from Kazakhstan. Local Environment, 23 (2). Published Online: 05 Nov 2017. https://doi.org/10.1080/13549839.2017.1397613

Komelina O.V., & Maksimenko, O. S. (2014). Modern factors for regional energy policy formation in housing and communal sector of Ukraine. Economics and Organization of Management, 1 (17)–2(18), 129-137 (in Ukrainian).

Korppoo, A., & Korobova, N. (2012). Modernizing residental heating in Russia: End-use practices, legal developments, and future prospects. Energy policy, 42, 213-220. DOI: 10.1016/j.enpol.2011.11.078.

Kubatko, O., & Kubatko, O., (2019). Economic estimations of air pollution health nexus. Environment, Development and Sustainability: A Multidisciplinary Approach to the Theory and Practice of Sustainable Development, Springer, 21(3), 1507-1517, June. https://doi.org/10.1007/s10668-018-0252-6.

Kubatko, O., & Kubatko, O. (2017). Economic estimations of pollution related cancer and nerves morbidity. International Journal of Ecology & Development, 32 (1), 33–43.

Kurbatova, T., Sidortsov, R., Sotnyk, I., Telizhenko, O., Skibina, T., & Hynek, R. (2019). Gain without pain: an international case for a tradable green certificates system to foster renewable energy development in Ukraine. Problems and Perspectives in Management, 17 (3), 464–476. https://doi:10.21511/ppm.17(3).2019.37.

Kurbatova, T., & Hyrchenko, Ye. (2018). Energy co-ops as a driver for bio-energy sector growth in Ukraine. IEEE 3rd International Conference on Intelligent Energy and Power Systems (IEPS), Kharkiv, September 10–14, P. 210–213. https://doi.org/10.1109/IEPS.2018.8559516.

Legendre, B., & Ricci, O. (2015). Measuring fuel poverty in France: Which households are the most fuel vulnerable? Energy Econ., 49, 620–628. doi: 10.1016/j.eneco.2015.01.022.

Li, R., Sineviciene, L., Melnyk, L., Kubatko, O., Karintseva, O., & Lyulyov, O. (2019). Economic and environmental convergence of transformation economy: the case of China. Problems and Perspectives in Management, 17(3), 233–241. http://dx.doi.org/10.21511/ppm.17(3).2019.19.

Lihtmaa, L., Hess, D. B., & Leetmaa, K. (2018). Intersection of the global climate agenda with regional development: Unequal distribution of energy efficiency-based renovation subsidies for apartment buildings. Energy Policy, 119, 327–338. https://doi.org/10.1016/j.enpol.2018.04.013.

Lin, B., & Jiang, Z. (2011). Estimates of energy subsidies in China and impact of energy subsidy reform. Energy Econ. 33, 273–283.

Lyeonov, S., Pimonenko, T., Bilan, Y., Štreimikiene, D., & Mentel, G. (2019). Assessment of green investments' impact on sustainable development: Linking gross domestic product per capita, greenhouse gas emissions and renewable energy. Energies, 12(20) doi:10.3390/en12203891.

MDL (2018). MDL opinion: Housing fund of Ukraine - statistics, problems, ways of recovery. Retrieved from https://mdl.kyiv.ua/dumka-mdl-zhitlovij-fond-ukrayini-statistika-problemi-shlyahi-ozdorovlennya/ (in Ukrainian).

Melnyk, L., Kubatko, O., Dehtyarova, I., Matsenko, O., & Rozhko, O. (2019). The effect of industrial revolutions on the transformation of social and economic systems. Problems and Perspectives in Management, 17(4), 381-391. doi:10.21511/ppm.17(4).2019.31.

Melnyk, L.G., Kubatko, O.V., & Kubatko, O.V. (2016). Were Ukrainian regions too different to start interregional confrontation: Economic, social and ecological convergence aspects? Economic Research-Ekonomska Istrazivanja, 29 (1); 573–582. doi: 10.1080/1331677X.2016.1174387.

Ministry of Social Policy of Ukraine (2018). Subsidies. Retrieved from https://www.msp.gov.ua/timeline/subsidii.html (in Ukrainian).

Ministry of Social Policy of Ukraine (2019). Minister of Social Policy Andriy Reva reports to the Verkhovna Rada on the start of the process of subsidies monetization and pensions indexation. Retrieved from https://www.msp.gov.ua/news/16770.html (in Ukrainian).

Ministry of Social Policy of Ukraine (2019a). Subsidy monetization was discussed in Kyiv by representatives of Ministry of Social Policy, World Bank and journalists from all over Ukraine. Retrieved from https://www.msp.gov.ua/news/17008.html (in Ukrainian).

Minregion (2018). Reform of the utility market and energy efficiency: current achievements and next steps. Ministry of Regional Development, Construction and Housing and Communal Services of Ukraine. Retrieved from https://dzki.kyivcity.gov.ua/files/2018/2/7/Minrehion.GKP.pdf (in Ukrainian).

Minregion (2019). Alyona Babak noted the irreversibility of the creation of condominiums in Ukraine. Retrieved from http://www.minregion.gov.ua/press/news/alona-babak-konstatuvala-nezvorotnist-protsesiv-stvorennya-osbb-v-ukrayini/ (in Ukrainian).

Mlaabdal, S. M. A., Chygryn, O., Kubatko, O., & Pimonenko, T. (2018). Social and economic drivers of national economic development: the case of OPEC countries. Problems and Perspectives in Management, 16(4), 155-168. doi:10.21511/ppm.16(4).2018.14.

Naftogas (2019). Dynamics of natural gas prices for Ukrainian consumers. Retrieved from http://www.naftogaz.com/files/Information/Dynamika-cina-2014-2018-Naselennya.pdf (in Ukrainian).

Naftogas (2019a). As a separate legal entity, Naftogaz received a net profit of 39.3 billion UAH in 2017, which is 48% more than in 2016. Retrieved from http://www.naftogaz.com/www/3/nakweb.nsf/0/2E5FC6B69F38A-444C225827A003F721A? (in Ukrainian).

National Energy and Utilities Regulatory Commission of Ukraine (2019). Retrieved from https://www.nerc.gov.ua/?id=15013 (in Ukrainian).

National Energy and Utilities Regulatory Commission of Ukraine (2019a). Report on the results of National Energy and Utilities Regulatory Commission's activity in 2018: decree No. 440, 29.03.2018. Retrieved from http://www.nerc.gov.ua/data/filearch/Catalog3/ Richnyi_zvit_NKREKP_2018.pdf (in Ukrainian).

Nova Poltava (2016). About 90% of multiapartment stores need thermo-modernization in Ukraine. Retrieved from http://www.nova.poltava.ua/v-ukra%D1%97ni-blizko-90-bagatopoverxivok-potrebuyut-termomodern-izaci%D1%97/ (in Ukrainian).

Oschadbank (2019). Loans for equipment that produce "green" energy. Retrieved from https://www.oschadbank.ua/ua/private/loans/kredituvannya-na-oblad-nannya-shcho-viroblya-zelenu-energ-yu (in Ukrainian).

Papada, L., & Kaliampakos, D. (2016). Measuring energy poverty in Greece. Energy Policy, 94, 157-165. https://doi.org/10.1016/j.enpol.2016.04.004.

Petrova, S., Gentile M., Mäkinen, I. H., & Bouzarovski, S. (2013). Perceptions of thermal comfort and housing quality: exploring the microgeographies of energy poverty in Stakhanov, Ukraine. Environment and Planning A, 45, 1240 – 1257. doi:10.1068/a45132.

Phimister, E., Vera-Toscano, E., & Roberts, D. (2015). The dynamics of energy poverty: evidence from Spain. Econ. Energy Environ. Policy, 4. doi: 10.5547/2160-5890.4.1.ephi.

Poputoaia, D., & Bouzarovski, S. (2010). Regulating district heating in Romania: legislative challenges and energy efficiency barriers. Energy Policy, 38 (7), 3820-3829. https://doi.org/10.1016/j.enpol.2010.03.002.

SAEE: State Agency on Energy Efficiency and Energy Saving of Ukraine (2019). State support for energy saving. Retrieved from http://saee.gov.ua/uk/consumers/derzh-pidtrymka-energozabespechenya. (in Ukrainian).

SAEE: State Agency on Energy Efficiency and Energy Saving of Ukraine (2019a). Information regarding capacity and amounts of electricity generated by renewable energy plants and sold at the feed-in tariff by owners of private households as of 01.01.2019. Retrieved from http://saee.gov.ua/sites/default/files/4_2018.pdf (in Ukrainian).

Saunders, M., & Schneider, K. (2000). Removing energy subsidies in developing and transition economies. In: 23rd Annual IAEE International Conference, International Association of Energy Economics, Sydney, 7–10.

Shih, J.-S., Burtraw, D., Palmer, K., & Liu, X. (2019). Water, storage and rates: the economics of electric residential water heating. Presentation October 9th, University of Maryland.

Siddig, K., Aguiar, A., Grethe, H., Minor, P., & Walmsley, T. (2014). Impacts of removing fuel import subsidies in Nigeria on poverty. Energy Policy, 69, 165–178.

Sineviciene L., Kubatko O. V., Sotnyk I. M. (2017). Determinants of energy efficiency and energy consumption of Eastern Europe post-communist economies. Energy & Environment. Prepublished October 9, 2017; DOI: 10.1177/0958305X17734386.

Sklyarov, R. (2019). Warm loans: should the government continue to pay. Retrieved from https://www.epravda.com.ua/rus/publications/2019/11/6/653367/

Solaymani, S. (2016). Impacts of energy subsidy reform on poverty and income inequality in Malaysia. Quality & Quantity, 50, 2707. https://doi.org/10.1007/s11135-015-0284-z.

Sotnyk I., Dehtyarova, I., Kovalenko, Y. (2015). Current threats to energy and resource efficient development of Ukrainian economy. Actual Problems of Economics, 11 (173), 137-145.

Sotnyk, I., Kurbatova, T., Dashkin, V., Kovalenko, Y. (2019). Green energy projects in households and its financial support in Ukraine. International Journal of Sustainable Energy. https://doi.org/10.1080/14786451.2019.1671389.

Sovacool, B. K. (2017). Reviewing, reforming, and rethinking global energy subsidies: towards a political economy research agenda. Ecological Economics, 135, 150-163. https://doi.org/10.1016/j.ecolecon.2016.12.009.

State Statistics Service of Ukraine (2019). Retrieved from http://www.ukrstat.gov.ua./.

Subsidies for Ukrainians: how much money was spent in 2014-2019 (2019). Retrieved from https://www.slovoidilo.ua/2019/08/06/infografika/suspilstvo/subsydiyi-ukrayinczіv-skilky-koshtiv-vytratyly-2014-2019-rokax (in Ukrainian).

Thomson, H., & Snell, C. (2013). Quantifying the prevalence of fuel poverty across the European Union. Energy Policy, 52, 563–572. http://dx.doi.org/10.1016/j.enpol.2012.10.009.

Thomson, H., Snell, C., & Bouzarovski, S. (2017). Health, well-being and energy poverty in Europe: a comparative study of 32 European countries. Int. J. Environ. Res. Public Health, 14(6). doi: 10.3390/ijerph14060584.

Thonipara, A., & Runst, P., & Ochsner, C., & Bizer, K. (2019). Energy efficiency of residential buildings in the European Union – An exploratory analysis of cross-country consumption patterns. Energy Policy, 129(C), 1156-1167. DOI: 10.1016/j.enpol.2019.03.003.

UKEEP: Ukraine Energy Efficiency Programme (2019). Retrieved from http://www.ukeep.org.

Ukrgasbank (2019). Loans for the purchase of solar power plants and heat pumps. Retrieved from https://www.ukrgasbank.com/private/credits/eco_energy (in Ukrainian).

UNECE (2009). Toward energy-efficient housing: prospects for UNECE member states. Committee on Housing and land Management, United Nations Economic Commission for Europe, Geneva.

UNIAN (2019). 2019 subsidies in Ukraine: how it works and who is waiting for the monetization of subsidies. Retrieved from https://www.unian.ua/society/10415382-subsidiji-2019-v-ukrajini-yak-pracyuye-i-na-kogo-chekaye-monetizaciya-subsidiy.html (in Ukrainian).

Urge-Vorsatz, S., & Tirado Herrero, D. (2012). Building synergies between climate change mitigation and energy poverty alleviation. Energy Policy, 49, 83–90.

USELF: Ukraine Sustainable Energy Lending Facility (2019). Retrieved from http://www.uself.com.ua.

Vasylieva, T., Lyulyov, O., Bilan, Y., & Streimikiene, D. (2019). Sustainable economic development and greenhouse gas emissions: The dynamic impact of renewable energy consumption, GDP, and corruption. Energies, 12(17). doi:10.3390/en12173289.

Verkhovna Rada of Ukraine (1997). Law of Ukraine "On electric power industry", No. 575/97. Retrieved from http://zakon3.rada.gov.ua/laws/show/575/97-%D0%B2% D1%80 (in Ukrainian).

Verkhovna Rada of Ukraine (2017). Law of Ukraine "On electricity market", No. 27-28 Retrieved from https://zakon.rada.gov.ua/laws/show/2019-19/stru/page4 (in Ukrainian).

Verkhovna Rada of Ukraine (2018). Law of Ukraine "On state budget of Ukraine". Annexes No. 1-9 to the State Budget of Ukraine for 2019. Retrieved from https: //zakon.rada.gov.ua/laws/show/2629-19 (in Ukrainian).

Waddams, C., & Deller, D. (2015). Affordability of utilities' services: extent, practice, policy. Centre on Regulation in Europe; Brussels, Belgium.

Yevdokimov, Y., Chygryn, O., Pimonenko, T., & Lyulyov, O. (2018). "Biogas as an alternative energy resource for ukrainian companies: EU experience". Innovative Marketing, 14(2), 7-15. doi:10.21511/im.14(2).2018.01.

Zanuda, A. (2018). Outcomes-2018: Ukrainian economy in five digits. Retrieved from https://www.bbc.com/ukrainian/features-russian-46633033 (in Ukrainian).

Chapter 8

Combating Inequality Via an Intercultural Strategy of The City: A Case Study of a Ukrainian City

Oksana Zamora[1], Svitlana Lutsenko[2]
[1]Department of International Economic Relations, Sumy State University, Ukraine
[2]Department of Pedagogy, Specific Education and Management, Sumy Institute for Regional Pedagogical Education After Graduation, Ukraine
E-mail: o.zamora@uabs.sumdu.edu.ua

ABSTRACT
The chapter is devoted to the analysis of a comprehensive approach of equality promotion via designing and implementing the intercultural strategy of the city. A multicultural city of Sumy at the North-Eastern part of Ukraine is used as a case study. Currently the city has just adopted the intercultural strategy incorporated within the city development strategy. It presents the best Ukrainian practices and challenges of the solution search for the issues of unequal access to the city infrastructure and municipal services in key spheres of the community life of the city visitors, migrants, and minorities of all kinds. The paper focuses on the empirical analysis of the cultural and educational components which appear to be essential in building the proper mentality and understanding of the need for the equality within the local community. There are a number of outlined strategic solutions based on the best world practices of the intercultural cities and adapted to the reality of a Ukrainian provincial city.
Key words: equality, economic inequality, inequality in education, migration, minorities, active citizenship.

1. Introduction

The history of humanity has always been marked by a strong migration trend which has been making a sensible impact on the population structure of strategically important locations. Economically strong cities are of a bigger interest for the people searching for better lives and, thus, they benefit more from the constant inflow of economically active and talented human resources. Globalization and equalization trends bring the immigration inflows to provincial cities as well, however, their further attractiveness depends on the internal policies, infrastructure and the local community abilities to embrace the newcomers as an additional resource rather than a challenge or even a threat. The cities who are aware of the effectiveness of a multicultural environment and who "use the difference of different cultures as potential for opportunities for the development of the local community" incorporate these principles and values into their strategies (Brunson, 2013). The Goal 10 among the 17 Sustainable Development Goals (SDGs) adopted by the United Nations Organization within the 2030 Agenda for Sustainable Development (United Nations, 2015) focuses on the inequality reduction. The issue of the inequality is perceived within and between the counties in income growth, all types of inclusion, equal opportunities and outcomes, equality-considering policies adoption, migration and mobilities, treatment of developing countries, financial and other kinds of assistance, etc. (Goal 10, 2015). Consequently, the relevant strategies developed at the local level are expected to correspond to the principles set up by this Goal.

A big number of researchers have explored the topic of intercultural cities and the ways and strategies for the equality promotion within the cities (Gabaix, Lasry et al., 2016; Osberg, 2001; Hall, 1979). Until Recently, the ways of reducing inequality have not been in the special focus of the researchers. However, since the UN has adopted the SDGs, scholars begin to give special focus to minimize the inequalities(Deaton, 2003; Langer, 2007, Lefranc, Pistolesi, Trannoy, 2008, Atkinson, 2015; Bhandari 2020; Bhandari & Shvindina 2019). Relevant strong attention to the cultural issues were supported by such researches as Lamont, Beljean, and Clair (Lamont, Beljean, Clair, 2014) and the group of the scientists who went deeper into this dimension of the inequality topic, while their research has proved that "human experience of inequalities is strongly grounded in cultural process" (Massey, McCall et al., 2014). This chapter is focused namely on the issue of inequality counteraction via the strategically supported culture and education environments favorable both for the hosting community and the city visitors, migrants, minorities of all kinds. We go deeper into the analysis of the cultural and partly educational components which appear to be essential for combating the inequalities and thus we answer the question of how to build the favorable culture-related conditions and to promote the proper mentality and understanding within the open local community. More serious attention to the educational sphere within the city development strategy based on the European intercultural educational policy had been paid in our recent paper (Vasylieva, Petrushenko, Mayboroda, Zamora, 2018), offering a set of recommendations for such peripheral cities like Sumy.

There was also a number of intercultural cities policy briefs and publications by the joint action of the Council of Europe and the EC which has presented the insights from intercultural cities throughout Europe and Asia: from Russia, Ukraine, Serbia to Italy, the Netherlands, etc. (About Intercultural Cities, 2018; Intercultural cities, 2009; Language policies, 2014; Urban citizenship, 2019; Urban policies, 2014). These examples demonstrate diverse solutions found by the cities that recognize themselves as intercultural: for example, among different urban policy approaches like guestworker, assimilationist, multicultural and intercultural policies, the last one demonstrates the most benefits both for the incomers and the local community. Positive success stories reveal that equality and togetherness are not enough; the migrants need not only the rights but also the opportunities and responsibilities to feel fully included. Nevertheless, there are examples where the immigrant minority searches or agrees to some kind of an isolated life-style within a hosting community only for the sake of the relatively decent economic situation: Roma communities in Macedonia (Spitalszky, 2018) and Ukraine (Bocheva, 2019), Moroccan and Turkish communities in Belgium (Gsir, Mandin, Mescoli, 2015), more rarely—Syrian communities in Germany (Ragab, Rahmeier, Siegel, 2017), etc. Lack of effort from the migrants' side to integrate, hope for settling without the need to accept the local rules and traditions, their lack of professional skills and language for communication, keeping up to some traditional values and principles that do not fit into the democratic systems of the hosting countries—all this adds to the challenges of being accepted in a foreign country. These cases sometimes result in complete destruction of the cooperation potential, strongly diminishing the chance of bridging between the minorities and the local community.

In addition, economic issues are usually accompanied by political and cultural problems creating challenges for religious practices, language use and development, ethno cultural practices and expressions (Langer, Brown, 2007). These are the signs of comprehending the migrants as a potential threat. Consequently, these challenges might contribute to unbearable conditions not only for the daily life of the immigrant minority but also destroy its economic potential. Yet, there are countries that manage to build at least some kind of neutrality policies, for example Belgium, where a number of rules and institutions were established for the sake of "equal treatment and recognition of the major ethno linguistic groups' languages and practices" (Langer, Brown, 2007). It is important to note, that the EU in overall is strongly focused on inter-religious and inter-faith dialogue on "international, European, national, and most of all—local level" (Vareikis, 2018).

There is a wide range of equality promotion experiences offered by communities of different kinds and characters, however, the empiric data state that each of them "should include proposals touching upon all aspects of the social structure relevant for the societal outcome in terms of economic inequality" (Atkinson, 2015). Deaton states that inequality in income leads to the reduction in "investments in health and education" (Deaton, 2003), which automatically reflects on the future of the upcoming generations lacking these two important components of a relevant quality. The circle closes up when skill prices change, which, in turn, reflects on the rising income inequality (Gabaix, Lasry, Lions, Moll, 2016). Theories on the causes of poverty refer to the concept of culture of poverty (Sameti, Esfahani, Haghighi, 2012) which should be recognized as one of the factors for counteraction. Children that grow up within the poor migrant or minority family must be aware of the prospects and accessibility of their good education, career chances, and inclusion into the hosting society.

Different opportunities, treatment and negative perception of the differences which are reflected in the inequality term may be classified as inequality of opportunity and inequality of outcome (Osberg, 2001), the last one is often used within the political discourse. We support this distinction as it includes social, cultural, and economic dimensions of the issue and provides a better understanding of its comprehensiveness. Treating the minority representative with his/her differences as a normal part of the community is the right approach, which is successfully practiced in some countries. However, an interesting example of this politically declared approach in 1960s in the USA, turned this country, though, to be nowadays recognized as one of "the most unequal countries in terms of both outcome and opportunity" (Lefranc, Pistolesi, Trannoy, 2008). We consider that one of the main reasons for this phenomenon is that the policies were focused on strengthening the racial and gender equalities, while social background cannot be diminished. This returns us somehow to the theory of poverty grounds.

The European Union, however, especially the Scandinavian countries (Collins, 2016), demonstrate a comprehensively positive experience of a successful strive toward the equality strengthening. The integration processes

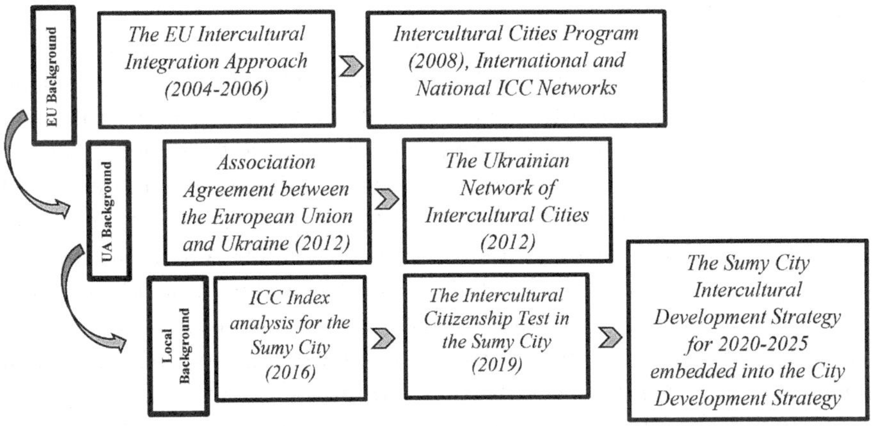

Figure 8.1. The Sumy City Roadmap toward the Interculturality *Source:* Authors.

of Ukraine with the EU make a sensible impact on the trends in social, cultural, and economic spheres. The main values which have been promoted in the country since the independence date include the notions of interculturality (Association Agreement, 2012). The Council of Europe develops the Intercultural Integration Approach basing on the principles of "cultural diversity, intercultural competence, multilingualism, intercultural education, hate prevention, combating racism and xenophobia, the role of the media, the interaction of migrants and host communities" (Brunson, 2013). Thus, the Intercultural Cities program was established involving the voluntarily joining cities, which further created relevant national networks.

In 2016, the Sumy city had joined this movement within the Intercultural Cities of Ukraine network and consequently had undergone the Intercultural Index analysis according to the ESC methodology (Sumy, 2016). The process became more organized after the Intercultural Citizenship Test in 2019 which laid the grounds for building the Intercultural City model and further development of the Sumy City Intercultural Development Strategy for 2020–2025. The Strategy was adopted at the LXVII City Council session (Decision, 2020) and became an integral part of the Sumy City Development Strategy till 2030 (Baranov, 2019). The local self-government has realized the need for that policy paper basing on the surveys data and European trends analysis.

The city annually receives thousands of students, tourists and hosts residents of different ethnic and religious backgrounds. Sumy is a regional centre for the administrative region neighbouring Russian Federation. The long multinational history and rich cultural heritage of the city obliges the community to start working in a more comprehensive way toward the better synergy with its minorities and guests. At the same moment it is not ready to represent the best of interculturality in its social and economic development, which was proven by the Interculturality Test (Intercultural Development, 2020). The national minorities require more active participation in the city governance, as well as more funding should be directed to strengthening the main areas of the intercultural policy. On the other hand, the minorities' and guests' religions and cultures exercising is expected but this exercising should be integrated into the existing lifestyle of the welcoming community. For the sake of peace and good neighbouring, the respect and understanding should be demonstrated by the guests and minorities representatives as well. That is why the Sumy City Intercultural Development Strategy for 2020–2025 is a conceptual document that was incorporated to the Sumy City Development Strategy till 2030. This allowed to outline not only the activities planned, but also required financing and responsible institutions and targeted audiences.

Yet, design of such a strategically important document faced some challenges, typical to Ukrainian cities, these are:

- there is no permanent statistics on residents by national, religious, or ethnic principle;
- some of the local minorities are quite closed both for cooperation and information revealing for the officials;
- the local authorities' initiatives may be perceived negatively; and
- some groups suffer from the lack of Ukrainian state language proficiency.

Thus, a number of different nations' representatives inhabiting Sumy may be determined only according to the last official census organized on December 5, 2001 (Statistic data, 2001). It is stated that out of 294,852 citizens, 84.6% were occupied Ukrainians and the rest were the nationals of: Russia (12.4%); Byelorussia (0.4%); Azerbaijan, Armenia, Georgia, Jewish, Poland, and Roma each (0.1%); and 1.8% did not admit their nationality. Greeks, Bulgarians, Moldovans, Lithuanians, Avars, Latvians, Karelians, etc. do not constitute 0.1% of the city population. The majority of them are Christians with a mixture of Islam and Judaism. In total, Sumy hosts 40 ethnic groups, which use Ukrainian and Russian languages for communication within the community. Most representatives of national minorities study Ukrainian in educational institutions, the incoming foreign students are obliged to take the language course during their preparation studies. The city's multilingualism is confirmed by the Intercultural Citizenship Test which states that more than 40% of respondents use

2–3 languages and another 25% speak 4 or more languages (Intercultural Development, 2020).

There are a number of registered NGOs which are nationally oriented, e.g., Sumy Regional Association of Armenian Culture “Artsakh,” Jewish Community Charitable Center “Hesed Khayim,” Sumy City Community Organization unions of Polish culture, Sumy Regional Non-Governmental Organization “Roma National Association,” etc. These organizations help their minorities to exercise their constitutional rights for the use and demonstration of their national cultural traditions, symbols, holidays, religion and so on. The most often form of these are the celebrations of national holidays and events where the national attire, symbols, arts, and traditions are presented. Each NGO provides the opportunity to study the national language and cultural heritage.

The ICC Index report on the assessment results of Sumy (Sumy, 2016) has revealed that in 2016–2017 comparing to other 81 cities participating in the ICC Network, Sumy was the 19th in overall and the 13th among the cities ranking with more than 200,000 inhabitants and the 14th among the cities where more than 15% were foreigners. However, ICC Index report shows that the cumulative intercultural index rate was the 67th. It indicates that that Sumy was more behind than ahead the average line. For example, the Sumy community demonstrated higher or average scores in the areas of intercultural competences, education, neighbourhood policy, public spaces, media, business, and employment. However, there was a sensible layback in the domains of social services, cultural and civic life, commitment, language, and competences development. It was also commented that the City Council was reporting on local intercultural events and even has launched a relevant page on its official website in 2016. To present the ICC background, including the designed Strategy, a webpage (Figure 8.2) was launched within the official website of the Sumy City Council (Sumy—An Intercultural City, 2016).

For now, the expert’s recommendation on using the social networks for nearing to the foreigners has not been fulfilled by the City Council. There is a Facebook page which is, however, maintained only in Ukrainian. This makes it useful only for the newcomers and residents who speak Ukrainian. The city authorities do not also have the diversity recruitment plan as according to the Ukrainian legislation only its citizens can serve as civil servants. Yet, there are some steps undertaken by the local authorities toward the Ukrainian language training for the vulnerable groups (e.g., unemployed, retired women).

As for the educational domain, there were some very progressive efforts that are worth mentioning: an increased parental involvement into the schools

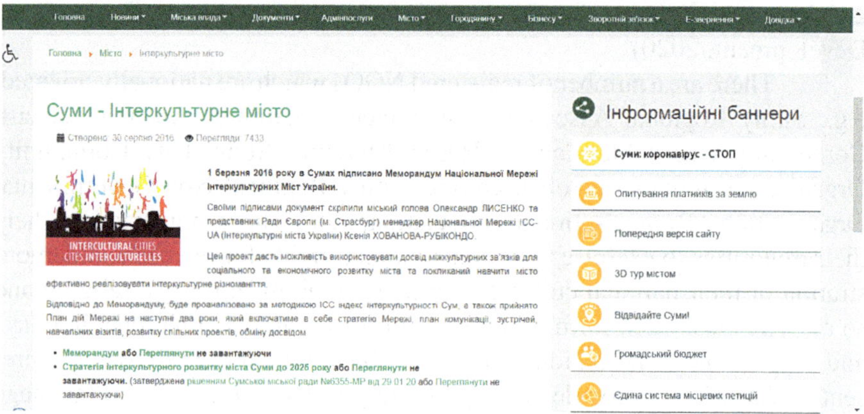

Figure 8.2. The Sumy City Council Webpage Dedicated to Interculturality
Source: The Sumy City Council Official Information Portal, a screenshot from (Sumy, 2016).

conflict prevention; professionalism of the school psychologists who acted as intercultural mediators and a bridge for communication between children and adults; intercultural projects for raising the awareness of pupils' different cultures and ethnic backgrounds; special surveys to identify any aspects of intolerance or radicalization; and information of sources of legal and social assistance. The experts recommended making an effort of communication with parents in their native languages wherever it's possible in order to increase their engagement and build understanding. The other recommendation was aimed at diversifying the school dishes, providing special spaces for religious rituals, and encouraging cultural mixing. It should be noted that the last recommendation is also valid for the university management, as well as for the urban management and cultural policies as there are city areas that are mainly inhabited by certain ethnic minorities (e.g., Roma community in the Baranovka neighborhood). Such public areas as parks, libraries or cafes, cultural venues are promoted to be interculturally prepared—e.g., the names of the streets and institutions are copied in English, the ATMs offer English language menu, once in a while the cinema offers the movies in English.

The experiences of festivals and trade fairs where traditional culture and food were presented have proved to be very positively accepted by the citizens. A relatively big number of arts and cultural events demonstrate the local community readiness to embrace other cultures if they are positively expressed (e.g., the annual international festival of brass music "Surmy Ukraiiny," where different cultures show musical and theatrical performances; the Jewish community, Aviv vocal studio). There are a number of ways how foreign nationals organize themselves into the groups and maintain their culture and exercise their traditions; however, they do not follow any systematic approaches.. This works in the opposite way: foreigners who come to study or live will not become isolated in their groups, but they are also not well understood by the Sumy community. That is why the most often issue is that foreign newcomers' behaviour is not understood/accepted/allowed simply because the locally established rules and traditions should have been explained from the very beginning. This step would help to build a harmonic coexistence in the Sumy community.

The survey has revealed that there is no discrimination issues in the field of entrepreneurial development, on contrary, there are areas in the city centre that are densely occupied by the people of foreign origin (mostly from the Central or Eastern Asia). There are also not only small but medium and big enterprises owned by foreigners from India, Azerbaijan, Poland, Italy, etc.

The Intercultural Index of the Sumy City demonstrated that there was a range of grant opportunities at the local level and others; however, there were no intercultural priority when distributing the funds. Yet, there are evidences that the city administration considers the proposals for improving the city interculturality positively having a Special Council that is obliged to present its decision for public through online and media consultation. Another positive experience is the local Cultural Outstanding People award that is conducted in cooperation with the local newspaper and a cultural-artistic centre. The expert's recommendation offered the City Council support to a local minority newspaper or radio program and makes more efforts in promoting a positive image of minorities or incoming foreigners. The city may organize anti-fake awareness raising campaigns in order to promote dialogue and understanding between different ethnic groups. Yet, Sumy has a municipal institution named the Sumy Promotion Agency that monitors and develops the city's openness to international relations. The local authorities have adopted an economic program "Sumy Open Space" aimed at promoting the international cooperation with the partner cities (e.g., organization of international exchanges, working visits, exhibitions, joint projects, etc.). In addition, the community is interested in international projects aimed at expanding the jointly fruitful economic relations.

The report also states that specialized institutions (e.g., hospitals, police, youth clubs, mediation centres, churches, etc.) are engaged into the intercultural mediation. For example, there are two organizations that are focused on inter-religious bridging: the United Church of Sumy and the Youth Friendly Clinic.

As a consequence of the war that is happening in the East of Ukraine, there are a steadily growing number of internally displaced persons who either seek for job opportunities or come to study. Anyhow they need relevant integration into the local community and all kinds of support.

The city promotes intercultural competence indirectly—via attracting foreign tourists and students, and directly via organization of interdisciplinary seminars, courses, and trainings with the involvement of foreign experts. The State Migration Service in the Sumy region provides the newly arrived migrants with a comprehensive road map that yet needs language diversity. It is also necessary to enhance the intercultural competence among civil servants, who provide services to foreign guests.

The results of the Intercultural Citizenship Test (Intercultural Development, 2020) revealed the stereotypical attitude of the Sumy citizens toward other ethnic origin or religion. Nevertheless, they believe that there is no inequality or barriers in the local community, but most of the foreigners do not share the same values as the respondents. Along with this it is interesting to note that most of the respondents do not attend events where foreign languages and other cultural backgrounds are gathered (Question 9, Appendix B to the Strategy). Of course, this leads to the situation that they will be unaware of the problems and challenges that foreigners face. On the other hand, the Sumy residents (83.4%) have demonstrated a tolerant attitude to other cultures assessing negatively, disrespect to other people's beliefs, and positively, the access to the health care system of illegal migrants. More than that, more than 60% of respondents are ready to assist a foreigner to protect his or her rights (Question 19, Appendix B to the Strategy).

The low grading of the question if the respondents contribute to the foreigners' welcome demonstrates a passive attitude to the newcomers. However, the people of other cultural backgrounds are welcomed to be involved in the decision-making and work in the social system (for example, teachers). This answer combined with the statement that Sumy citizens do not feel threatened (25.8%) or feel a low level of threat of the other cultures' influence on the local identity points out to a very low level of xenophobia in the city.

Up to today the city can boast of absence of the social conflicts except the traditional misperception of some of Azerbaijan and Armenia representatives. However, the following **major issues must be admitted as those that need greater attention**:

- low level of civic and legal education that leads to low activity in the city governance, misunderstanding of the legal norms, etc.;

- low level of awareness of the grants and local budget financing mechanisms, as well as projects design and management, which result in a very low use of the budgetary funds by the minorities;
- employment, medical services, and social protection issues when it comes to visiting foreigners and local residents with a foreign citizenship;
- not easily understandable behaviour of some nationalities by the local community combined with the absence of integration incentives from these nationalities' representatives which leads to the increasing gap;
- the foreign students coming for studies face the absence of English language knowledge, challenges related to public services, limited abilities for participation in the city entertainment, etc.

The number of the international students coming to enter the Sumy universities has been increasing during the last 15 years, while some of them stay to live in the city after the graduation. This means new businesses, employees, and mixed marriages. They have become a significant income source not only for the universities, but also for the local entrepreneurs who rent the apartments to them, sell the goods and services. According to ICC Index report the share of foreign nationals who possess a temporary residence status and who came to study in 2018 was 83%, and 15% were for family reunification with Ukrainian citizens and up to 0.02% come for the religious needs, respectively. The permanent residence permit is usually obtained via the marriage (42%), 41% acquire Ukrainian citizenship by the right of the territorial origin, and less than 1% restore the citizenship. The number of refugees or internally displaced people is quite low, e.g., in 2018 only 15 foreign nationals requested the refugee status (Migration Service, 2019).

However, despite the mentioned challenges, Sumy demonstrates a long-standing experience of peaceful coexistence and satisfactory cohesion within the city. These experiences will be improved by the newly adopted Strategy and can serve well for sharing with other cities.

2. Methods

This material is based on the combination of several sources of data that allowed building the problem understanding from different angles. The author was involved into the Sumy Intercultural Development Strategy development (Intercultural Development, 2020) and all the preparatory data collection works executed by the working group appointed by the city mayor. The Strategy paper is based on relevant legal background, both Ukrainian and European, and is strongly built upon the strategic planning methodologies. Thus, it has its own vision and mission declared as a separate chapter, as well as identified monitoring and evaluation indicators.

The data used for the study were obtained from the publicly available State Statistics of Ukraine, specifically requested data of the Main Department of Statistics in the Sumy region and of the State Migration Service of Ukraine in the Sumy region. In terms of the interculturality topic, the needs of the community were analysed using the results of the Intercultural Profile of the city prepared by the ICC experts (Sumy, 2016), Intercultural Citizenship Test (a survey), and the SWOT-analysis of all major areas of the city intercultural development conducted by the authors (Intercultural Development, 2020). The Test survey in Sumy involved 524 respondents, its questions and results were also included into the Intercultural Strategy paper.

There were also three meetings of focus experts groups conducted in order to get the written feedback which was used by the authors. Especially valuable was a face-to-face meeting with the representatives of minorities and communities, including migration and police services officials and the local university researchers and students with their verbal feedback on the Strategy draft presentation, this was supported by an audio recording and a further transcript of the meeting. These discussions allowed to reveal the qualitative data and the issues which was related to the foreign students challenges that were not discussed before.

The scientific side of the process was based on eight meetings of the working group, which were supported by the consultations with the public and EU experts in intercultural strategies development. A final public presentation and engagement of the wider audience to the Strategy paper discussion have proved that the format and approaches chosen were correct for the moment. The Strategy authors confirm that all active and interested community members' participation in the dialogue with the authorities and the working group allowed to make the Strategy activities relevant and realistic as it reflects the current needs and interests of the target audience.

The process of the Strategy design followed the Community Based Results Accountability Approaches methodology proposed by the EU experts and independent consultants-developers, as well as the recommendations provided by the Council of Europe (The Intercultural City, 2019). The systematic literature review allowed us to build the general framework of activities that must be reflected in the Strategy paper. P. Brunson names the most of them (Brunson, 2013):

1. Forming a positive attitude to cultural diversity.
2. Adjusting the main public services under the intercultural context.
3. Considering mediation and conflict resolution.
4. Migrants and minorities languages and Ukrainian state language training.
5. Building a positive media perception of the interculturality.
6. Civic engagement and intercultural governance approaches.
7. Training of the policy makers and local authorities/self-government administrative staff.
8. Hospitality programs/activities.

Basing on these theses that have created the framework of the city's mission, vision and the main strategic directions of future actions, the sets of operational goals were designed for each local life areas.

To sum up, the methods which were used for the data processing and development included the statistical analysis of quantitative data and the content analysis of qualitative data, an expert method, and the SWOT analysis followed by the deductive and inductive analysis methods.

3. Results

The SWOT analysis results (Intercultural Development, 2020) **in the field of education** present that it is characterized by openness, the local community is aimed at the lifelong learning using modern educational facilities and services of the local highly qualified personnel. However, the issues of intercultural and multicultural education remain unresolved, e.g., in the field of non-formal intercultural education. There are almost no extracurricular educational institutions, non-governmental minority or volunteers organizations who would be specifically focused on tolerance, interculturalism, xenophobia, and mediation. Educational activities aimed at shaping the intercultural competences do not involve minorities and foreigners in the joint development of a city "friendliness." This is reinforced by the fact that despite growing number of incoming foreign students, the young people who live in the city hardly meet the different cultures while studying. There is no open non-acceptance of the diversity or discrimination; however, this is rather because of the relatively low numbers of foreigners who study in the local universities.

The other issue that has been recently discussed a lot is that Sumy educators are not fully prepared to work with the diversity, vulnerability and, adding to this in the times of COVID-2019, online teaching. Teachers are more trained in the fields of teaching methodology, pedagogy, but not in intercultural dialogue, tolerance, mediation, and intercultural development. Language training is mostly directed to English but yet is not at the satisfactory level to provide fluent and relaxed communication with a foreign student.

That is why there are the following opportunities in the field of education for the Sumy city:

1. The city encourages cultural and educational organizations to work with the diversity issues and an intercultural dialogue.
2. Sumy City Council promotes the intercultural competences development through interdisciplinary seminars, courses and trainings on the basis of educational institutions, as well as language trainings.
3. The city takes measures to encourage intercultural mixing in public places, including schools and higher educational institutions. A wide range of cultural events are held in public places (public squares, parks and other cultural centres, museums and theatres) aimed at encouraging intercultural mixing.
4. Existence of the leisure facilities, schools, and courses where languages, cultures and traditions of national communities are studied.
5. Introductory legal conversations and practical classes for foreigners are held in educational institutions.
6. Openness of the local authorities for the proposals targeted at interculturality increase in the city.

In the field of culture, the SWOT analysis revealed the presence of citizens with a high level of cultural awareness and intercultural competence, which is the main argument for the successful implementation of a Strategy. If the culture considered as an iceberg according to Hall's model (Hall, 1979), then the part we see is a small manifestation of what we cannot see: the most invisible elements of culture, such as values, statuses, rules, traditions, and beliefs. During the interaction in intercultural situations, lack of awareness can lead to conflict situations or bad decisions. Thus, cultural awareness and tolerance, positive communication skills, intercultural and inter-religious mediation, and other intercultural competences are essential for both professional and social lives.

An important point to emphasize is that acceptation and respect of another culture does not mean that one must compromise one's own culture or sacrifice one's values or identity for the sake of "others." In fact, over time, it can cause even more problems and conflicts. Being tolerant to other cultures can be translated as trying to understand how to adapt our approaches, thoughts and decisions in a best way in order to deliver positive results. It is all about respecting other people's cultures as well as our own.

For a better understanding of the trends and the state of intercultural development of the city's cultural field, its weaknesses will be presented below. Thus, the reduction of the population, emigration of the active labor force abroad, including representatives of national communities, can be pushed by the war in the Eastern Ukraine. There is also a lack of clearly defined priorities for supporting intercultural policy in the city's strategic documents. Under certain circumstances the financial support for it may be reduced or fully eliminated that puts in danger the whole process of the Strategy implementation. The lack of training programs for public services, media, and educational institutions employees in multilingual and intercultural competence causes not only challenges in embracing foreign or minority residents but also adds to low support of the need to fund the Strategy activities. The lack of comprehensive statistical information on the number and composition of representatives of national communities will not allow to rationally plan financial support for intercultural strategy activities, as well as to take into account the needs and interests of all groups and communities of the city.

A low level of awareness of the city population of its intercultural advantages and, as a consequence, insufficient use of the intercultural potential of the city may be threatened by the inter-ethnic conflicts. The other weakness in this regard is—an insufficiently developed infrastructure of intercultural nature combined with the lack of information in foreign languages.

Basing on this, the following opportunities in the field of culture for the Sumy city may be proposed:

1. Increasing the level of interculturalism support and participation in international projects, including on the development of tourism and human capital. Informing and training representatives of communities and minorities in the city to receive grants, in particular through EU programs.

2. Readiness of local authorities of the city to support intercultural priorities financially and organizationally will help to implement the majority of the proposed activities within the Strategy.

3. Preservation and promotion of the historical heritage of minorities and communities, and its mapping in order to get acquainted with its cultural potential. Promotion of deeply traditional manifestations of cultures through cultural and art events. Increasing the level of identification of cultures of minorities and foreigners living in the city by the community.

4. Carrying out educational actions directed on formation of intercultural competences of the city residents, public servants and the newcomers. Strengthening the quality of initial adaptation of foreign guests of the city who came, in particular, for studies. Establishing good neighbourliness between residents and representatives of different groups.

5. The use of ethnic resources as a springboard for the development, modernization, and improvement of urban policy; involvement of the minorities and communities' representatives in the local decision-making. Recognition of the contribution of minorities and foreigners living in the city to the development of Sumy.

6. Creating a circle of agents of changes within the minorities and communities. Active involvement of their representatives in the cultural and artistic life of the city. Increasing the emotional connection of the citizens of the city with the representatives of minorities and foreigners living in the city on a personal level.

The Intercultural Development Strategy of the Sumy City (2020–2025) covers the generalized and somehow integrated areas of education, culture and sports, business and employment, public places and administrative services, security and safety, communications and media, intercultural governance, and civic engagement. The values that underpin the Sumy City Intercultural Strategy are aimed to unite all residents regardless of their age, gender, origin, beliefs, and aspirations.

Based on the analysis conducted in the framework of this study, the following basic elements mandatory for the first launch of the Intercultural Development Strategy were defined:

1. Identification of the main strategic goals (based on weaknesses and strengths), including mission and vision that should be presented to the target audiences and publicly discussed.

2. Decomposition of strategic directions for operational purposes that will allow defining the priority areas and the steps to success in them.
3. The ways of financing must be pre-identified as this determines the set of assumptions and limitations for the proposed activities. It is challenging but even the first draft of the Strategy paper requires also definition of the executing agencies/institutions as then their resources and potential can be exploited in the most suitable and effective way.
4. Indicators for monitoring the implementation of the operational objectives must be proposed at least for the first year of the Strategy activities implementation. This makes the set goals more realistic and measurable, and, thus, more understandable for the target audience, taxpayers, and the donors.

4. Discussion

Below the proposed solutions in the fields of education and culture will be presented for the discussion. They have embraced the results of the city analysis presented above as well as the European experiences in the field.

In the domain of education, the following main solutions were planned for the next 5 years for the Sumy city:

1. Equal access to education:
   - Language training;
   - Implementation of own and joint educational projects outside the educational process;
   - Establishment of cultural and educational centres of ethnic communities with the involvement of volunteers—representatives of these communities;
   - Introducing mentorship for representatives of minorities.
2. Social identity level improvements and social adaptation and intercultural awareness:
   - Annual educational events (International Tolerance Day, Cultural Diversity Day, Let's Be Friends, etc.);
   - Organization of non-formal intercultural education for community members;

- Intercultural social educational projects.

3. Intercultural competences formation:
   - Elective courses on intercultural education;
   - Regional innovative educational projects and experiments in educational institutions;
   - Didactic and methodological materials on problems of formation of intercultural competences;
   - Intercultural training of teachers.

4. Social cohesion trough diversity acceptance:
   - Development and implementation of a tolerance strategy at educational institutions;
   - Youth social events, flash mobs, quests, sports games with elements of national attributes with community/minority involvement;
   - Empowering citizens to implement their own cultural, artistic and sports projects and their functional support;
   - Ethno-cultural educational events;
   - Introducing Honorary Citizenship Award for Change Makers among different community groups in Sumy.

In terms of culture the following solutions were planned for the next 5 years basing on the needs analysis and researches of the successful European experiences:

1. The domain of popularization of the culture (heritage) of different minorities and communities through art and sports, examples of the proposed activities:
   - Annual "Bridging Cultures" festival, World Dance Day, Multicultural Food Fest;
   - An annual sports event with engagement of all communities and minorities of the city, with a demonstration of traditional sports and historical reconstruction;
   - Trainings, craft workshops, promotion of hobby groups that support the cultural heritage of communities;

- Creation of a students' ethno-cultural theatre on the basis of Theatre of Young Spectator, organization of crafts courses "Modern Fashion with Elements of National Decor," etc.;
- Development of non-expensive tourist routes and excursions taking into account the cross-cultural heritage, tourist trips to twin cities and other communities in the network of intercultural cities of Ukraine/ Europe.

2. Cultural diversity mapping for the sake of awareness of the community potential, examples of the proposed activities:
   - Creating an Intercultural Guide with a newsletter function for informing about the upcoming events;
   - Debunking myths and fears about particular cultures and religions through events with the involvement of relevant communities and minorities.
3. Preservation and popularization of the historical heritage of the minorities and communities that have lived or are currently living in Sumy, examples of the activities:
   - Establishment of the Intercultural Museum of the city with the audio guiding in relevant languages;
   - Setting up a space for the local minorities and communities arts from the past and those who are active now;
   - Creation of the "Book of Pride" by prominent communities and minorities who have lived in Sumy and made history through their achievements;
   - Establishment of memorials for historical persons whose activities are related to the intercultural development of the city;
   - A public project with the working title "How Did I Happen to Be Here."
4. Activists and volunteers' engagement toward promotion of the cultures and heritage of different communities, of the proposed activities:
   - Ambassadors Project (Agents of Changes) for intercultural dialogue in the field of culture and sport—their search, support, and training;
   - Prioritization of interculturalism among local grants and projects (needed a priority area: "Cultural and Sports Events Promoting the

Intercultural Character of the Sumy City"), information dissemination about the granting opportunities;

- Buddy Programme for/by the universities;
- Training for community and minority representatives to receive grants (EU's Creative Europe, House of Europe programs, etc.).

Below the sources and indicators are presented both for the education and the culture domains. However, except the set of these approved by the Sumy City Council, as a case study, we outlined the working materials related to the culture and education domains. Each of them was offered to have a specific success indicator in order to make the actual measurement possible and thus could be engaged additionally for monitoring and evaluation.

The working group discussion of the monitoring and evaluation tools which have been proposed but not adopted for the Sumy City Strategy was using an argument that there were no specifically indicated funding source for each planned activity, as well as the executors were not known yet. These lack of information was strongly limiting the ability to set up clear indicators to evaluate the implementation success of the Intercultural Development Strategy.

5. Conclusion

The City's Intercultural Development Strategy must be an integral part of the city's overall Development Strategy. Thus, assumption leads to the conclusion that both strategies must be designed simultaneously and for the same time period. Therefore, all measures defined within the Intercultural Strategy should help to implement the SDGs, overall City Development Strategy and be integrated into its philosophy. On the other hand, the intercultural component will be an indicator of the full implementation of the overall city development strategy as social priorities are usually tempted to be skipped under the conditions of crisis, political changes, etc. This complementarity and interconnectedness of these two strategies determine the framework of limitations and assumptions and risks for a City's Intercultural Development Strategy.

The Sumy city has followed this complementarity principle and namely after the City Council approval of the general development strategy for the Sumy city for the 2020—2030 period, the stage of implementation of this strategy began. For now the authorities are engaged in programming, development

Table 8.1. Monitoring and Evaluation Tools Proposed for Inequality Counteraction in Education and Culture Domains of the Sumy City Life

#	Domain	Monitoring and Evaluation Tools Proposed Within the Strategy	Monitoring and Evaluation Tools Proposed but Not Adopted For The Strategy
1	*Education*	• Comparative statistics on minority involvement and willingness to interact, increasing of intercultural awareness of minorities and the majority (number of joint intercultural projects, interviewing, testing).	• A number of change agents identified from each identified community, group of foreign nationals living in the city for a long time. • The procedure for officially assigning this status to a person and transferring it to a successor in case of change is defined. • The number of trainings each agent undergoes during the year of his/her work; 3 of them are obligatory in the first 2 months of his/her work (topics of intercultural dialogue, mediation and communication). • The number of agents included in the relevant working groups and structural units of local self-government corresponds to the number of key areas of work of the local self-government of Sumy. • Change agents participate in a minimum set of the meetings and meetings of the relevant local government working groups/units.
2		• Polling, testing for social adaptation and civic integration.	• Quantitative statistics of the city's senior foreign students trained to support newly arrived foreign students. • Quantitative statistics of the foreign students of each university who will give a positive feedback about the support from senior foreign students, the work of counselling centres and the effectiveness of the Buddy Program.

Table 8.1. (***Continued***)

#	Domain	Monitoring and Evaluation Tools Proposed Within the Strategy	Monitoring and Evaluation Tools Proposed but Not Adopted For The Strategy
3		• Quantitative statistics on the majority involvement in the creation of a national character of the institution and mentoring.	• Raising awareness of the Sumy community about deeply traditional manifestations of culture and traditions of minorities and foreigners living in the city.
4		• Introducing a competent approach to intercultural education (quantitative indicators).	• Launch of at least 1 "Modern Fashion With Elements Of National Decor" course on the basis of the relevant Sumy technical school/college. • Statistics of master classes on making traditional craft items and souvenirs held annually.
5		• Increasing minority involvement in the social life of the city (a survey).	• Statistics of representatives of communities and minorities of Sumy will annually take part in the scheduled city events. • Statistics of artists involved in the promotion of deeply traditional manifestations of culture and traditions of minorities and foreigners living in the city. • The establishment of a student ethno cultural theatre, the number of events in it. • The establishment of an intercultural museum, the number of events in it. Provision of the museum's expositions with translation/audio guide in several languages, including English during the first year of its existence. During the year, the minimum number of excursions from each school, kindergarten, and university is organized for representatives of communities and minorities and the majority.

#	Domain	Monitoring and Evaluation Tools Proposed Within the Strategy	Monitoring and Evaluation Tools Proposed but Not Adopted For The Strategy
6		• Reducing the number of unresolved minority issues due to their incompetence (quantitative statistics).	• On the basis of each university that teaches foreigners, an advisory point for socio-cultural adaptation of foreign students has been established. • Statistics of newly arrived foreign students who use the services of established help points. • Statistics of consultations the Sumy City Council Project Office conducts with community and minority representatives on the possibilities of obtaining grants for their needs. • Statistics of events aimed at debunking myths and fears about certain cultures and religions held annually and the number of citizens having attended each event.
7		• Introducing optional courses in formal and non-formal educational institutions (schools, colleges, and universities).	
8	*Culture*	• Increasing the number of visitors to networking events.	• Statistics of youth groups with intercultural orientation/ character. • Statistics of activities aimed at promoting interculturalism in the city community. • A developed intercultural guide and the number of its issues published.

Table 8.1. (*Continued*)

#	Domain	Monitoring and Evaluation Tools Proposed Within the Strategy	Monitoring and Evaluation Tools Proposed but Not Adopted For The Strategy
9		• Increasing the number of foreigners involved in the projects of the Sumy City Youth Leisure Centre (municipal youth organization).	• Statistics of training events held before each grant competition under EU programs for representatives of the city's communities and minorities. • The priority "Cultural And Sports Events To Promote The Intercultural Nature Of The City Of Sumy" to be introduced into local grant calls and projects. The minimum share of all applications submitted to the call for grants addressed to this priority is set up. • Each donor who has introduced the relevant priority conducts at least one explanatory event for potential applicants before the start of the competition. • The minimum number of representatives of the city's communities and minorities shall be introduced to the expert commission of the competition with the right of an advisory vote. It changes every year.
10		• Increasing the number of foreign visitors to the entertainment venues (theatres, cafes, etc.).	

#	Domain	Monitoring and Evaluation Tools Proposed Within the Strategy	Monitoring and Evaluation Tools Proposed but Not Adopted For The Strategy
11	*Culture (additional tools/indicators)*		• Control check-up of the elements of the Interculturality Index in the context of the effectiveness of the proposed measures—at least once a year.
12			*Popularizing the Sumy heritage:* • Statistics of commemorative signs been installed in a certain period of time. • Statistics of "The Book of Pride" copies in a minimum number of languages set, including Ukrainian and English. The number of its downloads from the website. • Statistics of budget tourist routes and excursion programs developed and used annually. • Increase of the presence of traditional craft items and souvenirs among the products sold during city events. • An art zone in the ethno museum opened for the arts of communities and minorities of the city.
13			*Visibility in media and information dissemination:* • Statistics of publications and video reports in local media about the activities of this section annually. • Statistics of views of the webpage of the Intercultural Guide per month. • Statistics of people included in the public project with the working title "How I Happened to Be Here" in the first year of its existence. The number of public events within this project organized annually.

of the implementation scheme and informing stakeholders about roles and responsibilities. Of course, it is clear that to achieve a strategic goal, one or several target programs must be designed as a basis for the development of appropriate budget financing documents. At the same time, it should be noted, that interculturalism as an approach (e.g., toward the gender equality) should become an integrated feature of all targeted urban development programs.

Discussed within this paper, Intercultural Development Strategy for a provincial Ukrainian city embraces systemic changes in the seven main areas of public involvement in Sumy city life: "Education," "Business and Employment," "Public Places and Administrative Services," "Security and Safety," "Culture and Sports," "Intercultural Governance," and "Communications and Media.". Basing on the presented data analysis and best suited European experiences in the interculturality and equality promotion within the Education and Culture, there are a number of steps needed to be taken by the local authorities in these domains. For example, in the field of education there is a need to carry out information and educational activities on intercultural integration, raising awareness, mixing, and intercultural competences training. These are expected both for the teachers and learners, hosting community representatives, and the newcomers. All these, consequently, require appropriate strategies, human and financial resources and infrastructure that will contribute to bridging between the people's groups, community and the authorities, and the city and its partners.

The most questionable issue in the process of the Intercultural Development Strategy design is how to prove to the target audience that the proposed activities will be fully implemented, and the results will be visible and feasible. This requires not only civic engagement and public discussions, but also careful designing of the monitoring and evaluation tools for each domain of the public life. However, this process will be strongly challenged by the absence of the clear understanding of the financial sources, funds amounts, and executors.

References

Bhandari, Medani P. (2020) Second Edition- Green Web-II: Standards and Perspectives from the IUCN, Policy Development in Environment Conservation Domain with reference to India, Pakistan, Nepal, and Bangladesh, River Publishers, Denmark / the Netherlands. ISBN: 9788770221924 e-ISBN: 9788770221917

Bhandari, Medani P. (2020), Getting the Climate Science Facts Right: The Role of the IPCC, River Publishers, Denmark / the Netherlands- ISBN: 9788770221863 e-ISBN: 9788770221856

Bhandari, Medani P. and Shvindina Hanna (2019) Reducing Inequalities Towards Sustainable Development Goals: Multilevel Approach, River Publishers, Denmark / the Netherlands- ISBN: Print: 978-87-7022-126-9 E-book: 978-87-7022-125-2

Brunson, P. R. (2013). Building Intercultural Strategies with Citizens: The Community Based Results Accountability Approach. Handbook. EU: Center for the Study of Social Policy, 32.

Statistic data of the Official Register of the Population (December 5, 2001). The Main Department of the Statistics of the Sumy Region: the Official Register of the Population. Retrieved from: http://sumy.ukrstat.gov.ua/?menu=301&article_id=758 [Online Resource].

Migration Service of the Sumy city (2019). Data, received for the inquiry of the Department of Communication and Information Policy of the Sumy City Council. Sumy, 2019.

Sumy: Results of the Intercultural Cities Index (October 2016). A comparison between 81 cities. Intercultural Cities. Council of Europe. Retrieved from: https://rm.coe.int/16806b2488 [Online Resource].

Hall, E. T. (1979). Foreword: Cultural models in transcultural communication. In Nonverbal Behavior, 11-17. Academic Press.

The Intercultural City Step By Step (2019). A practical guide for applying the urban model of intercultural inclusion. Revised edition, ed. I. D'Alessandro. Council of Europe. Retrieved from: https://rm.coe.int/the-intercultural-city-step-by-step-practical-guide-for-applying-the-u/168048da42 [Online Resource].

Atkinson, A. B. (2015). Inequality: What Can Be Done? Harvard University Press, 2015, 384 pages. Croatian Economic Survey, 17(2), December 2015, 113-118. doi:10.15179/ces.17.2.4

Langer, A., Brown, G. (2007). 'Cultural Status Inequalities: An Important Dimension of Group Mobilization', CRISE Working Paper no. 41, Centre for Research on Inequality Human Security and Ethnicity, Queen Elizabeth House, Oxford, 1, 6. Retrieved from: https://pdfs.semanticscholar.org/8dae/dbc5f247b7875b7736d46d4692abdff32c7f.pdf [Online Resource].

Massey, D., S., McCall, L., Tomaskovic-Devey, D. et al (2014). Understanding inequality through the lens of cultural processes: on Lamont, Beljean and Clair 'What is Missing? Cultural Processes and Causal Pathways to Inequality'. Socio-Economic Review, 12(3), July 2014, 609–636. Retrieved from: https://doi.org/10.1093/ser/mwu021

Lamont, M., Beljean, S., Clair, M. (2014). "What is Missing? Cultural Processes and Causal Pathways to Inequality". Socio-Economic Review, 1-36. Retrieved from: https://scholar.harvard.edu/lamont/publications/what-missing-cultural-processes-and-causal-pathways-inequality [Online Resource].

Deaton, A. (2003). Health, Inequality, and Economic Development. Journal of Economic Literature, XLI, March 2003, 113-158. doi:10.1257/002205103321544710. Retrieved from: https://pdfs.semanticscholar.org/e1e1/ace8783c06b9abcb7adef66125519bd49a36.pdf?_ga=2.247494343.1539128563.1577968856-1084006431.1577968856 [Online Resource].

Gabaix, X., Lasry, J., Lions, P., & Moll, B. (2016). The Dynamics of Inequality. Econometrica, 84(6), November 2016, 2071–2111. doi:10.3982/ECTA13569. Retrieved from: https://www.princeton.edu/~moll/dynamics.pdf [Online Resource].

Osberg, L. (2001). Inequality. International Encyclopedia of the Social & Behavioral Sciences, 2001, 7371-7377. https://doi.org/10.1016/B0-08-043076-7/01898-2

Intercultural Cities (2009). Ed. P.Wood. - Council of Europe Publishing, France, 124.

About Intercultural Cities (2018). Intercultural cities Programme, Council of Europe. Retrieved from: https://www.coe.int/en/web/interculturalcities/about [Online Resource].

Urban citizenship and undocumented migration (2019). Policy Brief "Urban Citizenship: Making places where everyone can belong". The Council of Europe, 8. Retrieved from: https://rm.coe.int/urban-citizenship-and-undocumented-migration-policy-brief/1680933628 [Online Resource].

Urban policies for intercultural education (2014). Intercultural Cities Policy Briefs. The Council of Europe. Retrieved from: https://rm.coe.int/CoERMPublicCommonSearchServices/DisplayDCTMContent?documentId=0900001680493bd3 [Online Resource].

Language policies for the intercultural city (2014). Intercultural Cities Policy Briefs. The Council of Europe. Retrieved from: https://rm.coe.int/CoERMPublicCommonSearchServices/DisplayDCTMContent?documentId=0900001680493bd5 [Online Resource].

Sameti, M., Esfahani, R. D., Haghighi ,H. K. (2012). Theories of Poverty: A Comparative Analysis. Kuwait Chapter of Arabian Journal of Business and Management Review, 1(45). February, 2012, 47.

Lefranc, A., Pistolesi, N., Trannoy, A. (2008). Inequality Of Opportunities Vs. Inequality Of Outcomes: Are Western Societies All Alike? The review of Income and Wealth, 54 (4), December 2008, 513-546. Retrieved from: https://doi.org/10.1111/j.1475-4991.2008.00289.x

Vasylieva, T.A., Petrushenko, Y.M., Mayboroda, T.M., Zamora, O.M. (2018). The Issues Of The City Development Strategy Design Basing On The European Intercultural Educational Policy. Visnyk of SSU, Economy Series (3), 2018, 38-43. doi: 10.21272/ 1817-9215.2018.3-6.

Baranov, A., Łątka, K. et al. (2019). Sumy City Development Strategy till 2030. Draft of the Sumy City Development Strategy 2030. Edited 04.12.2019. Retrieved from: http://minayev.com.ua/wp-content/uploads/2019/12/%D0%A1%D1%82%D1%80%D0%B0%D1%82%D0%B5%D0%B3%D1%96%D1%8F-%D1%80%D0%BE%D0%B7%D0%B2%D0%B8%D1%82%D0%BA%D1%83-%D0%BC%D1%96%D1%81%D1%82%D0%B0-%D0%A1%D1%83%D0%BC%D0%B8-2030.pdf [Online Resource].

Decision of the Sumy City Council (2020). № 6355-MP (VII Season, LXVII Session) "About the Adoption of the Sumy City Intercultural Development Strategy", January 29, 2020. (https://docs.google.com/viewer?url=http%3A%2F%2Fsmr.

gov.ua%2Fimages%2Fdocuments%2FRishennia%2FSesii%2F2020%2F29.01.2020%2F6355-MR%2F6355-MR.docx&embedded=true&fbclid=I-wAR0YN2suvp7pITwwuTP-sZFGUV1XxVebIRn2e_3Kla62rAwBXnEX-qHt_DnY , accessed 1 May 2020). [Online Resource].

Intercultural Development Strategy of the Sumy City for 2020-2025 (2020). An attachment to the № 6355-MP Decision of the Sumy City Council (VII Season, LXVII Session) "About the Adoption of the Sumy City Intercultural Development Strategy ", January 29, 2020. (https://docs.google.com/viewer?url=https://www.smr.gov.ua/images/documents/Rishennia/Sesii/2020/29.01.2020/6355-MR/6355-MR_dodatok_copy.docx&embedded=true, accessed 1 May 2020). [Online Resource].

Sumy – An Intercultural City (2016). A Sumy City Council Informational Portal, August 30, 2016. (https://smr.gov.ua/uk/misto/interkulturni-mista.html, accessed 1 May 2020). [Online Resource].

Goal 10: Reduce inequality within and among countries (2015). Sustainable Development Goals. New York (NY): United Nations; 2015 (https://www.un.org/sustainabledevelopment/inequality/, accessed 1 May 2020). [Online Resource].

United Nations (2015). Transforming our world: the 2030 agenda for sustainable development. Sustainable Development Goals Knowledge Platform. New York (NY): United Nations; 2015 (https://sustainabledevelopment.un.org/post2015/transformingourworld, accessed 1 May 2020). [Online Resource].

Spitalszky, A. (2018). Roma in the Republic of Macedonia: Challenges and Inequalities in Housing, Education and Health. Briefing, UK: Minority Rights Group International, November 2018. Retrieved from: https://minorityrights.org/wp-content/uploads/2018/11/MRG_Brief_Mac_ENG_Nov18.pdf [Online Resource].

Bocheva, H. (2019). Roma in Ukraine – A Time for Action: Priorities and Pathways for an Effective Integration Policy. Report, Budapest: Minority Rights Group International, 2019. Retrieved from: https://minorityrights.org/wp-content/uploads/2019/05/MRG_Rep_Ukraine_EN_Apr19.pdf [Online Resource].

Gsir, S., Mandin, J., Mescoli, E. (2015). Corridor Report on Belgium – Moroccan and Turkish Immigration in Belgium, INTERACT RR 2015/03, Robert Schuman Centre for Advanced Studies, San Domenico di Fiesole (FI): European University Institute, 2015.

Ragab, N.J., Rahmeier, L., Siegel, M. (2017). Mapping the Syrian diaspora in Germany: Contributions to peace, reconstruction and potentials for collaboration with German Development Cooperation, MGSoG: Maastricht University, 23 January 2017, 65.

Vareikis, E. (2018). Prevention of violence and discrimination against religious minorities amongst refugees in Europe. Parliamentary Assembly, Committee on Migration, Refugees and Displaced Persons report: Council of Europe, Strasbourg. Provisional version, Doc. 14429, Reference 4351 of 22 January 2018. Retrieved from: http://www.assembly.coe.int/LifeRay/MIG/Pdf/

TextesProvisoires/2019/20191202-ViolenceDiscriminationReligiousMinorities-EN.pdf [Online Resource].

Collins, Ch. (2016). We Should Take a Lesson from the Nordic Countries on Inequality, Institute for Policy Studies, July 18, 2016. (https://ips-dc.org/take-lesson-nordic-countries-inequality/, accessed May 1, 2020). [Online Resource].

Association Agreement between the European Union and Ukraine (2012). Government Portal: official website. Retrieved from: https://www.kmu.gov.ua/storage/app/media/uploaded-files/ASSOCIATION%20AGREEMENT.pdf [Online Resource].

Chapter 9

Causes and Ways to Overcome Socio-Economic Inequality in Ukraine

Tetiana Semenenko[1], Volodymyr Domrachev[2], Vita Hordiienko[3]
[1]Department of International Economic Relations, Sumy State University, Ukraine
[2]Department of Applied Information Systems, Taras Shevchenko National University of Kyiv, Ukraine
[3]Vita Hordiienko, PhD, Department of Management, Sumy State University
E-mail: t.semenenko@uabs.sumdu.edu.ua, domrachev@univ.kiev.ua, v.hordiienko@crkp.sumdu.edu.ua .

1. Introduction

The state of Ukraine's economy is characterized by the sharp impoverishment of the population contrasting with oligarchs' accumulation of fortunes through corruption. It results in political instability in the country. It is considered that the poverty of our time is different from the poverty of the past, because it is not caused by the natural scarcity of resources, as it was the case in the previous periods. Instead, its origins lie in the set of priorities of the rich. Therefore, today's poor can hardly expect any mercy as they are simply tossed aside. This situation is not uncommon—it is experienced by some of the developed economies of today. The Ukrainian society is facing a dilemma: either to reduce income inequality or to remain a third-world country with low levels of consumption.

The financial risks are increasing amid the worsening political relations in the world, falling oil prices, the spread of the COVID-19 pandemic, and the trade conflict between the United States and China. In its report (World Economic Outlook, April 2020) the IMF presented an analysis and forecast of changes in the economic situation both at the global and country level. The main economic challenge today is the pandemic that provoked a simultaneous struggle of countries with the spread of COVID-19 (Bhandari 2020a;2020b). The IMF predicts a 3.0% decline in the world's real GDP in 2020 and a sharp increase of unemployment. The Fitch report (Global Economic Outlook, 2020) indicates that the global economy will fall by 3.9% in 2020. It is predicted that restrictions associated with the Coronavirus pandemic will lead to the recession of unprecedented scale in the post-war period. This recession will be twice as severe as the recession after the global crisis of 2009.

Today it can be stated that the Coronavirus pandemic has become a historic challenge. In mid-February, as market participants began to fear that an epidemic would turn into a global pandemic, stock prices plummeted from previously high levels. Spreads have increased tremendously in the credit markets, especially in high-risk segments such as high-yield bonds, credits with low or no ratings, and private debts, the issuing of which has virtually ceased. Oil prices have fallen sharply amid the weakening global demand and the inability of OPEC + countries to reach an agreement to reduce production. This fact provoked a further risk aversion.

Given the lack of access to capital markets, it becomes evident that poor countries will become poorer. This will exacerbate economic inequality.

> *Poverty generates shame.*
> *He who is ashamed loses his courage.*
> *The coward suffers from humiliation.*
> *The humiliated are all despised.*
> *The humiliated are given to despair.*
> *The despondent loses his mind.*
> *The crazies die.*
> *Yes, poverty is the source of all misery.*
> *(Ancient Indian Wisdom)*

Due to the limited data, the authors do not claim to have conducted an extended analysis of the situation. Their full attention is given to the fundamental aspects that affect the future of the country.

The purpose of the article is to explore the problem of income inequality in Ukraine (corruption, inconsistency, absence of strategy of development, and absence of legislative base of punishment for crimes) and to outline the ways of fighting it.

The immediate goals facing Ukraine include:

- achievement of sufficient levels of economic growth;
- overcoming of surplus inflation;

- stabilizing of the level of external debt;
- stabilizing of budget deficits;
- solving the problem of demographic dynamics, etc.

2. Literature Review

In the first volume of Capital (Marx, 1952) K. Marx pointed to the inequality on the market of the owners of the commodity "labor" and the owners of capital. The latter can dictate their conditions, since the hired worker is deprived of property and, consequently, the means of subsistence, and cannot wait for a long time for something better to appear. The distribution of goods is a function of institutionalized power relations rooted in industrial relations.

Contrary to Marx, our contemporary Thomas Piketty (Piketty, 2014) argues that there is a tendency for hyper-concentration of goods. He also speaks about the emergence of a new class of "super managers." No economic laws determine the distribution of income and goods. In the aspect of our study, it is also important to note that the growth of the capitalist economy inevitably slows down. Capital gains are 3–5 bigger than economic growth. When the industrialized capitalist economy in its prime begins to stagnate, the previously accumulated goods become more important and powerful, while inequality increases rapidly.

Using the example of Britain, Danny Dorling (Dorling, 2019) demonstrates that the cost of the super-rich is too high for society. After all, since the Great Recession of 2008 only 1% has become richer, while for the rest of people it is getting increasingly harder. The gap between the haves and have-nots has turned into a chasm. While the rich found new ways to protect their wealth, everyone else was punished by austerity. A mere accident of being born outside the 1 percent will have a dramatic impact for the rest of one's life: it will shorten one's life expectancy, education and work prospects, as well as one's mental health.

According to the study conducted by Ukrainian scientists at the Institute for Demography and Social Research (Libanova et al., 2012), income inequality can (and usually does) lead to inequality in access to basic public goods (quality health care, quality education, and comfortable housing), and resources (financial, property, infrastructure, information, etc.) to the realization of their fundamental rights in general (the right to health, to life, to safety).

According to the data (World Employment and Social Outlook, 2020), the crisis caused by the pandemic has affected around 2.7 billion working people worldwide. All these people have experienced the negative effect of complete or partial termination of businesses. The COVID-19 pandemic crisis is projected to reduce 6.7% of working hours worldwide in the period April–June 2020. The losses for the global economy will be akin to 195 million full-time workers stopping work for the entire quarter. In Ukraine, unemployment is projected to rise to 9.4% in 2020 (The draft law on amendments to the Law of Ukraine "On the State Budget of Ukraine in 2020," 2020).

3. Independent Ukraine: without the Soviet Union, but with the Oligarchs

At the end of 1991, the Ukrainian Republic of the USSR, the second in importance, made its choice in favor of leaving the Union, thereby rendering its further existence impossible. This event opened the "door" for a new phenomenon when such words as "oligarch," "oligarchization," "deoligarchization" entered the lexicon of Ukrainians. Oligarchs are businessmen who have an active influence on politics and are the owners of the media. The key interest of this type of businesspersons is to restrict competition through the imposition of import duties and quotas, low rents for the extraction of minerals, special tariffs for supplying consumers with electricity and gas, low prices for the use of state railways and pipelines, operational control over state-owned enterprises, etc. As a result, Ukrainians are paying more and more for utilities, but the profits are not used to modernize production or to create energy independence but end up in offshore accounts of oligarchs. The state does not have enough money in the budget for social items—but it does not burden the oligarchs with high taxes on the extraction of raw materials and continues to subsidize state-owned enterprises instead of making them earn money for public good. Instead of the state budget, profits go to the same offshore accounts of oligarchs. At the same time, the country is losing investments that have not come to Ukraine due to artificial barriers built by oligarchs in order to protect themselves against undue competition. Such investments would mean new jobs and higher salaries, additional tax revenues to the budget, in particular, higher salaries for teachers, higher pensions, etc. According to the Head of the Delegation of the European Union to Ukraine Hugo Mingarelli, "… after reaching a certain level of social inequality it is impossible to achieve sustainable growth. In this country (Ukraine) the level of inequality is simply unacceptable. You cannot be a country where the majority of the population tries to survive on less than 200 dollars when

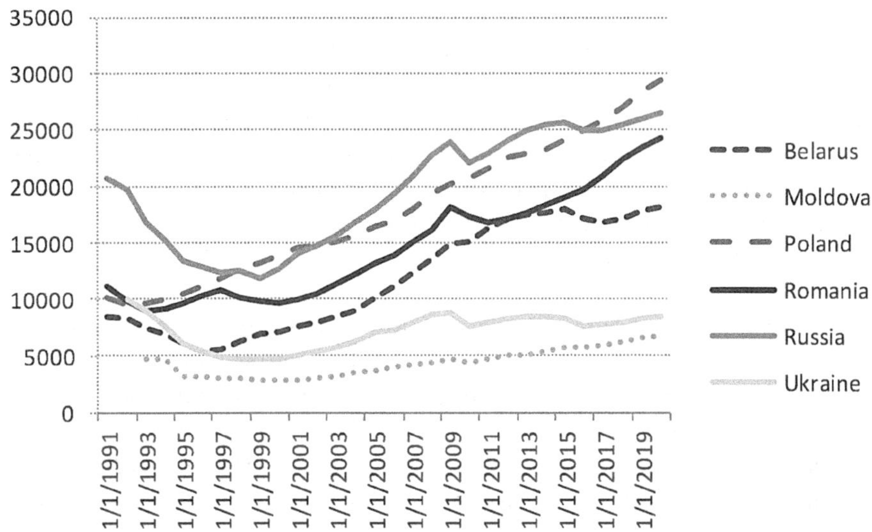

Figure 9.1. Comparative dynamics of per capita GDP according to purchasing power parity (international dollars) of individual countries.
Source: Authors (Constructed, using International Monetary Fund database) (World Economic Outlook, April 2020).

you have hundreds of Porsches, Ferraris and Lamborghinis on the streets. This should not continue. Social cohesion is very important" (News agency Unian, 2018).

The result of Ukraine's evolution over the past two decades is the sad fact that the country has lost its international ranking according to a number of indicators. The country, which in the beginning of its independence had a higher per capita GDP (purchasing power parity) than many of its neighboring countries, has become one of the poorest ones. The scale of these processes is illustrated by the data presented in Figure 9.1. It should be noted that except Ukraine in each of the countries in this sample—Belarus, Moldova, Poland, Romania, and Russia—the standard of living has increased.

Today, in most countries' inequalities in terms of capital, availability and value of real estate far outweigh the current income inequality. According to the study of Thomas Piketty (Piketty, 2014), in the last 30–40 years it has been the capital acquired due to speculative processes, not value-added capital, which has become a major cause of growing inequality. In addition, since the 1980s in the Western countries, assets were transferred into private hands and the capital was privatized. In the absence of opportunities to ensure an equitable distribution of economic results, including incomes, the authorities were forced to reduce excessive inequalities through their redistribution, in particular, using fiscal instruments. According to the report of the UK Oxfam charity called "Time to Care": today the fortunes of 2153 billionaires are greater than those of 4.6 billion people, or 60 percent of the world's population. The International Inequalities Institute considers inequality as one of the major problems of our time. It is a major factor in social instability, war, social tensions, and social conflicts.

> *Plagues, revolutions, wars, the collapse of the states bring back justice and make the rich poor, that is, reduce economic inequality. The mention of the "leveling of inequality" due to the collapse of the state dates back thousands of years: the last Roman aristocrats lined up for help from the Pope, the highest-ranking Mayans had to eat the same food as the common people. More recently, the anarchy in Somalia has narrowed the gulf in income of the general population and the kleptocrats in power [Walter Scheidel, The Atlantic]*

Social inequality is closely linked to the rise in poverty. The inability to get a quality education condemns children from poor families to smaller chances of finding decent-paying jobs and, consequently, to misery in adulthood. It also affects inequality in life expectancy. People from poor families are at higher risk of premature death or illness due to lack of funds for quality medicines and health care. Social inequality also has a negative impact on the prospects of economic growth, as the working-age population migrates to higher-income countries. Migration, the war in eastern Ukraine had an impact on the quantity of the population. Ukraine has decreased by almost 10 million of its citizens (please see Figure 9.2).

As for wages, according to the United Nations Organization (Ukrainian Journal « konomist», 2019), the general situation is as follows: the average wage of Ukrainians in dollar terms had increased since 1999 reaching a peak in December 2013—452.77 US dollars. Due to the military conflict in eastern Ukraine in 2014, it rapidly declined to 148.7 US dollars by February 2015. Later this indicator showed some growth, but in comparison with other countries the situation remains bleak (please see Figures 9.3 and 9.4).

At the same time, the uneven distribution of the average per capita equivalent income (please see Table 9.1) has an impact on the overall situation of the low average wage levels. The data show that there is a gradual redistribution of the population by the average per capita income towards an increase in the proportion of the population with higher income. This growth has been accelerating especially in the last two years.

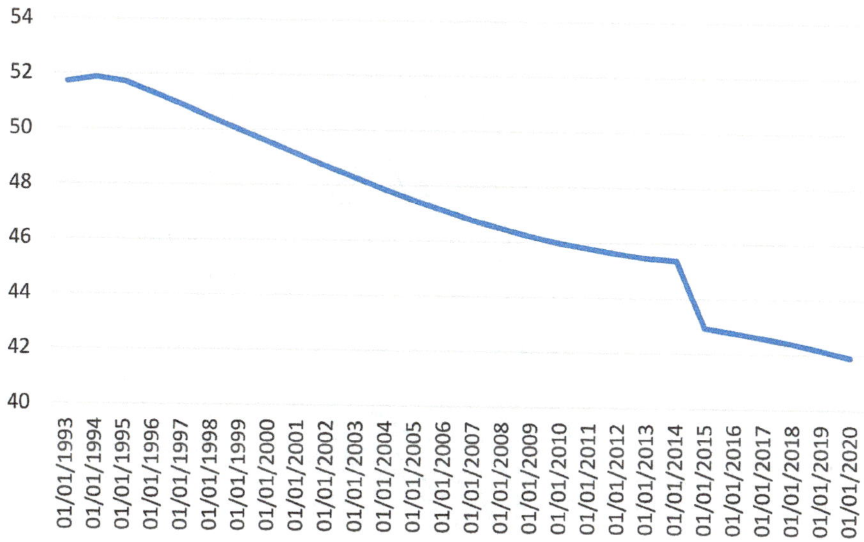

Figure 9.2. Population, million people, compiled by the authors.
Source: State Statistics Service of Ukraine, 2020.

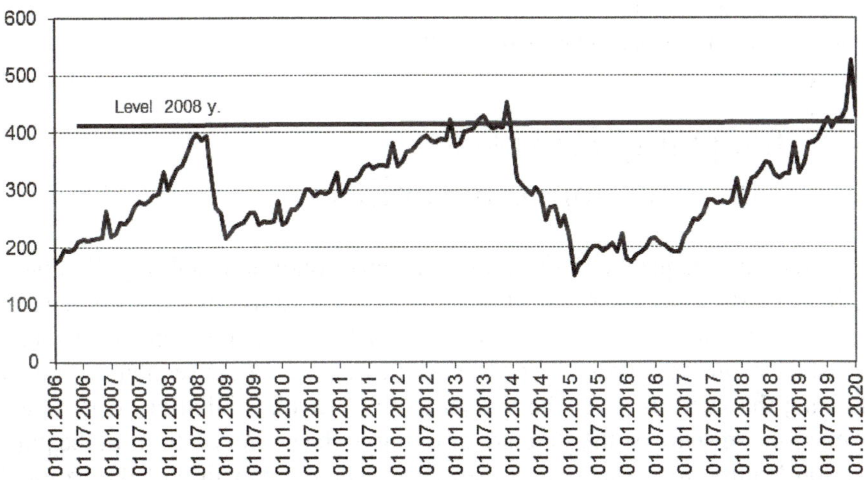

Figure 9.3. Average monthly wages, Ukraine, US $, compiled by the authors.
Source: State Statistics Service of Ukraine, 2020.

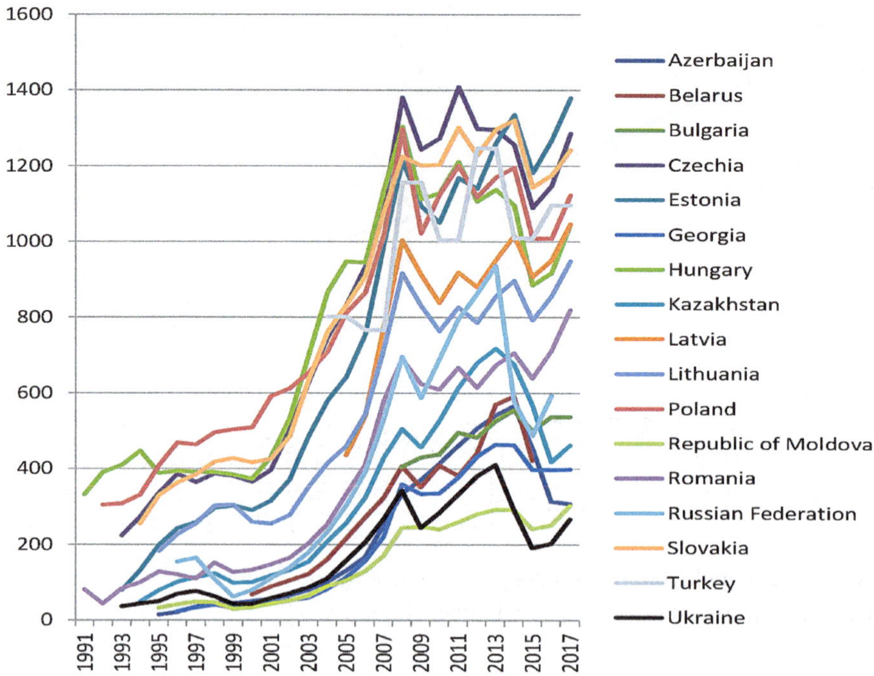

Figure 9.4. Gross average monthly wages by indicator, Country and Year, US $., compiled by the authors.
Source: UNECE Statistical Database, 2020.

4. Inequality Provokes Macroeconomic Instability

Income inequality often leads to macroeconomic instability. To counterbalance the imbalance, the state applies stabilization measures which, in turn, provoke a chain reaction: income redistribution and large-scale social transfers result in high budget expenditures and, as a consequence, budget deficits; the need to cover the budget deficit provokes inflation, which puts even more pressure on the standard of living of the poorer sections of the population leading to the increased inequality. The depreciation of the national currency and the galloping inflation made a significant contribution to the impoverishment of a large part of the population and uneven distribution of the national income in Ukraine.

According to theory, the ratio between the incomes of 20% of the richest sections of the population in the country should not exceed the incomes

Table 9.1 Distribution of the population by average per capita equivalent total income[1] (excluding the temporarily occupied territories the Autonomous Republic of Crimea and the city of Sevastopol)

	2010	2011	2012	2013	2014[2]	2015[2]	2016[2]	2017[2]	2018[2]
Distribution of the population (%) by average per capita equivalent total income per month, UAH									
under 1920,0	78,1	70,5	61,2	54,6	51,7	35,3	18,4	6,9	2,6
1920,1–2280,0[3]	21,9	29,5	14,5	16,6	16,5	17,6	16,0	7,7	4,0
2280,1–2640,0	…	…	9,3	10,5	11,7	15,9	16,4	11,0	6,7
2640,1–3000,0	…	…	5,6	6,8	7,8	11,4	13,6	11,0	7,6
3000,1–3360,0	…	…	3,1	4,3	4,6	6,7	10,0	10,7	9,4
3360,1–3720,0	…	…	2,1	2,0	2,5	4,4	7,9	10,8	9,2
3720,1–4080,0[3]	…	…	4,2	5,2	5,2	8,7	17,7	9,3	8,7
4080,1–4440,0	…	…	…	…	…	…	…	6,3	8,9
4440,1–4800,0	…	…	…	…	…	…	…	6,4	7,0
4800,1–5160,0	…	…	…	…	…	…	…	5,1	6,9
over 5160,0	…	…	…	…	…	…	…	14,8	29,0

[1]Since 2011, while compiling average per capita indicators as well as indicators of the population (households) differentiation by level of material well-being a scale of equivalency was started to be used. To ensure the comparability of time series indicators, data for 2010 have been revised taking into account the scale of equivalency.
[2]Excluding a part of temporarily occupied territory of the Donetck and Luhansk regions.
[3]In 2010 and 2011 over 1920 UAH, In 2012–2016 over 3720 UAH

Source: State Statistics Service of Ukraine, 2020.

of 20% of the poorest by more than 10 times. There is no official statistics on property stratification in Ukraine, but experts say that the gap between the income of the richest and the poorest in Ukraine is at least 40 times. In the US, which has always been the model of "wild capitalism," this figure is now 9. In Sweden this ratio is 3.5 to 1 (Zanuda, 2012).

The income of 33.9% of Ukrainians (almost 14 million people) is less than the actual subsistence level. The income gap between 10% of the richest and poorest Ukrainians decreased by 0,2 times this year, or by 4,2%. The difference between the residents of cities is 4.9 times and among the residents of villages—4.5 times (Ukrainian Journal « konomist», 2019). It should be noted that in 2017, based on the values of the Gini coefficients and the Palma ratio, Ukraine was identified as a country with the smallest gap between rich and poor. According to the World Bank data, researchers say that South Africa, Namibia, and Haiti are some of the most unequal countries in terms of income distribution. At the same time, Ukraine, Slovenia, and Norway are classified as the most equal nations in the world (UNECE Statistical Database, 2020). The dynamics of values of Gini coefficient in Ukraine from 2009 to 2018 are presented in Figure 9.5.

It should be noted that as of January 1, 2020, the highest minimum wage in the EU countries was in Luxembourg—€2142, which is 12.3 times

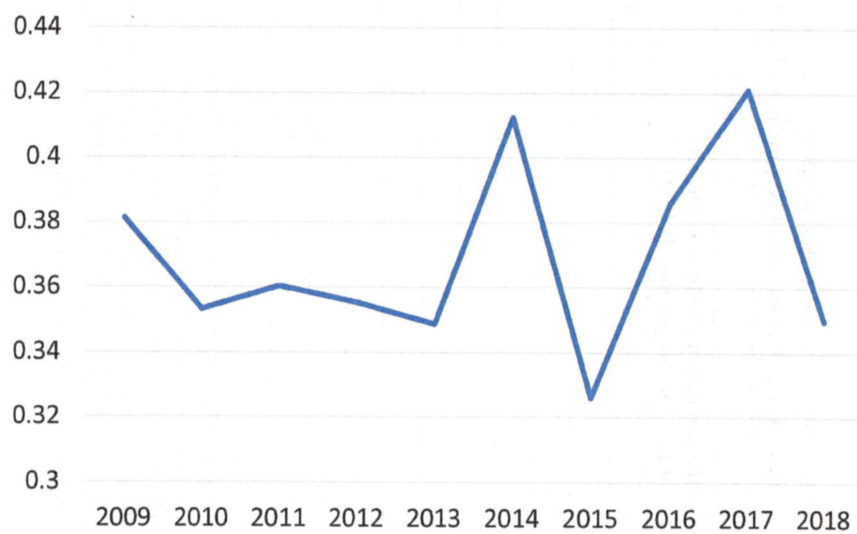

Figure 9.5. Dynamics coefficient Gini of household income reported in Gallup (Ukraine), compiled by the authors.
Source: The Word Happiness Report, 2019.

more than in Ukraine (4723 UAH or €174,6 at the rate of the National Bank of Ukraine on February 9) according to Eurostat. In general, the monthly minimum wage in the eastern EU countries was less than € 600, while in the northwest it reached more than €1500 (Berkal, 2020).

According to the data of the State Statistics Service of Ukraine (please see Table 9.2), it can be stated that there is a decrease in the population with average per capita equivalent total income per month, which is lower than the actual subsistence minimum.

According to (the Report on budget execution of the Pension Fund of Ukraine in 2019, 2020), in 2019 the Pension Fund ensured the payment of pensions to 11.3 million citizens. 378,8 thousand new pensions were allocated. The average amount of pension in 2019 grew by 437 UAH (16.5%)—from 2646 UAH to 3083 UAH. The minimum old-age pension increased from 1497 UAH to 1638 UAH. Most pensioners receive pension payments from 1501 to 2000 UAH—3949,5 thousand people (34,8%) and from 2001 to 3000 UAH—3872,8 thousand people (34,2%) (please see Table 9.3).

It should be noted that in Ukraine there is a significant difference between pension payments to different segments of the population. For examples, judges receive the highest pension payments. Their life-long cash allowances amount to an average of 28,7 thousand UAH. By a wide margin they are followed by pension payments to members of the military—the second position. Their average pensions in 2019 amounted to 4,6 thousand UAH. The lowest pension payments are received by persons with disabilities and persons who have been assigned social pension benefits—1,9 thousand UAH and 1,4 thousand UAH, respectively (please see Table 9.4).

It should be noted that the highest average pensions in Ukraine are received by residents of the Kyiv and Donetsk regions—UAH 4251 UAH and 4003 UAH respectively. The lowest pensions are received by the inhabitants of the Transcarpathian region (2463 UAH).

Income inequality contributes to the rise in criminalization of society. It is a direct cause of corruption as it provokes the meddling in politics and public administration by high-income individuals (oligarchs). The considerable difference in incomes makes it possible to use them for bribes. The corruption has a negative impact on the quality and quantity of public services, in particular, in education and health care, significantly reducing the effectiveness of public spending. The growing deficit of the state budget leads to a decrease in the wages of the budget sector employees. Along with the low level of financing of the entire social sphere this leads, in particular, to the charging of additional (off the books) parents' money for school repairs, to

Table 9.2 Differentiation of household living standards[1](excluding the temporarily occupied territories the Autonomous Republic of Crimea and the city of Sevastopol)

	2010	2011	2012	2013	2014[2]	2015[2]	2016[2]	2017[2]	2018[2]
Population with average per capita equivalent total income per month, lower than legally established subsistence minimum:									
million persons	3,6	3,2	3,8	3,5	3,2	2,5	1,5	0,9	0,5
% to the total population	8,6	7,8	9,0	8,3	8,6	6,4	3,8	2,4	1,3
Population with average per capita equivalent total income per month, lower than actual subsistence minimum:									
million persons	…	…	…	…	6,3	20,2	19,8	13,5	10,6
% to the total population	…	…	…	…	16,7	51,9	51,1	34,9	27,6
For information:									
Quintile ratio of differentiation of population total income, times	1,9	1,9	1,9	1,9	1,9	1,9	1,9	1,9	2,0
Quintile ratio of funds (by total income), times	3,5	3,4	3,2	3,3	3,1	3,2	3,0	3,3	3,3

[1]Since 2011, while compiling average per capita indicators as well as indicators of the population (households) differentiation by level of material wellbeing a scale of equivalency was started to be used. To ensure the comparability of time series indicators, data for 2010 have been revised taking into account the scale of equivalency.
[2]Excluding a part of temporarily occupied territory of the Donetck and Luhansk regions.

Source: State Statistics Service of Ukraine, 2020.

Table 9.3 The size of pensions in Ukraine

The amount of pension payment, UAH	The number of pensioners of all categories, persons	Share in total number, %	The average amount of pension payment, UAH
More than 10,000	259,519	2.3	12917.13
from 5001 to 10,000	1,052,454	9.3	6764.04
from 4001 to 5000	713,132	6.3	4446.51
from 3001 to 4000	1,407,255	12.4	3442.53
from 2001 to 3000	3,872,812	34.2	2361.63
from 1501 to 2000	3,949,512	34.8	1832.31
to 1500	80,046	0.7	940.93

Source: official website of the Pension Fund of Ukraine, 2020.

Table 9.4 Retirement benefits for different segments of the population

Types of pensions	As of 01.01.2019		As of 01.01.2020	
	Number of people	Average amount of pension, UAH	Number of people	Average amount of pension, UAH
Old-age pension	8,693,013	2648.24	8,534,558	3064.77
Disability pension	1,370,249	1994.94	1,405,704	2480.41
Survivor's pension	536,132	2254.57	539,097	2699.7
Superannuated pension	229,732	2476.92	220,594	2841.74
Social pensions	84,027	1499.59	77,632	1644.25
Pensions of members of the armed forces	553,958	4682.05	553,767	5303.01
life-long cash allowance of judges	3294	28,702.37	3378	45,874.27
Average retirement benefits		2644.66		3082.98

Source: official website of the Pension Fund of Ukraine, 2020.

purchase new equipment, to pay for additional services of schools, preschool and medical institutions.

5. What will COVID-19 Provoke—Equality or Inequality?

In the current COVID-19 pandemic situation most business structures face catastrophic losses that threaten their existence and solvency, especially for medium and small businesses. Millions of employees around the world are vulnerable to the loss of profits and production cuts. The harsh attitude of business owners towards the vulnerable groups of employees leads to increased unemployment.

Unprecedented employment cuts have begun in many countries [ILO Monitor 2nd edition, 2020]. As a result of the reduced production there is a decline in working hours and, consequently, wages of workers.

According to the estimates of the International Labor Organization [World Employment and Social Outlook—Trends, 2020] as of April 1, 2020, working hours will be reduced by 6,7 percent in the second quarter of 2020, which is the equivalent to the reduction of 195 million full-time employees. In the US these figures are impressive. Over 10.4 million unemployed people appeared in 2 weeks. It beat all the pessimistic forecasts from Morgan and Goldman Sachs, which forecast 5 and 7 million, respectively. The new and gloomier (and more realistic) forecasts from the same organizations give from 32 to 47 million unemployed people in the US at the height of the crisis.

In addition to the increase in the number of unemployed, many companies are also massively cutting their employees' salaries. Starting with the oil and aviation industry, and ending with the unexpected falling in the salaries of physicians not involved in the treatment of the Coronavirus disease (that is, those with other medical specialties).

The greatest losses of jobs and working hours will occur in the most affected sectors. The International Labor Organization (World Employment and Social Outlook—Trends, 2020) estimates that 1.25 billion workers representing nearly 38 percent of the global workforce are used in the sectors that are currently facing dramatic declines in production and high risks of job cuts. The key sectors include retail, services, power generation, and telecommunications. Most employees in these sectors are women.

Especially affected are workers in poor countries where these sectors have a high proportion of informal employees and workers with limited access to health care and social protection, particularly in Ukraine. Without the appropriate policy measures these countries are not capable of combating unemployment and, consequently, inequality.

Those employees, who continue to work during the pandemic, including health care workers, are exposed to significant risks. In the health sector these are mostly women. As we can see, new factors are emerging that provoke a further deepening of gender inequality.

Poor countries are unable to compensate for the losses of their entrepreneurs as it was done in the United States, for example. The Federal Reserve System has compensated more than half of the fall in profits—2 trillion US dollars.

A record number of Ukrainians returned to the country in March of this year—according to the former Cabinet Minister, Dmitry Dubilet, more than 270,000 Ukrainians came to Ukraine last month—this is the biggest homecoming of Ukrainian citizens in the last 4 years, not counting those who returned before the New Year.

According to the Deutsche Welle, from 5 to 10% of workers left Europe according to various estimates. This is mainly due to job cuts or reduction in working hours. Many found it unprofitable to work on the proposed terms and they decided to return to Ukraine. They have all lost their jobs in Europe and will increase the unemployment level in Ukraine. According to the Ministry of Agriculture of Germany, by the end of May this branch will need about 100,000 seasonal workers. However, due to the ban on entry to Germany from abroad only about 20,000 seasonal workers were able to enter the country. Farmers hope that by June the situation will change, borders will be reopened, and Ukrainians and Poles will return to Germany.

Accordingly, with the rising unemployment tens of millions of people who have recently lost their jobs (most of whom also have families) or whose salaries have been substantially reduced, in the near future will no longer be able to pay mortgages or rent. These people will be thrown out on the street or the banks that lent them money will go bankrupt, because they have nowhere to take the money from. The prognosis is bleak—it will exacerbate the crime situation in the country.

6. Gender Inequality in Ukraine

The study of the legal regulation of gender equality in Ukraine proves that it is central to the system of human rights and freedoms. Ukraine has adopted the Sustainable Development Goals, which state that gender equality is an important condition for achieving them. Our country acceded to the Beijing Declaration and Platform for Action on the promotion of women's rights, ratified the Convention on the elimination of all forms of discrimination against women, signed an Association Agreement with the European Union, which defines equal gender employment opportunities in various fields of activity, etc. The issue of gender equality is enshrined in the Constitution of Ukraine, the Law of Ukraine "On Equal Rights and Opportunities for Women and Men," the Human Rights Strategy, etc. Therefore, the current regulatory framework makes it possible to create a powerful mechanism for ensuring gender equality. At the same time, the Gender profile of Ukraine in 2019 confirms the existence of a considerable number of problems in this area demonstrating the declarative level of regulations in this area. For example, according to the Gender Inequality Index, Ukraine ranks 60th out of 160 countries in the world (Human Development Report, 2019).

Gender inequality is sharply manifested in the issues of employment on the labor market, levels of remuneration, career growth, occupation of managerial positions, representation in power, etc.

The indicator of labor force dynamics by gender in Ukraine (please see Figure 9.6) confirms that the employment rate of women is lower than the employment rate of men. Despite the dominant positions of women in

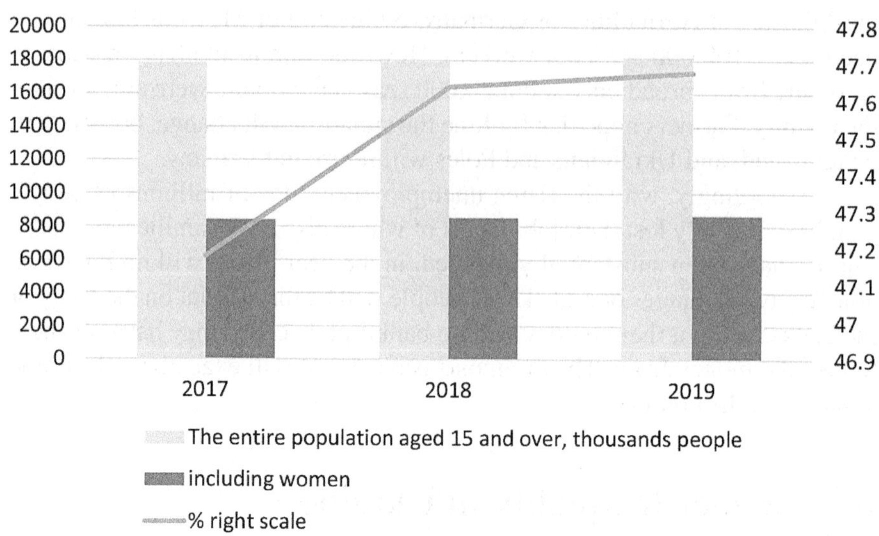

Figure 9.6. Labor force dynamics by gender in % to the total population (calculated by the authors).
Source: State Statistics Service of Ukraine, 2020.

the gender and age structure of Ukraine's population and in the structure of labor resources, women constitute only 46,7% of the total number of labor resources, whereas men—62,8% (Human Development Report, 2019).

The reasons of this inequality include: specific structure of the economic complex; gender segregation; income inequality of women and men; employment restrictions; different levels of domestic workloads; imperfect state policy relating to the labor market; the existence of false stereotypes in society; lack of affordable childcare; lack of flexible working modes, etc.

Gender inequality in the level of incomes in Ukraine is 21,2% as evidenced by the dynamics of the average monthly salary by gender (please see Figure 9.7).

Such a gap in average monthly salaries according to gender is related to the existing negative trends on the labor market, the main ones include: firstly, discrimination in employment; secondly, the existence of vertical segregation within a hierarchy of an enterprise or organization, and horizontal segregation between less paid and more highly paid positions within a certain cross-sectoral economic complex. In most cases, a woman who has been affected by gender inequality in relation to work payment continues to live in these conditions for the rest of her life—because her pension is dependent on the wages she received at working age. It should be noted that low

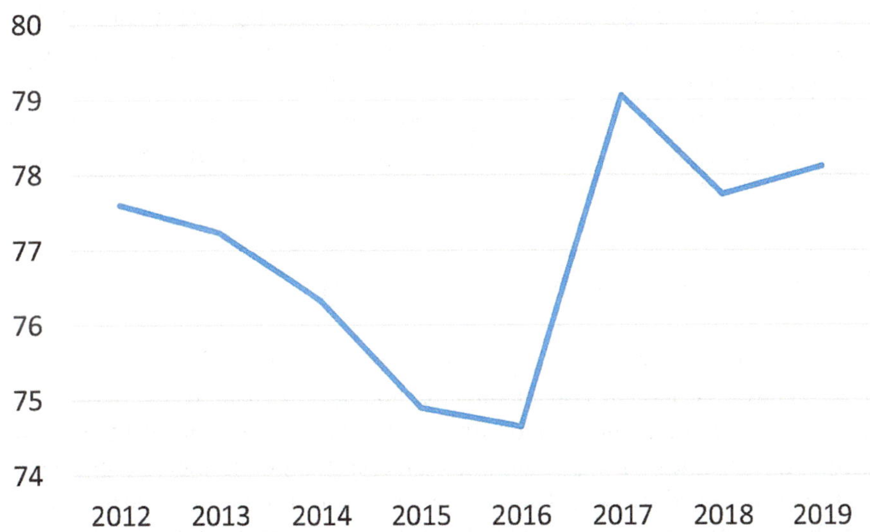

Figure 9.7. Percentage of the average monthly salary of women to the average monthly salary of men (calculated by the authors).
Source: State Statistics Service of Ukraine, 2020.

wages adversely affect both the work of an individual company (the motivation to work effectively is reduced) and the overall economic growth of the state (budgets are under-resourced).

Gender equality in career growth is a prerequisite on the labor market. There is a clear trend at the present stage of society's development: the higher the position or public agency, the less women it has. The analysis of the management structure of companies according to gender shows that most companies or organizations are managed by men (please see Figure 9.8).

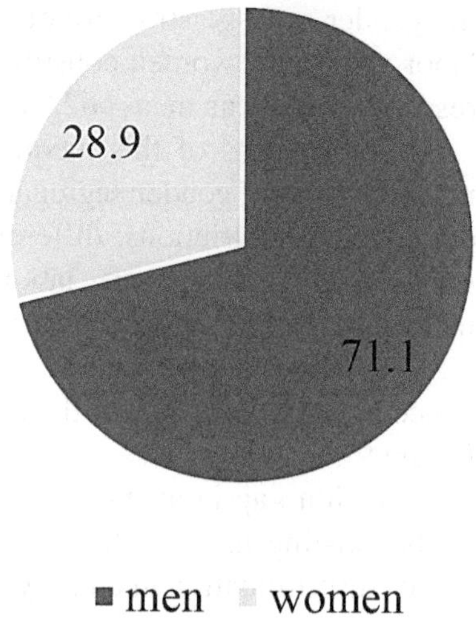

Figure 9.8. Management structure of companies by gender (calculated by the authors).
Source: State Statistics Service of Ukraine, 2020.

The solving of gender inequality on the labor market is possible provided that there is an effective state regulation to ensure citizens' right to work, protection against unemployment and earning of decent wages. In our opinion, the solution of this problem includes the following important methods and instruments of state regulation: improvement of the legal framework governing gender equality on the labor market; constant review of professions by gender; monitoring of wages; setting gender quotas in hiring; development of a strategy (program) that will stimulate employers to support gender equality of employees, etc.

7. Conclusions and Discussion

Ukraine is currently going through difficult times of its development. The country is facing considerable challenges measured by significant human, economic, financial, and infrastructural losses. The problem of social inequality occupies an important place among the total number of problems that need to be addressed. The areas of the state policy in ensuring social equality are the creation of conditions for overcoming poverty, the fight against excessive property stratification in society, as well as the bringing of social standards closer to the level of EU member states. This involves inclusive growth, according to which the achievement of good indicators is not an end in itself, while economic growth should be aimed at improving the quality of life of Ukrainian citizens, expanding their access to public goods, increasing the national welfare and its equitable distribution.

The main tools of state regulation of social inequality should be: legal support of the social sphere; direct state payments from budgets of different levels to finance the social sphere; social transfers in the form of social subsidies; establishment of social guarantees (minimum wage, subsistence level, consumer basket, minimum size of old-age pensions, scholarships, etc.); state regulation of prices for basic necessities; social insurance; implementation of state programs to solve certain social problems; establishment of social and environmental norms and standards, and control over their observance.

The implementation of systemic reforms in all spheres of activities should contribute to solving the problem of social inequality of the population. The Sustainable Development Strategy "Ukraine–2020" was approved by the Decree of the President of Ukraine of 2015, the purpose of which was the implementation of the European living standards in our country and Ukraine's taking up the leading positions in the world. The main prerequisite for the implementation of the Strategy is a social contract between the authorities, businesses and civil society, in which each party has certain responsibilities. The main reforms in the field of social equalization are the reforms of de-oligarchization, demonopolization, deregulation, the tax system, the pension system, the health care system, the renewal of political power, and anti-corruption reforms.

An important issue in the system of ensuring the rights and freedoms of citizens in Ukraine is the issue of gender equality, which unfortunately remains declarative. Gender inequality is acute in employment processes, wage levels, career advancement, management positions, government representation, etc. The solution to this problem is impossible without state intervention, which consists in the use of regulatory tools, as well as a wide range of administrative and economic methods.

References

Berkal, M. Nazvany minimal'nye zarplaty v stranah ES, Ukraina otstaet ot lidera v 12 raz (2020). [The minimum wages in the EU - Ukraine lags behind the leaders by a factor of 12]. Retrieved from: https://ubr.ua/labor-market/work-in-figures/sehodnja-ukraintsy-poluchajut-pochti-v-13-raz-menshe-chem-v-ljuksemburhe-3890336_[Online Resource].

Bhandari, Medani. P. (2020). In the Covid-19 Regime – What Role Intellectual Society Can Play. International Journal of Science Annals, 3(2), 5–7. doi:10.26697/

ijsa.2020.2.1 https://ijsa.culturehealth.org/en/arhiv https://ekrpoch.culture-health.org/handle/lib/71

Bhandari, Medani P. (2020), Second Thought- The Phobia Corona (COVID 19): What We Can Do, Journal of Global Issues & Solutions" - Institute for Positive Global Solutions - Bibliotheque: World Wide Society, VOL. XX, NO. 5ISSN#: 1544-5399 September-October 2020, https://bwwsociety.org/journal/current/2020/sep-oct/covid-19.htm

Bhandari, Medani P. (2020), The Phobia Corona (COVID 19) - What Can We Do, Scientific Journal of Bielsko-Biala School of Finance and Law, ASEJ 2020, 24 (1): 1-3, GICID: 01.3001.0014.0769, https://asej.eu/resources/html/article/details?id=202946

Derzhavna sluzhba statystyky Ukrainy (2020). [State Statistics Service of Ukraine]. Retrieved from: http://ukrstat.gov.ua [Online Resource].

Dorling, D. Inequality and the 1%. Published by Verso Books (2019). Retrieved from: http://www.dannydorling.org/books/onepercent/272pages [Online Resource].

Global Economic Outlook (2020). Retrieved from: https://www.fitchratings.com/research/sovereigns/global-economic-outlook-crisis-update-late-april-2020-coronavirus-recession-unparalleled-22-04-2020_[Online Resource].
https://www.imf.org/en/Publications/WEO/Issues/2020/04/14/weo-april-2020 [Online Resource].

Human Development Report 2019: United Nations Development Programme (2020) – New York. Retrieved from: http://hdr.undp.org/sites/default/files/hdr2019.pdf [Online Resource].

ILO Monitor 2nd edition: COVID-19 and the world of work Updated estimates and analysis (2020). Retrieved from: https://www.ilo.org/wcmsp5/groups/public/@dgreports/@dcomm/documents/briefingnote/wcms_740877.pdf [Online Resource].11.

Inequality index: where are the world's most unequal countries? (2019). Retrieved from: https://www.theguardian.com/inequality/datablog/2017/apr/26/inequality-index-where-are-the-worlds-most-unequal-countries?CMP=share_btn_fb [Online Resource].

International labor organization. World Employment and Social Outlook – Trends 2020 (2020). Retrieved from: https://www.ilo.org [Online Resource]. 108.

Libanova, E. (Eds.). Nerivnist v Ukraini: masshtaby ta mozhlyvosti vplyvu (2012) [Inequality in Ukraine: the scope and opportunities to influence it]. Kyiv: The M.V. Ptukha Institute for Demography and Social Studies of NAS of Ukraine. 120-198.

Marx, K. Kapital. Tom 1. Process proizvodstva kapitala (1952). [Capital. Volume 1 "The process of capital production"]. Retrieved from: http://istmat.info/files/uploads/59307/k._marks._kapital._tom_1._1952_g.pdf [Online Resource].

News agency Unian (2018). Riven nerivnosti v Ukraini prosto nepryiniatnyi - holova Predstavnytstva ES [The level of inequality in Ukraine is simply unacceptable

- Head of the EU Delegation]. Retrieved from: https://www.unian.ua/economics/finance/10026065-riven-nerivnosti-v-ukrajini-prosto-nepriynyatniy-golova-predstavnictva-yes.html [Online Resource].

Piketty, T. (2015). About capital in the twenty-first century. American Economic Review. Retrieved from: http://piketty.pse.ens.fr/files/PikettyAER2015.pdf [Online Resource].

Proekt Zakonu pro vnesennia zmin do Zakonu Ukrainy "Pro Derzhavnyi biudzhet Ukrainy na 2020 rik" [The draft law on amendments to the Law of Ukraine "On the State Budget of Ukraine in 2020"]. Retrieved from: http://w1.c1.rada.gov.ua/pls/zweb2/webproc4_1?pf3511=68570 [Online Resource].

The Word Happiness Report (2019). Retrieved from: https://worldhappiness.report/ed/2019/ [Online Resource].

UNECE Statistical Database (2020). Retrieved from: https://w3.unece.org/PXWeb2015/pxweb/en/STAT/STAT__20-ME__3-MELF/60_en_MECCWagesY_r.px/?rxid=0806c85a-23f8-4249-a4d0-10980df459d1 [Online Resource].

UNECE Statistical Database (2020). Retrieved from: https://w3.unece.org/PXWeb2015/pxweb/en/STAT/STAT__20-ME__3-MELF/60_en_MECCWagesY_r.px/?rxid=0806c85a-23f8-4249-a4d0-10980df459d1 [Online Resource].

World Economic Outlook. International Monetary Fund database (April 2020). Retrieved from:

Yakyi rozryv mizh dokhodamy naibahatshykh ta naibidnishykh ukraintsiv (2019). [What is the gap between the incomes of the richest and poorest Ukrainians]. Ukrainian Journal «Ekonomist». Retrieved from: http://ua-ekonomist.com/18020-yakiy-rozriv-mzh-dohodami-naybagatshih-ta-naybdnshih-ukrayincv.html [Online Resource].

Zanuda, A. Mainova nerivnist v Ukraini: hirshe, nizh na "dykomu Zakhodi", BBC Ukraina (2012). [Property inequality in Ukraine: worse than in the "Wild West", BBC Ukraine]. Retrieved from: https://www.bbc.com/ukrainian/business/2012/01/120119_inequality_usa_uk_ukraine_az [Online Resource].

Zvit pro vykonannia biudzhetu Pensiinoho fondu Ukrainy za 2019 rik (2020) [Report on the implementation of the budget of the Pension Fund of Ukraine in 2019]. The Pension Fund of Ukraine database. Retrieved from: https://www.pfu.gov.ua/2121043-zvit-pro-vykonannya-byudzhetu-pensijnogo-fondu-ukrayiny-za-2019-rik/ [Online Resource].

Chapter 10

Education as A Tool for Prevention Discrimination Against LGBTQ+ People

Shapoval Vladyslav, Medani P. Bhandari[1]*, Shvindina Hanna*

ABSTRACT

This chapter presents the Lesbian, Gay, Bisexual, Transgender and Queer or Questioning (LGBTQ+) people struggle around the world in general and in Ukraine in particular. Further, chapter gives the examples of how education can help to understand the LGBTQ+ people's rights and prevent from the discrimination against LGBTQ+ people's rights. In general, inequality can be seen in every sphere of political, social, economic, cultural, and religious systems; however, LGBTQ+ people's struggles are relatively new and painful. This chapter outlines the general history of such struggles and the current situation of LGBTQ+ people.

1. Introduction

The issue of self-identification as Lesbian, Gay, Bisexual, Transgender and Queer or Questioning (LGBTQ+) has a long history of struggles. The identity issue is associated with the history of the development of human civilization; however, the society have been dominated by the political, economic, and community elite; therefore, the marginalized people's identity remained under shadow. The minority struggle was always there throughout the history of human civilization. However, those struggles were always dominated or surpassed by the mainstream elites. It is important to note that we do not

[1] Correspondence—Prof. Medani P. Bhandari, medani.bhandari@gmail.com

have recorded documents or history of such domination. The reason for the absence of these records is that the marginalized people's voice was and has always been suppressed and dominated by the societal elites. Whatever history we have is primarily of those people who were in power, and nowadays history created is also about the people in power. Discrimination based on gender, age, origin, ethnicity, disability, sexual orientation, class, and religion were embedded in the society and still can be observed in various forms. The identity issue due to sexual orientation was considered as crime or deviant behavior in most parts of the world. The struggle of Lesbian, Gay, Bisexual, Transgender and Queer or Questioning (LGBTQ+) is the struggle of identity, and against the discriminative behavior by the mainstream societal elites.

It is also the fight of an individual's self-respect and dignity. Every person should have the right to live in his/her own way without harming rights of others. We think identity is how a person thinks about himself or herself, how society recognizes her or him, and how he or she presents or wants to present within and in the surrounding s/he lives. A person can have multiple identities, related self-portrait, and identity created by functions (relation, role, profession, orientation, duties, responsibilities, outlook, and situation). The struggle of Lesbian, Gay, Bisexual, Transgender and Queer or Questioning (LGBTQ+) is for the individual rights and freedom of how they want to present themselves in terms of sexual orientation. More than that, it is the struggle for self-respect and dignity.

> An "identity" refers to either (a) a social category, defined by membership rules and (alleged) characteristic attributes or expected behaviors, or (b) socially distinguishing features that a person takes a special pride in or views as unchangeable but socially consequential (or (a) and (b) at once). In the latter sense, "identity" is modern formulation of dignity, pride, or honor that implicitly links these to social categories (Fearon 1999:1).
>
> in political theory, questions of "identity" mark numerous arguments on gender, sexuality, nationality, ethnicity, and culture in relation to liberalism and its alternatives (Young 1990; Connolly 1991; Kymlicka 1995; Miller 1995; Taylor 1989 as in Fearon 1999:1)

Personal identity is a set of attributes, beliefs, desires, or principles of action that a person thinks distinguish her in socially relevant ways and that (a) the person takes a special pride in; (b) the person takes no special pride in, but which so orient her behavior that she would be at a loss about how to act and what to do without them; or (c) the person

> feels she could not change even if she wanted to. Most often, I will argue, the (a) meaning applies, so that for usage in ordinary language personal identity can typically be glossed as the aspects or attributes of a person that form the basis for his or her dignity or self-respect. Used in this sense, "identity" has become a partial and indirect substitute for "dignity," "honor," and "pride" (Fearon 1999:11).
>
> Identity is "people's concepts of who they are, of what sort of people they are, and how they relate to others" (Hogg and Abrams 1988:2; as in Fearon 1999:4).
>
> Identity "refers to the ways in which individuals and collectivities are distinguished in their social relations with other individuals and collectivities" (Jenkins 1996:4; as in Fearon 1999:4).

Person's identity is how she or he define, how she or he presents, how she or he wants to be seen from others' perspective. As noted earlier, the struggles of Lesbian, Gay, Bisexual, Transgender and Queer or Questioning (LGBTQ+) struggle for validation from others, struggle for inclusiveness, equity, and fight for the discrimination they have been facing throughout. *The lesbian and gay movement has battled for political and cultural change, and the role of identities within that pursuit has changed considerably over time. For the lesbian and gay movement, then, cultural goals include (but are not limited to) challenging dominant constructions of masculinity and femininity, homophobia, and the primacy of the gendered heterosexual nuclear family (heteronormativity). Political goals include changing laws and policies to gain new rights, benefits, and protections from harm* (Bernstein 2002:334).

> LGBT and intersex people face discrimination and human rights abuse in every country of the world, ranging from loss of jobs and housing to extreme violence and even murder. Approximately 80 countries still criminalize LGBT relationships in some way, and many other types of laws deny even the most basic rights and dignity to LGBT and intersex people. While legal rights are an important measure of how any country treats its LGBT and intersex citizens, it is only one measure of broader social acceptance and of realities faced by LGBTI men and women (GLAAD 2021).
>
> Conservative cultural and societal norms are reinforced by anti-LGBT religious forces—another major opponent in international work. …… "Very clearly, the main opponents of LGBTI rights are dogmatic religious forces (from all religious traditions and in all quarters of the world), either operating on their own, or

through state actors." ...."The root causes of [homophobia, lesbophobia, and transphobia] stem mostly from tradition and traditional interpretations of religion, which promote rigid models of social existence based on a reproductive foundation of humanity and sexuality." ..... religious leaders as major opponents (MAP 2020:15).

As lesbian, gay, bisexual, and transgender (LGBT) advocates have fought to gain rights and recognition around the globe, their shared endeavors and coordinated activism have given rise to an international LGBT movement. Over the past century, advocates have recognized common aims and collaborated in formal and informal ways to advance the broader cause of sexual equality worldwide. Advocates in different contexts have often connected their struggles, borrowing concepts and strategies from one another, and campaigning together in regional and international forums. In doing so, they have pressed for goals as diverse as the decriminalization of sexual activity; recognition of same-sex partnerships and rainbow families; bodily autonomy and recognition for transgender and intersex people; nondiscrimination protections; and acceptance by families, faith communities, and the public at large. At times, the international LGBT movement—or to be more accurate, LGBT movements—have used tactics as diverse as public education, lobbying and legislative campaigns, litigation, and direct action to achieve their aims. The result has been a gradual shift toward recognizing LGBT rights globally, with these rights gaining traction in formal law and policy as well as in public opinion and the agendas of activists working for human rights and social justice. The movement's aims have also broadened, being attentive to new issues and drawing common cause with other campaigns for bodily autonomy and equal rights. At the same time, gains have triggered ferocious backlash, both against LGBT rights and against broader efforts to promote comprehensive sexuality education, access to abortion, the decriminalization of sex work, and other sexual rights. Understanding this advocacy requires consideration of important milestones in global LGBT organizing; how LGBT rights have been taken up as human rights by domestic, regional, and international bodies; and some of the main challenges that LGBT advocates have faced in contexts around the globe (Thoreson 2021—Summary from—An International LGBT Movement).

The history of gay rights movements, initially LGBT and later LGBTQI (Lesbian, Gay, Bisexual, Trans, Queer and Intersex), can be understood only in light of the forms of persecution and oppression

> faced by individuals who had emotional and sexual relations with persons of their own gender and/or did not conform to the social expectations of their own gender. Their emergence dates back to the first half of the nineteenth century. There has been an increasing awareness of the demands made by the LGBTQI movement during the early twenty-first century, notably with regard to measures for combatting discrimination, which are at the foundation of the European Union's Charter of Fundamental Rights (2000) (Régis Schlagdenhauffen 2020—Digital Encyclopedia of European History).

The LGBTQ+ began to take place in the Western world from enlightenment era, mostly become visible, when a Swiss man *Heinrich Hössli (1784–1864) published in German the first essay demanding recognition of the rights of those who followed what he called masculine love. Following Hössli; the German jurist Karl-Heinrich Ulrichs (1825–1895) wrote twelve volumes between 1864 and 1879 as part of his "Research on the Mystery of Love Between Men" ("Forschungen über das Räthsel der mannmännlichen Liebe")....... A first gay liberation movement emerged in Berlin in 1897, revolving around the doctor Magnus Hirschfeld (1868–1935),co-founder of the Wissenschaftlich-humanitäre Komitee (WhK, Scientific-Humanitarian Committee). The committee took multiple actions: a petition in favor of appealing § 175, the publication of books and brochures on homosexuality, the publication of a review (Jahrbuch für sexuelle Zwischenstufen, Yearbook for Intermediate Sexual Types), and the circulation of an educational film on the damage caused by homophobia (Anders als die anderen, Different from Others, 1919). In the ensuing period, committee branches were created in a number of German cities and neighboring countries whose laws condemned homosexuality (Austria, the Netherlands, Sweden, etc.). ... The Union for the Rights of Men (Bund für Menschenrechte), founded in 1922 by Friedrich Radszuweit (1876–1932), was the first to be open to lesbians* (Régis Schlagdenhauffen 2020—Digital Encyclopedia of European History). This founding movement in Europe extended other part of the world and paved the pathway for the LGBTQ+ rights.

In the United States, in 1924, Henry Gerber founded The Society for Human Rights in Chicago. It was the first documented gay rights organization in the United States (the details of LGBTQ+, movements prepared by the CNN research team is included in the annex-one). There are several researches on the LGBTQ+ rights (Morris 2021; Bechdel 2006; Bornstein1994; Bronski 2011; Carbado and McBride 2002; Carter 2004; Cenziper and Obergefell 2016; Fadermam 2015 1999; Feinberg1996; Jacobs 1997; Johnson 2004;

Moraga and Anzaldua 1981; Scholinski 1998; Shilts 1987; Short 2013; Thoreson 2014). Similarly, there are hundreds of International and National Non-Governmental Organizations working for the rights of LGBTQ+ (MAP 2020). However, the identity and rights problems of the LGBTQ+ are still not solved. The major opponents of LGBTQ+ rights are the mainstream religious houses. "*Very clearly the main opponents of LGBTI rights are dogmatic religious forces (from all religious traditions and in all quarters of the world), either operating on their own, or through state actors." .... "The root causes of [homophobia, lesbophobia, and transphobia] stem mostly from tradition and traditional interpretations of religion, which promote rigid models of social existence based on a reproductive foundation of humanity and sexuality* (MAP 2020:15). The countries with the religious ground still do not accepts the rights of LGBTQ+ and still consider LGBTQ+ as a deviant behavior or even a crime. Every country has own story of struggle and unfolded chapters of discriminations, against the rights of LGBTQ+.

LGBTQ+ struggles are also struggling for the freedom to survive with the identity of LGBTQ+.

> The first is freedom of speech and expression—everywhere in the world. The second is freedom of every person to worship God in his own way—everywhere in the world. The third is freedom from want—which, translated into world terms, means economic understandings which will secure to every nation a healthy peacetime life for its inhabitants—everywhere in the world. The fourth is freedom from fear—which, translated into world terms, means a world-wide reduction of armaments to such a point and in such a thorough fashion that no nation will be in a position to commit an act of physical aggression against any neighbor—anywhere in the world. (F. Roosevelt, 1941, para. 73 as in Grant and Gibson 2013).
>
> [The UN members] believe that men and women, all over the world, have the right to live … free from the haunting fear of poverty and insecurity … They believe that science and the arts should combine to serve peace and the well-being, spiritual as well as material, of all men and women without discrimination of any kind. They believe that …the power is in their hands to advance … this well-being more swiftly than in any previous age. (UNESCO, 1949, p. 259 as in Grant and Gibson 2013).

This chapter is just one of them among many. This chapter presents LGBTQ+. Struggle in Ukraine, which holds the former USSR dominated power centric mentality.

2. LGBTQ+ Struggle in Ukraine

The collapse of the USSR and its regime led to the liberalization and democratization of Ukrainian society. Yet, the Soviet background and mentality reflected on a major group of the country that created conservatism and reluctance to new changes and open-mindedness. On the other hand, Ukrainian society is considered a traditional and community one. If we examine the traditionalism part, it means that Ukrainians appeal to their traditions and customs, and sustained practice. If we would look at the community part, the Ukrainian society ranks the family values and relationships with other people ones the most important. All these factors have an impact on readiness to accept new changes, reforms and challenges that especially have actively become implemented after the Revolution of Dignity.

The LGBTQ+ rights started being publicly discussed after the Euromaidan. In 2014, the pro-European government was formed as a desire of the Ukrainian nation. That meant a new hope for Ukrainians as well as LGBTQ+ people in Ukraine. The integration with the European Union requires to have similar laws and regulations of an applicant country. It should be mentioned every country in the EU has its own laws toward the LGBTQ+ community that sometimes might have broader laws and protection, but there is an umbrella of treaties and laws combating discrimination based on sexuality. It is the Treaty on the Functioning of the European Union [1]. The articles 10 and 19 are the main driving forces for protecting this community and legalize the same-sex activity in all members of the European Union. Ukraine should adopt a similar law that would allow LGBTQ+ people to feel protected and indiscriminate. Yet, the society remains intolerant and hostile toward this community, which makes creating the law more complicated.

We believe education is the right tool that can help with forming a more inclusive and open-minded society in Ukraine. The issue is not discussed formally at any governmental educational institutions; this creates an illusion of the absence of it and leads to misunderstanding and hatred of the majority group of people in Ukraine, especially in the young generation, which is a driving force of Ukrainian future. On the other hand, the majority of religious and right-wing organizations stand for the traditionalism and proceeding family values of Ukrainian society that cultivates the hate and intolerant attitude in their followers toward the LGBTQ+ community. One way to educate is to create lessons that would discuss the LGBTQ+ issue at school or include some topics into the school curricular plan. We believe the most efficient method to cultivate critical thinking and a desire is to question everything. In such a way, a student would get the information from a first source and could decide by himself if it should be trusted or not. First, we need to understand the current situation with LGBTQ+ rights in Ukraine.

3. Key Milestones in Ukrainian LGBTQ+ History and Modern Situation

The development of the LGBTQ+ community became possible after Ukraine received its independence from the USSR. During the Soviet era, the Criminal Code illegalized homosexuality and any same-sex activities. In 1991, the process of creating and writing the new constitution started. In 1996, by the majority, the new Constitution of Ukraine was adopted and ratified. It guarantees "*defense of constitutional rights and freedoms of man and citizen*" [5]; this legalizes same-sex activities and privacy. During Ukrainian history, we may see there were no intentions or interest from politicians toward the LGBTQ+ community. The former president of Ukraine, Leonid Kravchuk said such words in 1999, "*We need to work in the independent state for 500 years, and I think that only then will we discuss the problems of sexual minorities.*" [12]. This example took other representatives and believed that there were more important things to consider. We state that the development of Human Rights and the freedom of Ukrainians is one of the most important topics that should be discussed. It will give positive outcomes in the future like the satisfaction of daily life and the rise of living standards as people would be happier and content with their basic needs. After the Revolution of Dignity, there was a shift in Ukrainian society toward Human Rights. Nowadays, the LGBTQ+ issue is discussed more.

Let's see what we have as of 2019. According to the Interim Report of the LGBT Human Rights Nash Mir Center, "*Since the beginning of 2019 Ukrainian legislation on the interests and rights of LGBT people has not been changed. The implementation of the LGBTI components of the Action Plan on Human Rights, the deadline of which had already expired in previous years, moved no farther beyond its stalling point.*"[2] As we see, the interests toward the LGBTQ+ people of the Ukrainian government have not increased, and it did not satisfy the needs of the LGBTQ+ community. During the political campaign, the parties and presidential nominees did not show any interest neither. Mostly, all of them appealed to the majority group that consists of senior and middle-aged people by the populist slogans and problems that "seemed" to be more important such as low salaries, pension reform, peace. In addition, the government failed to carry out the Action Plan on Human Rights that includes LGBTQ+ components [2].

About the legislative part, it should be said there is still no anti-discrimination law that would give proper protection to the people of the community. One of the ways to legally sue someone based on the act of discrimination by its sexuality or gender identity is referring to Article 161 of the Criminal Code of the Republic of Ukraine, which states, "*Willful actions inciting national, racial or religious enmity and hatred, humiliation of national honor and dignity, or the insult of citizens' feelings in respect to their religious convictions, and also any direct or indirect restriction of rights, or granting direct or indirect privileges to citizens based on race, color of skin, political, religious and other convictions, sex, ethnic and social origin, property status, place of residence, linguistic or other characteristics.*"[3] As we see this article does not directly include the terms "sexual orientation," "LGBTQ+," "gender identity" that makes it vaguer and may be misinterpreted.

Also, there is no possibility to have same-sex marriage in Ukraine [4]. Article 51 of the Constitution of Ukraine states, *"Marriage is based on the free consent of woman and man. Each of the married couple has even rights and duties in marriage and family"* [5], and Article 21 of the Ukrainian Family Code stands, *"A marriage is a family union between a woman and a man, duly registered in a public civil status act registration authority"*[6]. These articles make it impossible to legally register same-sex unions in the country. To add, the Ukrainian Family Code bans same-sex couples from the adoption in Article 211, *"The spouses, as well as persons referred to in paragraphs 5 and 6 of the present Article may be adopters. Persons of the same sex may not be adopters"* [6].

Transsexuality in Ukraine is considered as a psychiatric disorder [7], while the World Health Organization excluded transgenderism from the mental illness category in 2018 [8]. It means that Ukraine has an outdated list of disorders and actions should be taken to establish a new variant that would increase inclusiveness in society.

On the other hand, there are some victories and opportunities for LGBTQ+ people in Ukraine. After the revolution, the Pride March has become a way to cultivate tolerance, inclusiveness and acknowledge diversity in Ukrainian society. The year of 2019 became memorable for Ukrainian LGBTQ+ history. The capital of Ukraine held the largest Pride March ever. Also, big cities such as Kharkov, Odesa had their smaller ones. The number of various pro-LGBTQ+ events and projects has increased too. For instance, the conference in Kyiv "Transgender issues: challenges and perspectives in modern Ukraine and world", or the LGBTQ+ photo exhibition "We Were Here" that was dedicated to the organizer and other LGBTQ+ veterans and volunteers of the ATO[13].

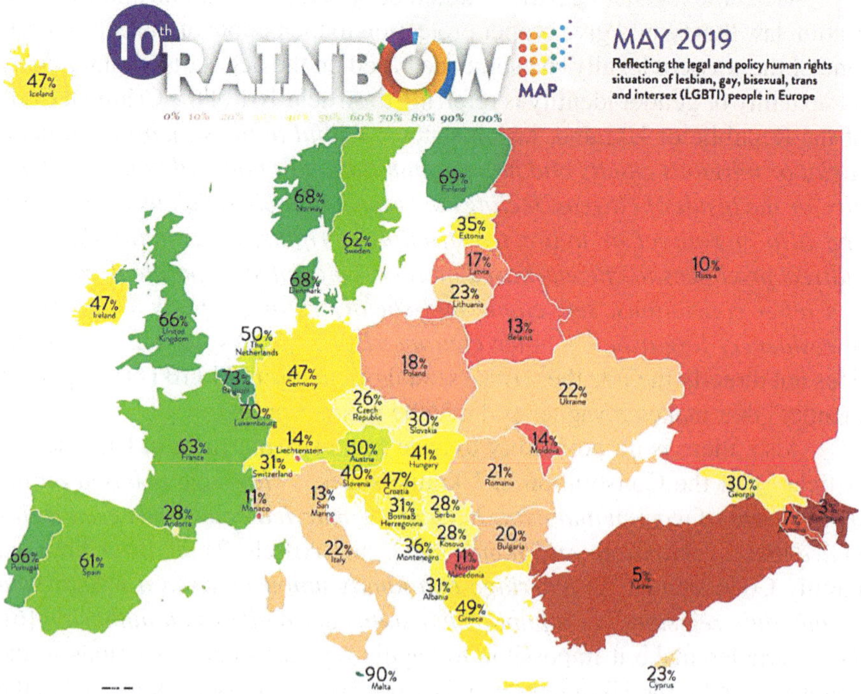

Figure 10.1. General European Score on achieved LGBTQ+ human rights in 2019
Open Source: Rainbow Europe

In 2016, The Ministry of Health of Ukraine adopted a new regulation where states gay and bisexual men may donate their blood [9]. In 2015, the Ukrainian Parliament approved an amendment to the Labor Code [10]; this amendment bans discrimination based on gender identity and sexual orientation at work. It was a requirement of the European Union for Ukraine to proceed with the application of free-visa travel. The summary made by Rainbow Europe puts Ukraine with 22% of achieving of LGBTQ+ human rights [11]. It vividly differs from the scores of the neighboring countries. For instance, the Russian Federation has 10%, Poland has 18%, and Moldova, 15%. It shows a good tendency of Ukraine if we compare to the Eastern European countries. Also, the LGBTQ+ people may openly serve in the Ukrainian army and homosexuality is not an excuse for exemption. Still, there are a lot of things to do.

4. LGBTQ+ and Ukrainian Education

> Education shall be directed to the full development of the human personality and to the strengthening of respect for human rights and fundamental freedoms. It shall promote understanding, tolerance, and friendship among all nations, racial or religious groups, and shall further the activities of the UN for the maintenance of peace. (The Universal Declaration of Human Rights, Article 26.2, as in Grant and Gibson 2013)

According to Article 53, "*Everybody has a right to education. Complete universal secondary education is obligatory*" [5]. This guarantees and obligates to have at least full secondary education for every citizen of the country. Referring to UNESCO, the literacy rate in Ukraine is 99.7%. It makes Ukraine one of the leading states across the world. It should be mentioned that 70% of Ukrainians have a secondary or higher education. The country ranks 4th on the World Bank Enrollment Index (tertiary education) [14]. In addition, this sector always receives one of the biggest parts of the GDP and is considered well-developed. Are the instruments of anti-discrimination also well-developed in the Ukrainian legislature and are educational institutions safe and inclusive places for studying?

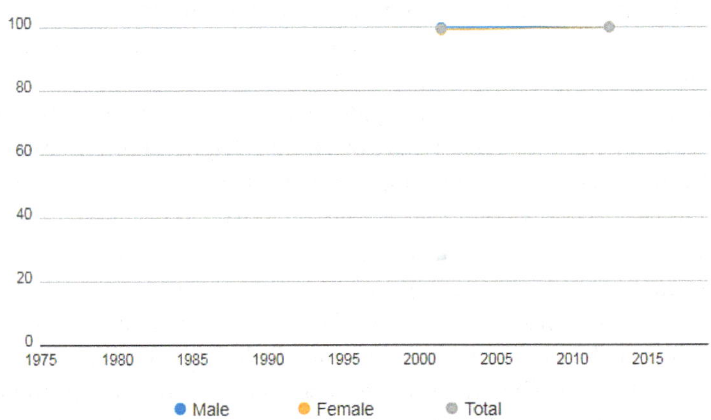

Figure 10.2. Literacy rate among the population aged15 years and older
Open Source: UNESCO http://uis.unesco.org/country/UA

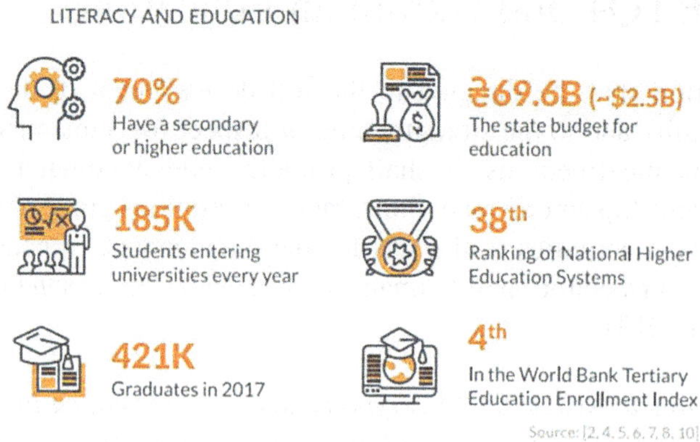

Figure 10.3. Highlights of Education in Ukraine
Open Source: The Country That Codes. IT Industry in Ukraine. 2019 Market Report: https://s3-eu-west-1.amazonaws.com/new.n-ix.com/uploads/2019/08/29/Ukr-report-29.pdf

The Law of Ukraine "On the principles of prevention and combating discrimination in Ukraine" was established in its final version in 2014. This law prohibits any kind of discrimination on different aspects of daily life for every citizen of Ukraine, and in education too. The law establishes a list of non-discriminatory grounds. However, there are no terms such as "gender identity" or "sexual orientation," but there is a phrase "or other grounds." This means that the list in the provision of the Law is non-exhaustive and illustrative. This interpretation is confirmed as an analysis of international law, the practice of its application by various international judicial and quasi-judicial authorities, and the practice of the European Court of Human Rights. For example, the European Court of Human Rights finds that the list of non-discriminatory grounds includes sexual orientation, gender identity (transgender or transgender), and others [15]. As we see the students have juridical protection against discrimination. They may appeal to court and police based on this law. Another good alternative is the new law "On Amendments to Certain Legislative Acts of Ukraine on Counteracting Bullying (Harassment)" that was signed in 2019 [15]. The law provides a new ground for protection in the educational institution at all levels and includes different participants to be fined. According to the law, a new role at Ukrainian schools includes "social teachers." They are in charge of prevention of bullying (harassment), support students who were bullied or witnessed wrongdoers as well as give parents advice [15]. Based on laws, the educational environment in Ukraine is safe and protected from discrimination. Is it true in reality?

According to the National School Environment Survey, that was conducted by the "Tochka Opory" an LGBT organization, in 2019, concluded, "*Schools across the country are hostile to LGBT students, most of whom hear negative statements on a regular basis, and are subjected to victimization (victim blaming) and discrimination. As a result, many LGBT teens avoid school activities and school in general*" [18], There are some statistics:

- Almost 50% of respondents do not feel safe due to their sexual orientation at school;
- About 37% of interviewed students missed at least a school day during the previous month due to their insecurity at school;
- About 90% of LGBT students have been verbally harassed at school (in the form of names or threats) and;
- 54% were subjected to physical abuse because of their sexual orientation or gender identity;
- 66% of students, who were abused, never reported about harassment or bullying to school staff, because they were sure that school staff would not interact or their actions would not be efficient.
- 55% of those who reported to school personnel about harassment, were instructed not to pay attention;
- Almost 73% of the interviewed students mentioned that had heard homophobic statements from school staff [18].

We may conclude that the school environment is not safe for LGBTQ+-students even having anti-discrimination and anti-bullying laws. We believe the lack of the most important factor causes harassment toward the sexual minority. It is an educational component. The lack of educational programs about human sexuality, sex education for schools, and general inferiority of LGBTQ+ issues in Ukraine make violation of laws and harassment in different forms toward LGBTQ+ people happen in the country. The Ukrainian Government should provide an action plan on educating its citizens about the issue and cultivating a tolerant attitude that would increase inclusiveness in society.

4.1. Education as a tool of prevention of the discrimination against LGBTQ+ people

> Teaching for social justice might be thought of as a kind of popular education—of, by, and for the people—something that lies at the heart of education in a democracy, education toward a more vital, more muscular democratic society. It can propel us toward action, away from complacency, reminding us of the powerful commitment, persistence, bravery, and triumphs of our justice-seeking forebears—women and men who sought to build a world that worked for us all. Abolitionists, suffragettes, labor organizers, civil rights activists: Without them, liberty would today be slighter, poorer, weaker—the American flag wrapped around an empty shell—a democracy of form and symbol over substance. (Ayers, 2008, para. 1, as in Grant and Gibson 2013)

We believe education is the best way to cultivate inclusiveness and tolerance in the society. Education can boost the essence of critical thinking which allows students to check and question all the information that is given to them and search on their own origin or background. The education may show the difference; people and students may learn how to be empathic and

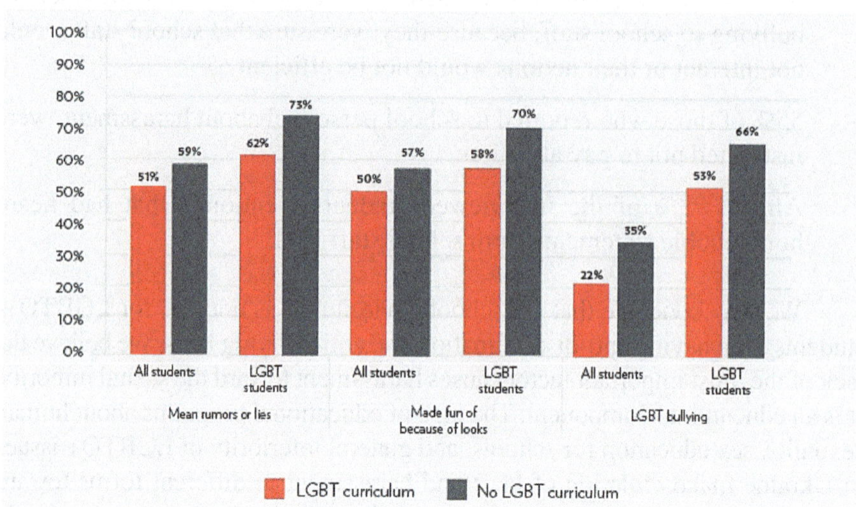

Figure 10.4. Students who learn about LGBT Issues in the curriculum report less harassment

Source: California Safe School Coalition http://casafeschools.org/

Developed by Century Foundation

tolerant to others who may differ from them. Teaching about the LGBTQ+ issue can help people better understand the issue and people of the community [21]. In addition, the classroom will learn how to respect others and how to solve any problems peacefully without using violence [20].

According to the Convention on the Rights of the child, *"...Education of the child shall be directed to the preparation of the child for responsible life in a free society, in the spirit of understanding, peace, tolerance, equality of sexes, and friendship among all peoples, ethnic, national and religious groups and persons of indigenous origin."*[19]. Tolerance and inclusion are also set as aims in the Human Rights Convention that all State Parties agreed to follow and implement nationally. That is so important to auditing if the national governments implant this policy.

The reports published by GLAAD, have realized the more people are familiar with the LGBT issue the more understanding and comfortable they would with them. That reduces the harassment and discrimination toward the community [21]. Also, the results of surveys conducted by the California Safe School Coalition have shown that students who study the issue lesser abuse and bully LGBTQ+ people. For instance, the percentage of bullying toward LGBTQ+ students was reduced by 13% if students have a LGBTQ+ curriculum (Figure 10.4). That is why education can be a great tool for preventing discrimination against LGBTQ+.

4.2. Educational tools that exist in Ukraine to combat discrimination toward the LGBTQ+ community

In the international practice, the best examples are the USA and The European Union, reducing discrimination against LGBTQ+ by educational tools are aggregated by all members of the country such as government, public, businesses, and educational institutions. This comprehensive approach involves people getting to know more about the LGBTQ+ community on different levels as well as guaranteeing protection and assistance from different factors of society. If we consider Ukrainian reality, the best and more efficient tools were created by private initiatives such as NGOs, international organizations, and businesses. Unfortunately, there is no governmental support in this direction, except, the legal basis which remains undeveloped.

One of the best tools that were established is sex education. Most European countries include not only biological aspects of sex difference or STI, HIV/AIDS, but also sexuality and gender identity in their curriculum. If we look at the "Sexuality Education in Europe and Central Asia" report conducted by German Federal Centre for Health Education

Main topics dealt with	How extensive? Extensive	Briefly	Not
Biological aspects and body awareness	X		
Pregnancy and birth		X	
Contraception (including at least three effective methods)		X	
HIV/AIDS	X		
STIs	X		
Love, marriage, partnership	X		
Sexual pleasure			X
Sexual orientation			X
Gender roles		X	
Online media and sexuality		X	
Access to safe abortion in the framework of the national law			X
Mutual consent to sexual activity		X	
Sexual abuse/violence		X	
Domestic violence		X	
Human rights and sexuality			X

Figure 10.5. Curriculum content about sexual education in Ukrainian schools
Source: "Sexuality Education in Europe and Central Asia" Report https://www.ippfen.org/sites/ippfen/files/2018-05/Comprehensive%20Country%20Report%20on%20CSE%20in%20Europe%20and%20Central%20Asia_0.pdf

(BZgA) in close cooperation with the International Planned Parenthood Federation—European Network (IPPF EN), we may see sexual orientation is not included into any curriculum to study at any grades of school in Ukraine. Furthermore, there is no information in the school curriculum even about human rights. It makes students more vulnerable and ignorant about their rights (Figure 10.5) [22].

The best assistance at raising awareness about LGBTQ+ is made by non-governmental organizations. For instance, the LGBT Center "Nash Svit," National LGBT-Portal of Ukraine, NGO "The Alliance. Global," All-Ukrainian NGO "The Gay Alliance. Ukraine," NGO "Insight," NGO "Tochka Opory," Charity Fund "GenderZ," NGO "Parental Initiative," TERGO, NGO "KyivPride," and many others that almost cover all big cities and oblast centers of Ukraine. Let's examine some of them and their activities.

"Alliance. Global '' works nationally with the headquarter in Kyiv. It provides a paralegal assistance for LGBTQ+ people and Men who have sex with men (MSM). Also, anyone can get tested on different diseases, including HIV in Kyiv, Kharkov, and Dnipro. They have a PrEP info-camping

Figure 10.6. Picture of "Friedlt Pill" with the slogan
Source: http://www.gweek.com.ua/2019/09/blog-post_16.html

that aims to promote the drugs that may reduce the likelihood of reviving HIV during unsafe sex or other ways. The NGO holds various supporting groups, especially, for people who have HIV positive status. There is also a shelter for MSM and transgender people who are going through a difficult life situation [23].

"GenderZ" works also in the human rights direction in Zaporizhia. It works with different issues that include raising awareness about sexuality and gender identity, reducing hostility toward LGBTQ+, gender inequality, and gender stereotyping, preventing HIV and other STI diseases by organizing various campaigns and events [24]. The most vivid one is the "Friendly Pill" project that aims to "heal" homophobia and transphobia. It is an InfoCampaing which uses an educational tool to raise awareness about LGBTQ+. In order to get a pill, a person needs to take a test that includes questions about the issues and explanations to them. The organizers tell that pills are not a medicine, but just a sweet dragee (contains sugar and flavors). It is a trick that they use to involve more people. The project is supported by the German Federal Foreign Office [25]. "It is a medicine against homophobia and transphobia"

LGBT Center "Nash Svit'" is one of the oldest LGBTQ+ organizations in Ukraine. Now it is based in Kyiv. The main goals are to provide primary legal assistance and counseling for victims of discrimination and hate speech, monitoring of human rights respect toward LGBTQ+ community, and preparation of reports about human rights LGBTQ+ in Ukraine. They also provide legal education for LGBTQ+ people by organizing seminars and workshops. The Center has different textbooks and methodologists about LGBTQ+, trainings about social work with homosexuals, police, and LGBTQ+, and advocacy and monitoring human rights for activists that are free and publicly available. The organization has established credibility and a lot of foreign and Ukrainian media and organizations use its reports, for instance, the US Department of State, OSCE, and many others.

Tochka Opory is a non-governmental organization that has a few directions to work in. First is the assistance of LGBTQ+ and marginalized teenagers. They organize workshops, trainings for teachers, official guardians and parents, psychologists, and social workers and inform them about the issue. The second direction is advocacy focusing on the Sustainable Development Goals and LGBTQ+. The last one is about creation of an inclusive environment at schools. It also should be mentioned that the organization became the first LGBTQ+ initiative that established cooperation with other non-LGBTQ+ ones. The result of the cooperation was the project «Ukrainian Corporate Equality Index» [27]. The Ukrainian Corporate Equality Index is a national survey of corporate policies, rules, and practices of private companies to support equality and diversity and to prohibit discrimination in the workplace. The index includes studies on combating discrimination on the grounds of gender, disability, sexual orientation, and gender identity. The study is a joint development of the NGO Support Point of the UA, the NGO Social Action, and the National Assembly of People with Disabilities of Ukraine with the support of the International Renaissance Foundation [28].

The next organization is "KyivPride." It does the Pride Week and the Equality March during the global Pride Month. In 2019, the biggest Pride March was held in Ukraine organized by this organization [30]. Around 8000 participants across Ukraine and representatives of Foreign Institutions took part in the event. It is twice higher than the previous year [29]. The memorable moment of the march was LGBTQ+ military column that also participated in the march [31]. Besides this, there are other workshops and activities that are conducted by the organization.

There is another organization, Tergo. It is a parental initiative of LGBTQ+ children that supports in different ways non-traditional sexual minorities. One of their work was a social advertisement that includes videos against homophobia and bullying at schools and coming out stories. They provide different informational materials and assistance about the LGBTQ+ community and help parents and other people their children [32].

Figure 10.7. Picture from the Kyiv Pride March 2019. It says, "Love will not hurt anyone"
Source: Radio Liberty https://www.radiosvoboda.org/a/kyiv-pride-2019-photos-of-eaquality-march/30015203.html

5. Conclusion

The case study of Ukraine shows the process of developing Human Rights for LGBTQ+ and creating an inclusive society. It became possible after the collapse of the USSR and the Revolution of Dignity that boosted the Human Rights activities. We believe education is an essential tool for fighting with discrimination and harassment against the LGBTQ+ community. Results showed that students that have an LGBTQ+ curriculum at school are less likely to become wrongdoers and LGBTQ+ students suffer less bullying. That is why it is so vital to have governmental support and assistance that may provide not only a legal basis but comprehensive protection and support on different levels. Including topics of sexuality and gender identity to the school curriculum can increase the acceptance and tolerance in the society that would start to cultivate since childhood. There is still a lot to do but with such NGOs' support and other foreign institutions, Ukraine may become one day an LGBTQ+-friendly country.

Lesbian, Gay, Bisexual, Transgender and Queer or Questioning (LGBTQ+) is the identity issue—which has a long history of struggles. The identity issue is associated with the human civilization; however, the society have been dominated by the political, economic, and community elite; therefore, the marginalized people's identity remained under a shadow. Discrimination based on gender, age, origin, ethnicity, disability, sexual orientation, class, and religious belief, minority, language, immigration status etc. are in the society throughout the history of human civilization. The LGBTQ+ struggle is to have an individual right and identity in a way that one wants to have a dignified life whatever that person choses to be recognized as that person's sexual choice. The struggle against the discrimination of LGBTQ+ rights is a global movement. The case study presented in this chapter is one of them; however, almost every country can have similar type of struggles one way or another. It is also about individuality, and the validation of the ways of life an individual wants to survive. To survive with the LGBTQ+ identity is a great challenge, not only in the external environment, but also the challenge of within. The mainstream society, family, community, and nation have to accept the notion that, every person should have the right to live a validated life, in a way that an individual wants to live. A person's identity is how she or he defines, how she or he presents, how she or he wants to be seen from others' perspective. Therefore, LGBTQ+ wants to be as they want to be and seek the validation of themselves as they feel, believe, and maintain their lives.

References

Appiah, Kwame Anthony and Henry Louis Gates Jr., eds. (1995), Identities. Chicago, IL: University of Chicago Press.

Bechdel, Alison(2006), *Fun Home: A Family Tragicomic*, Houghton Mifflin

Bernstein, Mary (2002), Identities and Politics: Toward a Historical Understanding of the Lesbian and Gay Movement, Social Science History, Vol. 26, No. 3 (Fall, 2002), pp. 531-581 (51 pages) Cambridge University Press.

Bloom, William. (1990), Personal Identity, National Identity, and International Relations. Cambridge: Cambridge University Press

Bornstein, Kate (1994), *Gender Outlaws: On Men, Women, and the Rest of Us*, Routledge

Bronski, Michael (2011), *A Queer History of the United States*, Beacon Press

Carbado, Devon and Dwight McBride, eds (2002), *Black Like Us: A Century of Lesbian, Gay and Bisexual African American Fiction*, Cleis Press

Carter, David (2004), *Stonewall: The Riots That Sparked the Gay Revolution*, Macmillan

Cenziper, Debbie and Jim Obergefell, (2016), *Love Wins: The Lovers and Lawyers Who Fought the Landmark Case for Marriage Equality*, Harper Collins Publishers

Chauncey, George. (1994), Gay New York: Gender, Urban Culture, and the Makings of the

Connolly, William E. (1991), Identity/Difference: Democratic Negotiations of Political Paradox. Ithaca, N.Y.: Cornell University Press.

Erikson, Erik H. (1968), Identity: Youth and Crisis. New York: Norton.

Faderman, Lillian (2015), *The Gay Revolution: The Story of the Struggle*, Simon & Schuster, Faderman, Lillian (1999), *To Believe in Women: What Lesbians Have Done for America – A History*, Houghton Mifflin

Fearon, J. D. (1999), What Is Identity (As We Now Use the Word)? California: Stanford University. http://www.stanford.edu/~jfearon/papers/iden1v2.pdf

Feinberg, Leslie (1996), *Transgender Warriors*, Beacon Press

Gay Male World, (1890-1940), New York: Basic Books.

GLAAD (2021), LGBT Rights and Discrimination to LGBT and intersex people, https://www.glaad.org/vote/topics/global-lgbt-rights

Gleason, Philip. (1983), "Identifying Identity: A Semantic History." Journal of American History 6:910–931.

Carl A. Grant and Melissa L. Gibson (2013), "The path of social justice": A Human Rights History of Social Justice Education, Equity & Excellence in Education, Vol. 46, No. 1 (2013): 81-99. DOI. © 2013 Taylor & Francis (Routledge) https://epublications.marquette.edu/cgi/viewcontent.cgi?article=1427&context=edu_fac

Hogg, Michael and Dominic Abrams. (1988), Social Identifications: A Social Psychology of Intergroup Relations and Group Processes. London: Routledge.

Jacobs, Sue-Ellen (1997), *Two-Spirit People: Native American Gender Identity, Sexuality and Spirituality*, University of Illinois

Jenkins, Richard. (1996), Social Identity. London: Routledge.

Johnson, David (2004), *The Lavender Scare: The Cold War Persecution of Gays and Lesbians in the Federal Government*, University of Chicago Press Books

Kymlicka, Will. (1995), Multicultural Citizenship: A Liberal Theory of Minority Rights. Oxford: Clarendon Press

Levy, M. (2020, June 15). *Gay rights movement. Encyclopedia Britannica*. https://www.britannica.com/topic/gay-rights-movement

MAP (2020), International LGBT Advocacy Organizations and Programs, An Overview- LGBT Movement Advancement Project (MAP), Denver, CO 80205, USA www.lgbtmap.org

Miller, David. (1995), Nationalism. London: Oxford University Press.

Moraga, Cherrie and Gloria Anzaldua, (1981), *This Bridge Called My Back: Writings by Radical Women of Color*, Persephone Press

Morris, Bonnie J. (2021), History of Lesbian, Gay, Bisexual and Transgender Social Movements, 2021 American Psychological Association, USA- https://www.apa.org/pi/lgbt/resources/history

Perry, John, ed. (1975), Personal Identity. Berkeley: University of California Press.

Ringmar, Erik. (1996), Identity, Interest, and Action: A Cultural Explanation of Sweden's Intervention in the Thirty Years War. Cambridge: Cambridge University Press.

Scholinski, Daphne (1998), *The Last Time I Wore a Dress*, Riverhead Books

Shilts, Randy (1987), *And the Band Played On: Politics, People, and the AIDS Epidemic*, St. Martin's Press

Short, Donn (2013), *Don't Be So Gay! Queers, Bullying, and Making Schools Safe*, UBC Press

Taylor, Charles. (1989), The Sources of the Self: The Making of the Modern Identity. Cambridge, MA: Harvard University Press

Thoreson, Ryan (2021), An International LGBT Movement, Oxford University Press 2021, https://doi.org/10.1093/acrefore/9780190228637.013.1214 https://oxfordre.com/politics/view/10.1093/acrefore/9780190228637.001.0001/acrefore-9780190228637-e-1214

Thoreson, Ryan (2014), *Transnational LGBT Activism*, University of Minnesota Press University Press.

Wendt, Alexander. (1994), "Collective Identity Formation and the International State." American Political Science Review 88:384–96 www.international-lgbt-advocacy-organizations-and-programs.pdf

Young, Iris Marion. (1990), Justice and the Politics of Difference. Princeton, N.J.: Princeton

Web References

1. Treaty on the Functioning of the European Union: https://eur-lex.europa.eu/LexUriServ/LexUriServ.do?uri=CELEX:12012E/TXT:EN:PD
2. Interim Report on the situation of LGBT people in Ukraine in 2019: https://gay.org.ua/en/blog/2019/09/04/lgbt-situation-in-ukraine-in-2019-january-august/
3. The Criminal Code of the Republic of Ukraine: https://www.unodc.org/res/cld/document/ukr/2001/criminal-code-of-the-republic-of-ukraine-en_html/Ukraine_Criminal_Code_as_of_2010_EN.pdf
4. Same-sex marriage in Ukraine: accept or deny: https://www.unian.info/society/2395054-same-sex-marriage-in-ukraine-accept-or-deny.html
5. Constitution of Ukraine: https://www.justice.gov/sites/default/files/eoir/legacy/2013/11/08/constitution_14.pdf
6. Ukrainian Family Code: http://jafbase.fr/docEstEurope/FAMILY_CODE_OF_UKRAINE.pdf
7. Transgender life in Ukraine officially labeled as 'disorder': https://www.kyivpost.com/lifestyle/transgender-life-in-ukraine-officially-labeled-as-disorder-394156.html?cn-reloaded=1
8. WHO takes transgenderism out of mental illness category? https://www.theguardian.com/society/2018/jun/20/who-takes-transgenderism-out-of-mental-illness-category
9. Ukraine has lifted the ban on blood donation for homosexuals: http://gaynewseurope.com/en/2016/04/27/ukraine-has-lifted-the-ban-on-blood-donation-for-homosexuals/
10. Rada pushes through non-discrimination amendment to Labor Code: https://www.unian.info/politics/1181197-rada-fails-second-attempt-to-pass-bill-on-non-discrimination-in-labor-relations.html

11. Rainbow Europe https://rainbow-europe.org/#8665/0/0
12. Andriy Maymulakhin; Olexandr Zinchenkov; Andriy Kravchuk (2007). "Ukrainian Homosexuals and Society - Report 2007" (PDF). Kyiv: LGBT Human Rights NASH MIR Center. Retrieved 12 March 2016 https://www.gay.org.ua/publications/gay_ukraine_2007-e.pdf
13. Annual Review: Ukraine. Rainbow Europe: https://www.ilga-europe.org/sites/default/files/ukraine.pdf
14. THE COUNTRY THAT CODES. IT Industry in Ukraine. 2019 Market Report: https://s3-eu-west-1.amazonaws.com/new.n-ix.com/uploads/2019/08/29/Ukr-report-29.pdf
15. PREVENTION AND ANTI-DISCRIMINATION IN UKRAINE. International Organization for Migration (IOM), Representation in Ukraine: http://iom.org.ua/sites/default/files/iom_booklette-06_1kolonka_screen.pdf
16. Poroshenko signs anti-bullying bill into law https://www.unian.info/society/10411221-poroshenko-signs-anti-bullying-bill-into-law.html
17. Coming to school soon: is Ukrainian education ready for LGBT students?: https://life.pravda.com.ua/columns/2019/08/20/237930/
18. The national survey of the school environment in Ukraine on LGBT adolescents: https://issuu.com/fulcrumua/docs/doslidzhennya-2
19. Convention on the Rights of the child: https://www.unicef-irc.org/portfolios/general_comments/GC1_en.doc.html
20. Education For Peace: Top 10 Ways Education Promotes Peace: https://centralasiainstitute.org/top-10-ways-education-promotes-peace/
21. Can Education Reduce Prejudice against LGBT People?: https://tcf.org/content/commentary/can-education-reduce-prejudice-lgbt-people/
22. https://www.ippfen.org/sites/ippfen/files/2018-05/Comprehensive%20Country%20Report%20on%20CSE%20in%20Europe%20and%20Central%20Asia_0.pdf
23. NGO "Alliance. Global": http://ga.net.ua/
24. Charity Fund "GenderZ": http://genderz.org.ua/
25. "Friendly Pill". official Website of InfoCampaing: https://pigulka.org.ua/
26. LGBT Center "Nash Svit": https://gay.org.ua/
27. NGO "Tochka Opory": https://t-o.org.ua/
28. Ukrainian Corporate Equality Index. Official Website: http://cei.org.ua/
29. How many people attended the 2019 Equality March: https://www.the-village.com.ua/village/city/city-news/286437-skilki-lyudey-vzyalo-uchast-u-marshi-rivnosti-2019
30. NGO "KyivPride": http://kyivpride.org/
31. During the 2019 Equality March, the LGBT military column will be held for the first time: https://www.the-village.com.ua/village/city/city-news/286411-pid-chas-marshu-rivnosti-2019-vpershe-proyde-kolona-lgbt-viyskovih
32. NGO "Parental Initiative" TERGO": http://tergo.org.ua/

Annex-one

LGBTQ Rights Milestones Fast Facts in USA (CNN Editorial Research)
Source: CNN- 2021- LGBTQ Rights Milestones Fast Facts (CNN Editorial Research)- Updated 2:49 PM EST, Tue February 2, 2021 https://www.cnn.com/2015/06/19/us/lgbt-rights-milestones-fast-facts/index.html

Year	Action—Movements / Achievements
2021	President Joe Biden signs an executive order repealing the 2019 Trump-era ban on most transgender Americans joining the military. "This is reinstating a position that the previous commanders and, as well as the secretaries, have supported. And what I'm doing is enabling all qualified Americans to serve their country in uniform," Biden said, speaking from the Oval Office just before signing the executive order.
2021	Secretary of Transportation Pete Buttigieg becomes the first Senate-confirmed LGBTQ Cabinet secretary.
2020	The Supreme Court rules that federal law protects LGBTQ workers from discrimination. The landmark ruling extends protections to millions of workers nationwide and is a defeat for the Trump administration, which argued that Title VII of the Civil Rights Act that bars discrimination based on sex did not extend to claims of gender identity and sexual orientation.
2020	The 4th Circuit Court of Appeals rules in favor of former student, Gavin Grimm. in a more than four-year fight over restroom policies for transgender students. The ruling states that policies segregating transgender students from their peers is unconstitutional and violate federal law prohibiting sex discrimination in education. The decision relies in part on the Supreme Court's decision in June 2020, stating that discrimination against people based on their gender identity or sexual orientation violates Title VII of the Civil Rights Act of 1964.
2020	The general election results in three legislative firsts. Sarah McBride wins the Senate race for Delaware District 1 and will become the nation's first person who identifies as transgender to serve as a state senator. Ritchie Torres wins the House race for New York District 15 and will become the first Black member of Congress who identifies as gay. Mauree Turner wins the race for Oklahoma state House for District 88 and will become the first nonbinary state legislator in US history and first Muslim lawmaker in Oklahoma.

Year	Action—Movements / Achievements
2019	New York Governor Andrew Cuomo signs a law banning the use of the so-called gay and trans panic legal defense strategy. The tactic asks a jury to find that a victim's sexual orientation or gender identity is to blame for a defendant's violent reaction. New York follows California, Rhode Island, Illinois, Nevada and Connecticut as the sixth state to pass such a law.
2019	Billy Porter becomes the first openly gay Black man to win the Emmy for best lead actor in a drama series.
2018	The Pentagon confirms that the first transgender person has signed a contract to join the US military.
2018	Daniela Vega, the star of Oscar-winning foreign film "A Fantastic Woman," becomes the first openly transgender presenter in Academy Awards history when she introduces a performance by Sufjan Stevens, whose song "Mystery of Love" from the "Call Me By Your Name" soundtrack, is nominated for best original song.
2018	The Trump administration announces a new policy that bans most transgender people from serving in military. After several court battles, the Supreme Court allows the ban to go into effect in January 2019.
2018	Democratic US Representative Jared Polis wins the Colorado governor's race, becoming the nation's first openly gay man to be elected governor.
2017	The 7th Circuit Court of Appeals rules that the Civil Rights Act prohibits workplace discrimination against LGBTQ employees, after Kimberly Hively sues Ivy Tech Community College for violating Title VII of the act by denying her employment.
2017	District of Columbia residents can now choose a gender-neutral option of their driver's license. DC residents become the first people in the United States to be able to choose X as their gender marker instead of male or female on driver's licenses and identification cards. Similar policies exist in Canada, India, Bangladesh, Australia, New Zealand, and Nepal.
2017	The US Department of Defense announces a six-month delay in allowing transgendered individuals to enlist in the United States military. Defense Secretary Jim Mattis writes that they "will use this additional time to evaluate more carefully the impact of such accessions on readiness and lethality." Approximately a month later, President Donald Trump announces via Twitter that the "United States Government will not accept or allow Transgender individuals to serve in any capacity in the US Military..."

Year	Action—Movements / Achievements
2017	Virginia voters elect the state's first openly transgender candidate to the Virginia House of Delegates. Danica Roem unseats incumbent delegate Bob Marshall, who had been elected 13 times over 26 years. Roem becomes the first openly transgender candidate elected to a state legislature in American history.
2016	The Senate confirms Eric Fanning to be secretary of the Army, making him the first openly gay secretary of a US military branch. Fanning previously served as Defense Secretary Carter's chief of staff, and also served as undersecretary of the Air Force and deputy undersecretary of the Navy.
2016	Obama announces the designation of the first national monument to lesbian, gay, bisexual and transgender (LGBTQ) rights. The Stonewall National Monument will encompass Christopher Park, the Stonewall Inn and the surrounding streets and sidewalks that were the sites of the 1969 Stonewall uprising.
2016	Secretary of Defense Carter announces that the Pentagon is lifting the ban on transgender people serving openly in the US military.
2016	A record number of "out" athletes compete in the summer Olympic Games in Rio de Janeiro. The Human Rights Campaign estimates that there are at least 41 openly lesbian, gay and bisexual Olympians—up from 23 that participated in London 2012.
2016	Kate Brown is sworn in as governor of Oregon, a day after she was officially elected to the office. Brown becomes the highest-ranking LGBTQ person elected to office in the United States. Brown took over the governorship in February 2016 (without an election), after Democrat John Kitzhaber resigned amidst a criminal investigation.
2015	Secretary of Defense Ash Carter announces that the Military Equal Opportunity policy has been adjusted to include gay and lesbian military members.
2015	The US Supreme Court hears oral arguments on the question of the freedom to marry in Kentucky, Tennessee, Ohio and Michigan. On June 26 the Supreme Court rules that states cannot ban same-sex marriage. The 5-4 ruling had Justice Anthony Kennedy writing for the majority. Each of the four conservative justices writes their own dissent.
2015	Boy Scouts of America President Robert Gates announces, "the national executive board ratified a resolution removing the national restriction on openly gay leaders and employees."

Year	Action—Movements / Achievements
2014	The United States Supreme Court denies review in five different marriage cases, allowing lower court rulings to stand, and therefore allowing same-sex couples to marry in Utah, Oklahoma, Virginia, Indiana, and Wisconsin. The decision opens the door for the right to marry in Colorado, Kansas, North Carolina, South Carolina, West Virginia, and Wyoming.
2013	In United States v. Windsor, the US Supreme Court strikes down section 3 of the Defense of Marriage Act, ruling that legally married same-sex couples are entitled to federal benefits. The high court also dismisses a case involving California's proposition 8.
2012	In an ABC interview, Obama becomes the first sitting US president to publicly support the freedom for LGBTQ couples to marry.
2012	The Democratic Party becomes the first major US political party in history to publicly support same-sex marriage on a national platform at the Democratic National Convention.
2012	Tammy Baldwin becomes the first openly gay politician and the first Wisconsin woman to be elected to the US Senate.
2011	"Don't Ask, Don't Tell" is repealed, ending a ban on gay men and lesbians from serving openly in the military.
2010	The 9th Circuit Court of Appeals upholds a ruling that the state of Idaho must provide gender confirmation surgery for Adree Edmo, an inmate in the custody of the Idaho Department of Correction. The ruling marks the first time a federal appeals court has ruled that a state must provide gender assignment surgery to an incarcerated person. According to the court opinion, "the gender confirmation surgery (GCS) was medically necessary for Edmo and ordered the State to provide the surgery." In July 2020, Edmo receives her gender confirmation surgery and a May 2020 appeal by Attorney General of Idaho, Lawrence Wasden, is denied as moot by the US Supreme Court in October 2020.
2009	Milk is posthumously awarded the Medal of Freedom by President Barack Obama.
2009	Obama signs the Matthew Shepard and James Byrd Jr. Hate Crimes Prevention Act into law.
2008	The California Supreme Court rules in re: Marriage Cases that limiting marriage to opposite-sex couples is unconstitutional.
2008	Voters approve Proposition 8 in California, which makes same-sex marriage illegal. The proposition is later found to be unconstitutional by a federal judge.

Year	Action—Movements / Achievements
2006	The New Jersey Supreme Court rules that state lawmakers must provide the rights and benefits of marriage to gay and lesbian couples.
2005	The California legislature becomes the first to pass a bill allowing marriage between same-sex couples. Governor Arnold Schwarzenegger vetoes the bill.
2004	The first legal same-sex marriage in the United States takes place in Massachusetts.
2003	The US Supreme Court strikes down the "homosexual conduct" law, which decriminalizes same-sex sexual conduct, with their opinion in Lawrence v. Texas. The decision also reverses Bowers v. Hardwick, a 1986 US Supreme Court ruling that upheld Georgia's sodomy law.
1998	Martin Luther King Jr.'s widow, Coretta Scott King, asks the civil rights community to help in the effort to extinguish homophobia.
1998	Matthew Shepard is tied to a fence and beaten near Laramie, Wyoming. He is eventually found by a cyclist, who initially mistakes him for a scarecrow. He later dies due to his injuries sustained in the beating.
1998	Russell Henderson and Aaron McKinney from Laramie, Wyoming, make their first court appearance after being arrested for the attempted murder of Shepard. Eventually, they each receive two life sentences for killing Shepard.
1997	Comedian Ellen DeGeneres comes out as a lesbian on the cover of Time magazine, stating, "Yep, I'm Gay."
1997	DeGeneres' character, Ellen Morgan, on her self-titled TV series "Ellen," becomes the first leading character to come out on a prime-time network television show.
1996	President Clinton signs the Defense of Marriage Act, banning federal recognition of same-sex marriage and defining marriage as "a legal union between one man and one woman as husband and wife."
1996	Hawaii's Judge Chang rules that the state does not have a legal right to deprive same-sex couples of the right to marry, making Hawaii the first state to recognize that gay and lesbian couples are entitled to the same privileges as heterosexual married couples.
1995	The Hate Crimes Sentencing Enhancement Act goes into effect as part of the Violent Crime Control and Law Enforcement Act of 1994. The law allows a judge to impose harsher sentences if there is evidence showing that a victim was selected because of the "actual or perceived race, color, religion, national origin, ethnicity, gender, disability, or sexual orientation of any person."

Year	Action—Movements / Achievements
1993	President Bill Clinton signs a military policy directive that prohibits openly gay and lesbian Americans from serving in the military, but also prohibits the harassment of "closeted" homosexuals. The policy is known as "Don't Ask, Don't Tell."
1983	Lambda Legal wins People v. West 12 Tenants Corp., the first HIV/AIDS discrimination lawsuit. Neighbors attempted to evict Dr. Joseph Sonnabend from the building because he was treating HIV-positive patients.
1982	Wisconsin becomes the first state to outlaw discrimination based on sexual orientation.
1979	The first National March on Washington for Lesbian and Gay Rights takes place. It draws an estimated 75,000 to 125,000 individuals marching for LGBTQ rights.
1978	Harvey Milk is inaugurated as San Francisco city supervisor, and is the first openly gay man to be elected to a political office in California. In November, Milk and Mayor George Moscone are murdered by Dan White, who had recently resigned from his San Francisco board position and wanted Moscone to reappoint him. White later serves just over five years in prison for voluntary manslaughter.
1978	Inspired by Milk to develop a symbol of pride and hope for the LGBTQ community, Gilbert Baker designs and stitches together the first rainbow flag.
1976	After undergoing gender reassignment surgery in 1975, ophthalmologist and professional tennis player Renee Richards is banned from competing in the women's US Open because of a "women-born-women" rule. Richards challenges the decision and in 1977 and the New York Supreme Court rules in her favor. Richards competes in the 1977 US Open but is defeated in the first round by Virginia Wade.
1975	The first federal gay rights bill is introduced to address discrimination based on sexual orientation. The bill later goes to the Judiciary Committee but is never brought for consideration.
1975	Technical Sergeant Leonard P. Matlovich reveals his sexual orientation to his commanding officer and is forcibly discharged from the Air Force six months later. Matlovich is a Vietnam War veteran and was awarded both the Purple Heart and the Bronze Star. In 1980, the Court of Appeals rules that the dismissal was improper. Matlovich is awarded his back pay and a retroactive promotion.

Year	Action—Movements / Achievements
1974	Kathy Kozachenko becomes the first openly LGBTQ American elected to any public office when she wins a seat on the Ann Arbor, Michigan City Council.
1974	Elaine Noble is the first openly gay candidate elected to a state office when she is elected to the Massachusetts State legislature.
1973	Lambda Legal becomes the first legal organization established to fight for the equal rights of gays and lesbians. Lambda also becomes their own first client after being denied non-profit status; the New York Supreme Court eventually rules that Lambda Legal can exist as a non-profit.
1973	Maryland becomes the first state to statutorily ban same-sex marriage.
1973	First meeting of "Parents and Friends of Gays," which goes national as Parents, Families and Friends of Lesbians and Gays (PFLAG) in 1982.
1973	By a vote of 5,854 to 3,810, the American Psychiatric Association removes homosexuality from its list of mental disorders in the DSM-II Diagnostic and Statistical Manual of Mental Disorders.
1970	Community members in New York City march through the local streets to recognize the one-year anniversary of the Stonewall riots. This event is named Christopher Street Liberation Day and is now considered the first gay pride parade.
1969	Police raid the Stonewall Inn in New York City. Protests and demonstrations begin, and it later becomes known as the impetus for the gay civil rights movement in the United States.
1969	The "Los Angeles Advocate," founded in 1967, is renamed "The Advocate." It is considered the oldest continuing LGBTQ publication that began as a newsletter published by the activist group Personal Rights in Defense and Education (PRIDE) in 1966.
1961	Illinois becomes the first state to decriminalize homosexuality by repealing their sodomy laws.
1961	The first US-televised documentary about homosexuality airs on a local station in California.
1955	The first known lesbian rights organization in the United States forms in San Francisco. Daughters of Bilitis (DOB). They host private social functions, fearing police raids, threats of violence and discrimination in bars and clubs.
1953	President Dwight D. Eisenhower signs an executive order that bans homosexuals from working for the federal government, saying they are a security risk.

Year	Action—Movements / Achievements
1952	The American Psychiatric Association's diagnostic manual lists homosexuality as a sociopathic personality disturbance.
1950	The Mattachine Society is formed by activist Harry Hay and is one of the first sustained gay rights groups in the United States. The Society focuses on social acceptance and other support for homosexuals.
1924	The Society for Human Rights is founded by Henry Gerber in Chicago. It is the first documented gay rights organization

Source: CNN- 2021- LGBTQ Rights Milestones Fast Facts (CNN Editorial Research)- Updated 2:49 PM EST, Tue February 2, 2021 https://www.cnn.com/2015/06/19/us/lgbt-rights-milestones-fast-facts/index.html

Chapter 11

Hidden Inequalities in Education: Case of In-Service English Language Teachers Training in Ukraine

Alla Krasulia[1], Yevhen Plotnikov[2], Liudmyla Zagoruiko[3]
[1]Department of Germanic Philology, Sumy State University, Ukraine
[2]Department of Applied Linguistics, Nizhyn Gogol State University, Ukraine
[3]Department of Foreign Languages, Pavlo Tychyna Uman State Pedagogical University, Ukraine

ABSTRACT
The chapter aims at investigating the issue of access to the continuing professional development (CPD) of in-service English as foreign language (EFL) teachers in Ukraine in the context of current educational reform. The study reveals the barriers and challenges those teachers with different social and professional background face while trying to get involved into the high quality CPD activities. In addition, the chapter examines the existing ways of in-service EFL teachers training and the mismatch between the provided activities and those needed by teachers. The quantitative approach was used to generate research data and evaluate empirical evidence by means of descriptive statistics. The study utilized holistic description, analysis, and surveys to address teachers' experience in CPD activities, as well as their beliefs on forms and methods of such training, and provided an in-depth exploration of teachers' professional needs and the role of certain factors influencing the access of different teacher groups to quality professional development. Poor technological facilities, unclear goals and practices, as well as financial factors, such as funding to support travel or cost of attendance, were identified as key barriers to CPD access.

1. Introduction

Ensuring education for all has been one of the most challenging issues over the last centuries. The right to education is a basic human right (UNESCO, 2009). On the one hand, equal access to education contributes to gaining fundamental living skills by all people regardless their gender, age, and social status. On the other hand, equal educational opportunities have broader meaning as they reduce poverty, decrease social stratification, and promote cultural identity.

On the contrary of equality, the phenomenon of educational inequality could be followed. In education, inequality of opportunity and inequality of condition are distinguished (Breen & Jonsson 2005). The first type of inequality is seen in obtaining educational services without taking into consideration gender, age, class, and economic background. Inequality of condition grounds on the aspects related to financial rewards, working, and living conditions, and includes rights of the contributors and consumers of educational services. In the frameworks of our research, teachers take both places.

Due to a range of social factors and different roles that today's teacher plays in society (see Klasnic, 2019), teaching profession is considered one of the most demanding in the labor market. A constant upgrade of professional skills and competencies is not only an internal urge and necessity, but also a part of external requirements imposed on teachers by educational authorities. In our research, we attempted to highlight EFL teachers' continuing professional development (CPD) in Ukraine and show whether teachers have equal access to CPD in fact or this is simply a right proclaimed in official documents. From one perspective, education is just and objective; from the other perspective, it is unjust and biased (Somel, 2019). Attempts to avoid the dilemma are futile, since contradictions resurface in every instance and result in theoretical inconsistencies within the theories (Meyer, 1986).

2. Background of the Study and Literature Review

The idea of inequality in education is usually researched within the framework of two areas: educational attainment and levels of achievement (Blanden & MacMillan, 2016; Montt & Kelly, 2015). For a long period, such inequalities were mainly associated with primary or secondary education (see Duru-Bellat, 2015), but nowadays inequality in education move far beyond schools. For example, in EU countries, educational inequality often relates to tertiary and even postgraduate education (Bubbico & Freytag, 2018; Hall,

2012; Rodriguez-Pose and Tselios, 2009). At the same time, the level and scope of such inequalities can be rather different. Traditionally, distinction is made between direct and indirect inequality. Direct inequality refers to a complete lack of educational opportunities, while indirect

> "occurs when unfair treatment of a group (or groups) is not the explicit purpose of a policy or action, but still results in ... [educational] inequality" (Pachamama Alliance, n.d.)

Indirect inequality is also determined by the inability to access high-quality education or by the lack of conformity of forms and types of services to consumer requirements. One of the examples of indirect inequality is the barriers that teachers face while doing their professional development activities.

In the present study, the (in)equality in education is considered in the frameworks of lifelong or continuing learning, the concept which appeared almost a century ago in two influential works (Yeaxlee, 1926; Lindeman, 2013 [1929]), that formed an intellectual basis for a comprehensive view of education as a continuous process or as E. Lindeman put it, "education is life." Alongside this, the historical aspects of (in)equality in education were analyzed by a wide range of scholars, for example (Apple, 2011; Blanden & MacMillan, 2016; Jacob, 1981; Pfeffer, 2008; Schindler & Lörz, 2012) to mention just a few. An extensive amount of research focuses on inequality at school for both teachers and students (Ansalone, 2009; Batruch et al., 2018; Breen et al., 2010; Guclu & Onder, 2014; Hargreaves, 2014; Mehan, 1992; Owens, 2018). Theoretical approaches to educational inequality are also highlighted in different studies (Herrera, 2007; Howe, 1994; Meuret, 2011). Besides, access to education and equal educational opportunities is also the subject of research by international organizations (Bubbico & Freytag, 2018; Chzhen et al., 2018; Hippe et al., 2016; OECD, 2017; UNESCO, 2017; World Bank, 2010).

Analysis of the relevant research reveals the following obstacles to equal access to education: lack of financial support (Mongkolhutthi, 2019; Rubenson & Desjardins, 2009); inability to choose forms of the training (Dailey-Hebert et al., 2014; Levine & Sun, 2002; McConnell et al. 2013); insufficient knowledge of using teaching tools, for instance ICT technologies (Crimmins et al., 2017; Feist, 2003; Kopcha, 2012; Lee, 2000; Macià & Garcia, 2016); lack of motivation to study (Abramova, 2013; Badri et al., 2016; Broad, 2015; Kennedy, 2011; Bhandari 2020; Bhandari and Shvindina 2019), etc. Broadly speaking, there is no single cause of inequality. The economic fundamentals of educational inequality are relevant for all the countries of the world, but in many cases, institutional barriers can also play a significant role.

The object of our study is EFL teachers' professional development. There have been several attempts to define what "teachers' professional development" means. Some researchers define professional development as a process that goes beyond a purely informational framework and is aimed at helping teachers adapt their activities following current changes and requirements in such a way as to increase the effectiveness of student learning (Burke et al., 1990); aims to develop professional competencies of teachers, forming beliefs, views, and understanding of their professional functions and inclusion in active professional work (Joyce & Showers, 2002); occurs in an environment that promotes increased initiative, creativity, and reflection of specialists (Bredeson, 2002); it is the result of a systematic analysis of the teacher's own activities (Villegas-Reimers, 2003).

Accordingly, the authors consider the professional development of a teacher as a multidimensional process aimed at qualitative changes in the competence of a specialist, due to the general patterns of intellectual and cultural development of a person and occur in a favorable social environment. It begins at the stage of a person's training in the professional program of a special educational institution (professional education, initial training), continues at the workplace (induction, in-service training) and is supported by a wide range of formal, informal, and additional (formal, non-formal, and informal) education throughout professional life.

In addition, researchers of professional development of teachers, based on the analysis of various models of world practice, distinguish managerial and democratic professionalism (Day & Sachs, 2008). The first is more dependent on the needs of the official institution (school, college, or university), is planned by managers and supported by targeted measures: training for innovation, professional development to improve a certain area of activity, the creation of an initiative group to spread positive experience, training as part of the planned certification of teachers. The second is located in an alternative field, depends on the initiatives of the participants, their interests, needs, tastes, and is open to consumers (students, teachers, and lecturers). Educational research claims that these approaches exist in parallel, and sometimes conflict with each other: a top-down approach from the administration to the teacher: more regulations, more rewards for compliance, etc.; and the other, bottom-up approach, from the initiative of the employee working to improve the educational process, in which the administration should be interested (Fullan, 2012).

The relationship between these types is a defining prerequisite for centralized and decentralized approaches to organizing teacher training and development.

In our opinion, monitoring of inequality in education is crucial to understanding how, and how much, education contributes to more equitable societies. This necessarily calls for a series of choices to be made. First, educational inequality can be examined using indicators that capture different aspects of education ranging from resources to access, participation, and attainment. Second, different inequality measures can be used to summarize the degree of dispersion for a given educational indicator. Third, while it may be interesting to view the distribution of an education indicator throughout the population, policy-makers need to know how its value varies by individual characteristics if they are to address issues of inequality. Fourth, different data sources are available that measure different aspects of the education process and provide information on background characteristics (Antoninis et al. 2016). We try to use different educational indicators to define the barriers which EFL teachers face pre- or during participation in CPD activities. Thus, according to (Razumkov Center., 2002) the level of accessibility of education in Ukraine is determined by two indicators: physical accessibility of educational institutions and financial accessibility of educational services for the majority of citizens.

There are a number of reasons for the violation of access to education at different levels in Ukraine. The research by Lukina (2019) outlined the range of risks that threaten the success of public policies to ensure equal access to education. These are the following:

1. risks caused by political instability (military threat, military conflict in the East of the country, the appearance of internally displaced persons, the exact number of which remains uncertain;
2. demographic risks (demographic crisis, growth of internal migration of the population, reduction of the share of the rural population, problems of small schools, etc.);
3. risks of inequality in the quality of educational services based on the territory and population of the locality;
4. risks caused by insufficient competence of managerial personnel in the field of education;
5. cultural risks that restrict access to education (religious, ethnic, gender, moral, ethical, etc.);

6. risks of a social and economic nature (decrease in the financial viability of the population, low level of material and technical support for the educational process and scientific institutions, low wages in the field of education, the outflow of scientists and teachers abroad, etc.);
7. risks of a psychological (psychological and mental) nature, manifested in resistance to changes in the field of education, unwillingness to update the ways of organizing the educational process, change the conceptual approaches to the formation of educational systems, determining educational needs, etc.

In addition, the academician of the National Academy of Sciences of Ukraine M. Zhulynskyi stated that one of the factors that interferes in the implementation of competitive advantages of Ukrainian education is low social status of teachers and lecturers. At the same time, the insufficient level of professional training influences the educators' work in the conditions of the information society. According to the researcher, one of the priority measures that should be taken to solve the above problem is to radically change the methodology of training specialists in the field of education, who should have desire and need for constant self-studying (Razumkov Center, 2002).

In our opinion, decentralization in teacher education is directly related to the readiness of teachers to make their own choices of professional development form. The researchers asked the teachers to evaluate this readiness. Seventy-five people representing 19 regions of Ukraine and the city of Kyiv participated in the survey. 36% of the respondents rated their readiness highly (grades 8–10), 46% of the respondents consider themselves more ready than not (grades 5–7), and 18% of the respondents (grades 2–4) said that teachers are not ready to choose their own forms of professional development. The results of the survey also indicate that a large part of those who work in teacher education demonstrate their readiness for qualitative changes in professional development, namely, to overcome isolation, formalism in the design of training content, and to introduce interactive forms of work. The key to the effectiveness of this activity can be both official institutions of postgraduate pedagogical education, as well as non-governmental organizations, professional associations, and regional centers of pedagogical excellence (Tezikova & Plotnikov, 2019).

"Equal Access to Quality Education" Project

Ukraine collaborates with an array of international organizations on different educational issues. One of the projects analyzed in our study in terms of equal access of EFL teachers to CPD is the project "Equal access to Quality Education." It was conducted in 2006–2010 in Ukraine and financed by the *International Bank for Reconstruction and Development* (World Bank, 2011). The objective of the Project was to provide equal access to quality education and to improve the efficiency of the education system to prepare Ukrainian graduates for the knowledge society. Our attention was drawn by one of the Project's components—professional development of educators. This component was aimed at raising the awareness, professional knowledge, and competences of teachers and school directors to implement the reforms underway in teaching and learning in secondary school education. The above mentioned component had four sub-components. Henceforth, we would like to familiarize with the Project's results assessment considering the above-mentioned sub-components.

- Sub-component 1.1.—Training of trainers. As the Project assessment showed the target was reached: 110 (100%) master-trainers did the training course, so they could pass on knowledge to educators and school staff by giving lectures on implementation of innovative teaching methods, use of information and communication technologies, and curriculum design.
- Sub-component 1.2.—Improvement of teachers' qualifications. A total of 1637 educators were trained, although this is fewer than the target number (4000 teachers, 2200 school principals). According to the Project managers such poor achievement is caused by several reasons: only 10 of 28 regional in-service teacher training centers (RITTC) took part in the training; some of them stopped their participation at the beginning of 2010 as the Directorate did not pay for their services; lack of institutional support from the Ministry of Education and Science (MES) regarding training of trainers and improvement of teachers' qualification; problems with distant learning due to inadequate skills and lack of commitment among the teachers and educators to work with online, web-based learning environment. As a result, no changes were introduced to the certification system of learning competence following the piloting of the new educator testing schemes as had been anticipated by the project. However, it was an important step for the modification of the current certification system of competences.
- Sub-component 1.3.—Leadership training for school directors. An appropriate tutorial was developed and prepared for publication, but

this was not issued under the decision of the MES. Six hundred and thirty nine school directors had access to new management methods. However, it is difficult to evaluate the training efficiency as no changes were introduced to the certification system of learning competence by the pilot application of new attesting schemes for educators.

- Subcomponent 1.4.—Improvement of regional institutes of in-service teachers. At the mid-term review of the Project (2008), it was decided to cancel the study visits abroad.

According to the abovementioned results, the Project was not successful in some areas; it has largely failed to meet its planned goals. Teacher training was conducted properly, and Ukraine has made significant progress in implementing external assessment of knowledge as a tool for managing the quality of education.

TALIS Survey

The second project we would like to draw your attention to is the all-Ukrainian monitoring survey on teaching and learning among principals and teachers of secondary schools. It was carried out by the scientists of the Ukrainian Educational Research Association using the tools of Teaching and Learning International Survey (TALIS) 2013, conducted by the Organization for Economic Cooperation and Development (OECD, 2014). The survey took place between February and August 2017 in the framework of the project "Teacher and Education Reform: Quality Assessment in an International Context," which is implemented by the all-Ukrainian Foundation "Step by Step" with the support of the Ministry of Education and Science of Ukraine (Shchudlo *et al.* 2017). Three-thousand six hundred teachers from 201 secondary schools from all regions of Ukraine participated in the survey.

We are interested in the chapter on the experience of professional development and support of teachers. The researchers considered "professional development" as a series of activities aimed at developing individual skills, knowledge, qualifications, and other characteristics of teachers with the ultimate goal of improving their teaching practices.

The survey stated that this development can be achieved in many ways, from the most formal (such as courses or seminars) to more informal approaches (such as collaborating with other teachers or participating in extracurricular activities).

The main focus of TALIS was on the following types of professional development activities:

- Courses/workshops (e.g., on teaching methods and/or other educational challenges).
- Pedagogical conferences and seminars (where teachers and/or researchers present their research findings and discuss pedagogical issues).
- Observation visits to other schools.
- Observation visits to enterprises, public and non-governmental organizations.
- In-service training courses at enterprises, public institutions, and non-governmental organizations.
- Qualification programs (e.g., courses aimed at obtaining a higher qualification level).
- Participation in professional associations established for the purpose of professional development of teachers.
- Individual or collaborative research project on professionally oriented topics.
- Mentoring or/and mutual visits by colleagues as part of official practice at school.

The survey results showed some barriers to the teachers' professional development (percentage of the teachers who "agree" or "fully agree" with barriers to professional development) (Table 11.1) (Shchudlo *et al.* 2017).

Table 11.1. Barriers to professional development of teachers.

Barriers	**The average indicator for Ukraine (%)**
I do not have the appropriate conditions (e.g., lack of qualifications, experience, position)	4,9
Professional development is too expensive/financially unavailable	23,8
There is no employer's support	16,5
Professional development programs are implemented during my working time	54,0
I do not have time because of family responsibilities	11,5
There are no appropriate professional development programs	18,8
There are no rewards for participating in such events	50,2

Therefore, the results of the survey conducted by the Ukrainian Educational Research Association based on the TALIS methodology show that Ukrainian teachers cannot improve their professional skills due to the fact that training events are arranged during their working time. The second barrier is lack of rewards or incentives for professional development.

"New Ukrainian School" Reform

Currently, the reformation processes in Ukrainian education directly depend on the readiness of teaching staff to grasp the declared ideas, principles of activity, and promising results. In pedagogical education, these processes often require a broad discussion of the most rational ways of implementation and exchange of effective experience. The dialogue between the Ministry of Education and Science, and the ones who train teachers, who work in schools, and who is responsible for the further professional development of teachers is extremely relevant and timely. Thus, professional development and educational reforms are interconnected (Mukan *et al.*, 2019).

In Ukraine, the education sector is undergoing ambitious and long-overdue reforms that have the potential to radically transform this sector. New law *"On Education"* was adopted in 2017 for the entire education sector. This law, as well as the fiscal decentralization reform (2014), are a major change in the direction of devolution of power from the central government to the local and the expansion of the decision-making powers of local authorities and local institutions (such as schools and universities) (Institute of Educational Analytics, 2018).

Business entities in Ukraine report that it is difficult to find employees with relevant hard and soft skills. In particular, a 2014 World Bank survey of enterprises found that 40% of them operating in four key sectors (agriculture, food industry, information technology, and renewable energy) reported significant gaps between the skills currently available to their employees and the skills needed to achieve business goals (Del Carpio *et al.*, 2017). The reason for this is probably the outdated curriculum, textbooks and equipment, as well as lack of motivation and appropriate training of teachers.

Resolution of the Cabinet of Ministers of Ukraine No. 800 of August 21, 2019 approved the Procedure for improving the qualifications of teaching and research staff. According to point 7 of the Procedure, pedagogical and research staff independently choose specific forms, types, directions, and subjects of providing educational services for professional development. The subject of professional development may be an institution (its structural division), a scientific institution, or another legal entity or individual, including an individual entrepreneur who carries out educational activities in the field of professional development of teachers and/or researchers. Funding sources for teachers and research staff professional development include state and local government budgets, money of individuals and/or legal entities, own income of educational institutions and/or their founders, and other sources not prohibited by law. Self-financing of training could be conducted by teachers and research staff of educational institutions who work in these institutions at the main place of work and takes professional training outside the training plan of the institution (Postanova Kabinetu Ministriv Ukrainy № 800, 2019).

To improve the quality assurance system in education, the Ministry of Education and Science plans to introduce voluntary certification of teachers in 2019–2020 (part 3 of article 41 of the Law of Ukraine "On Education") (Zakon Ukrainy "Pro Osvitu", 2017). The Ministry is currently developing a model for teacher certification in the form of an external assessment of teachers' professional competencies through independent testing, self-assessment, and learning from practical experience. Voluntary certification is implemented as a tool for material incentives (according to article 4 and article 61 of the Law of Ukraine "On Education") and professional growth. Here are some suggestions (Sondergaard, 2018) for optimizing the implementation of teacher certification.

- *Focus on working with current teachers.* At present, when the MES is dealing with certification of teachers before as well as during their work, licensing of current teachers based on the needs of the "New Ukrainian School" concept may be more important, since the reforms will reduce the need for new teachers and there will be relatively few new teachers.

- *Introduce certification as part of a systematic strategy to improve indicators on teachers' activity.* Certification or re-certification of teachers who are already employed is carried out in educational systems when there is a need to radically improve the quality and/or achieve a new status. Ideally, re-certification should not exist by itself as a simple professional development program. It should rather be implemented as part of a systematic strategy that takes into account the entire continuum of teacher training and development, including high-quality initial teacher training, introduction to the course of duties, structured and flexible continuous development of professional skills (CDPS), and motivational career pathway. One of the key components of the modern teacher certification system is also the accreditation of educational institutions. Finally, the system as a whole should be based on transparent, recognized and agreed standards and competencies for all teachers.

- *Voluntary certification is appropriate, but it should be carefully monitored.* The approach to certification, which is provided voluntarily within the framework of the "New Ukrainian School" concept, not only has certain advantages (its participants are only really interested in it and motivated teachers), but can also create some challenges. First, there is an issue of harmonizing the status of certified teachers and non-certified teachers. Second, there is the question of how to build up the dynamics of certification, so that more and more teachers get certified. The fact is that the system requires a certain critical number of certified teachers.
- *To think carefully about the timing.* Is it realistic to meet the schedule required for full certification of all teachers? This is where a more comprehensive approach to the implementation of the reform of the "New Ukrainian School" can be useful, according to which the re-certification is consistent in time with the retirement of teachers and the formation of hub schools. When merging schools into centers based on reference schools and through the retirement of teachers, the reform will take advantage of the fact that the drop in the number of teachers, combined with the reduction in the number of teacher positions and the possibility for school leaders to refuse to sign a contract, to encourage existing teachers to re-register, which will give them a better chance to return to work or stay in the system (Sondergaard, 2018).

The analysis of existing professional development programs for EFL teachers of modern Ukrainian schools has shown their conventionalism, regulation, and control by the administrations of educational institutions and the lack of readiness of a significant number of teachers to independently choose ways for self-improvement. The first forms are associated with the activities of state institutions of professional development, the second—supported by professional organizations, associations, and voluntary associations of professionals who offer alternative models of teacher training and seek their official recognition.

3. Method

3.1. Research Questions

As shown above, the process of CPD does not seem to be a straightforward issue for many in-service teachers. In order to analyse the existing often hidden inequalities, we conducted a research addressing four main barriers that teachers face when in need or wishing to update their professional competence. For the purpose of the study, the notion of barriers was transformed into the following research questions:

1. *The financial barrier*: EFL teachers are reluctant to participate in CPD activities causing [additional] financial expenses. *RQ 1*. Would removing financial barriers ensure more equal access to EFL teachers' professional development?
2. *The spatial barrier*: EFL teachers are reluctant to participate in those CPD activities, which are conducted far from their place of living. *RQ 2*. Would giving opportunity to choose different modes of CPD activities ensure more equal access to professional development?
3. *The content barrier*: EFL teachers are reluctant to participate in pre-defined CPD activities. *RQ 3*. Would giving an opportunity to choose the content of CPD activities ensure more equal access to professional development?
4. *The psychological barrier*: Less trained EFL teachers are reluctant to participate in CPD activities due to the fear of showing occupational mismatch. *RQ 4*. Would using a differentiated approach based on participants' professional competence level while conducting CPD facilitate more equal access to CPD activities?

3.2. Instruments

While doing the research, we followed the strategies suggested by Noble and Smith (2015), namely, accounting for personal biases which may have influenced findings; meticulous record keeping; respondent validation; data triangulation, etc. Thus, we used two instruments to answer the research questions.

Questionnaire

In order to address the research questions, the researchers distributed a questionnaire. To check the reliability of the close-ended questions, the questionnaire was piloted to a group of 14 teachers. The results of Cronbach Alpha calculations were lower than 7 for both Part 2 and 3 of the questionnaire, which is not acceptable (Pallant, 2016). The questionnaire was reviewed by four experts in the field of foreign language teaching methodology and CPD at the authors' home universities. As a result, a number of ambiguous items were either changed or excluded. The revised version of the questionnaire was administered to another group of teachers (12 in total). Cronbach Alpha at that stage turned out to be 8 (Part 2) and 71 (Part 3). Thus, the final questionnaire (Appendix A) consisted of three parts with 19 questions in total. The first part was used to collect the participants' background information, namely gender, age, place of work (both type of settlement and type of institution), and teaching experience. The second, reflective part focused on the participants' reflective practices and critical appraisal of their own experience of CPD. The third part encompassed teachers' metacognitive reflection, professional views, and attitudes toward the nature of CPD, CPD providers, and frequency (Table 11.2).

Table 11.2. Distribution of items in the questionnaire.

Part of the questionnaire	Corresponding item
Background information	1, 2, 3, 4, 5, 6, 7
Reflective information	8, 9, 10, 11, 12, 15, 16
Views and attitudes	13, 14, 17, 18, 19

The questionnaire contained items of different types: 10% of the questionnaire items were a five-point Likert scale varying from (1) "not at all" to (5) "completely"; another 29% of the items were open-ended questions; and the rest 61% of the questionnaire comprised close-ended multiple choice/response items with an average of 4 alternatives. All the questionnaire items were in English.

Oral Interview

To address the study research questions, the researchers also created a semi-structured interview. The purpose of the interview was to obtain more specific information about the existing inequalities in access to quality CPD and the reasons behind teachers' reluctance to participate in CPD activities. A number of informal conversations with educators, as well as the conduction and observation of teacher training sessions, had taken place prior to the research phase and are reflected in pre-research notes. Alongside the questionnaire, the interview questions were also subject to piloting. First, two experts at the authors' home universities checked the questions to make sure that they explored the necessary areas of research. Second, five EFL teachers from rural and urban schools reviewed the questions for clarity and comprehensibility. Having made necessary changes, we developed an interview protocol that consisted of three sets of questions split into two categories (Appendix B). Both the questionnaire and the interview development were informed by the findings from an initial literature review.

3.3. Participants

Purposive sampling for this study was developed by keeping a representative mix of:

- teachers from rural and urban schools;
- teachers from schools of different types;
- teachers from different regions; and
- teachers with different professional background and experience.

The research area covered 16 out of Ukraine's 27 primary administrative units (59%). The study dealt with the following demographics of the sample population: of 66 teachers who completed the questionnaire, 64 (97%) were female, 2 (3%) were male. Their teaching experience ranged from 1 to more than 40 years. Table 11.3 indicates the profile of the participants.

Table 11.3. Participating teachers' demographic information.

Variable	Characteristic	Qnty	Percentage (%)
Gender	Male	2	3
	Female	64	97
Place of work	Village	19	28.8
	Small city	20	30.3
	Big city	27	40.9
Teaching experience	Less than 5 years	11	16.7
	6–10 years	14	21.2
	11–15 years	14	21.2
	16–20 years	12	18.2
	21–25 years	5	7.6
	26–30 years	8	12.1
	31–35 years	1	1.5
	36–40 years	1	1.5

Since the secondary education landscape is quite wide, the sampling included teachers from different types of educational establishments ranging from traditional state-owned institutions to private entities to self-employed tutors. Most of the participants (72%) represent primary,

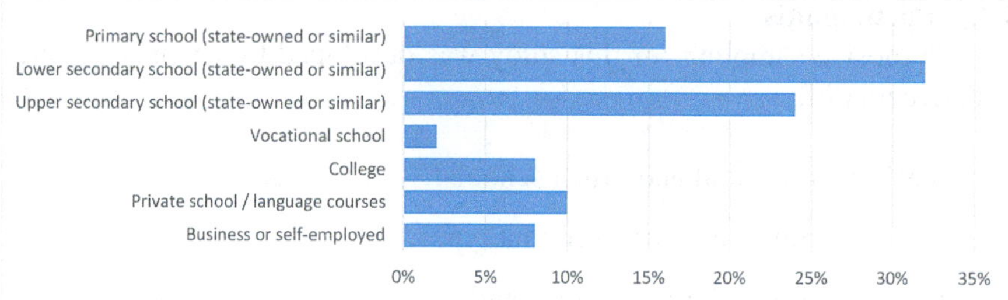

Figure 11.1. Distribution of participants according to their employment institution.

lower, and upper secondary schools because due to the peculiarities of the national educational system those institutions are usually bound together as one school and take the great majority of the levels 1–3 education sector according to the European Qualification Framework (Ministry of Education and Science of Ukraine, 2019). The distribution of the participants according to the type of educational institutions they work for is presented in Figure 11.1.

3.4. Data Collection and Analysis

Procedure

A mixed methods approach based on both quantitative and qualitative procedures was adopted for data collection and analysis. The administration of the research instruments took place in November/December 2019 (questionnaire) and January 2020 (interviews). The questionnaire was distributed online by means of the Microsoft Forms website. Sixty-eight (68) responses were obtained, which were then reduced to 66 copies after discarding the surveys that were incomplete. The average time for filling out the questionnaire about 16 minutes. The demographic information presented in the previous subsection was obtained at this stage. The purpose of the study was partially disclosed to the participants (i.e., teachers were informed that the focus of the project was CPD in education) to minimize any presumable bias and influence on the teachers' responses. Full anonymity of the responses was guaranteed and provided. After administrating the questionnaire, the descriptive statistics was calculated for two main questionnaire parts (Tables 11.4 and 11.5).

After administering the questionnaire, one of the researchers conducted semi-structured interviews lasting between 25 and 40 minutes with 23% (15 teachers) of the sample population. Informed consent was obtained from the interview participants and all of them were informed of their rights and

Table 11.4. Descriptive statistics of teachers' reflective practices on experience of CPD.

	Min	Max	Mean	Median	Mode	SD
Satisfaction with CPD activities	2	5	4.24	4	5	.8
Influence of CPD activities	1	4	2.96	3	4	.93
Participation in professional organisations	1	3	2.5	3	3	.63
Access to CPD	1	3	2.43	2.5	3	.61

Table 11.5. Descriptive statistics of teachers' professional views and attitudes.

	Min	Max	Mean	Median	Mode	SD
Frequency of CPD	1	4	1.92	2	1	1.11
Ability to choose forms and topics of CPD	1	4	3.56	4	4	.95
Readiness to build personal CPD schemes	1	5	3.33	3	3	1.15

assured of their anonymity. The interviews were conducted in a face-to-face mode, in English and/or Ukrainian depending on respondents' choice and were audio-recorded for transcription, translation, and coding. The audio files were content analysed in line with the guidelines by Cohen, Manion, and Morrison (2007) for analysing qualitative data.

After doing the initial transcription and translation (where necessary), the results were coded and grouped taking into account the similarity of the emerging topics. The transcripts and the emerged groups were subsequently submitted to other two researchers, who double-checked the analysis. Thus, the data were analysed cyclically throughout the whole period of fieldwork research, with each data collection cycle influenced by the analysis of the previous data. Summative data analysis was done in relation to each participating teacher and then across two or three cases.

Finally, the teachers were invited to comment on the findings, had an opportunity to correct errors and misleading interpretations, and expressed their opinion whether the analysis and conclusions adequately reflected the phenomena being investigated. Validation of this kind is commonly cited in qualitative research literature as a tool to enhance both the ethicality and trustworthiness of the findings (e.g., Anderson, 2017; Noble & Smith, 2015).

Limitations of the Study

The limitations of this study include the small sample size of teachers surveyed (66) and interviewed (15). Teachers volunteered for the survey in a current study so may have had more interest in the topic than teachers who did not participate. At the same time, the participating group may be slightly more active in their professional activities. This fact may influence the total scores.

Because of the limitations and the qualitative nature of the study, the authors do not generalize them to wider groups and circumstances. The quantitative part of the study covered a larger number of participants and thus may be of more interest to researchers. The insights provided into the practices of participating teachers and into the factors behind the obvious and hidden obstacles to professional development may be of use not only to educators, but also to other professionals in wider contexts worldwide.

4. Results

The data below is the interview excerpts and some generalized findings from the questionnaire. It is categorized according to the research questions and corresponding barriers.

4.1. The Financial Barrier

Twelve (80%) out of 15 interviewed teachers expressed the necessity in some additional funding for their professional development due to the lack of sufficient salary level.

[Int. 2] "Many useful resources are provided on a paid basis. But nobody gives any funds for this." "They reimbursed travel and lodging expenses [*for state designated CPD courses*]. But I have to pay for any other CPD activities, which are also necessary."

[Int. 3] [*Have you experienced any problems while having CPD?*] "Financial. I have to pay for everything myself. Even now [*at activities provided by a university*] I pay for my courses." "Let the state pay for our CPD, because, first of all, it is a vital necessity for it and the school."

[Int. 4] "You are asking what the problem is? Lack of funding for the courses. Not just travelling and accommodation expenses. I am ready to pay for some webinars myself but not all of them."

[Int. 10] "Money is the main barrier. The authorities should pay for our CPD."

[Int. 13] "It would be desirable to have a normal high-speed Internet at school, so that we can do something online."

4.2. The Spatial Barrier

[Int. 1] "The restriction is that you need to take courses with tight deadlines. Sometimes they say that we only have courses now and that's it. Hypothetically speaking, I am given a choice of time and place, but in fact I do not have it."

[Int. 2] "An obstacle is that due to family commitments, there is not always time to take online courses or trainings. And they can't be postponed. In my opinion, it is better to take CPD at home than to go somewhere. I mean, at local institutions that provide this opportunity."

[Int. 3] "It would be nice not to go anywhere, but to have our CPD at home."

[Int. 13] "The biggest problem was traveling to the place where the courses were held. I live 250 km from [central city of the region] and had to leave at 4 am, and even book a seat on the bus. The road there is very bad and it is a long drive. Therefore, it is not convenient for me to go there for courses. I wish it were closer."

The questionnaire also showed that 25% of the participating teachers are eager to have CPD sessions not very often (Figure 11.2). More interesting is that 78.8% out of this quantity are teachers from villages and small cities (Figure 11.3).

Another data to reveal the hidden inequality are teachers' responses to Question 14 of the questionnaire (Can you say that CPD activities are easily accessible for you?). Fifty percent of the respondents claimed that CPD activities are not always easily acceptable (Figure 11.4), notwithstanding the fact that in theory all the teachers are on equal grounds. Such results may consider both financial and spatial barriers.

4.3. The Content Barrier

[Int. 1] [*a person, who conducted formal CPD courses*] "I don't get any useful information from her. Really. Actually, you can

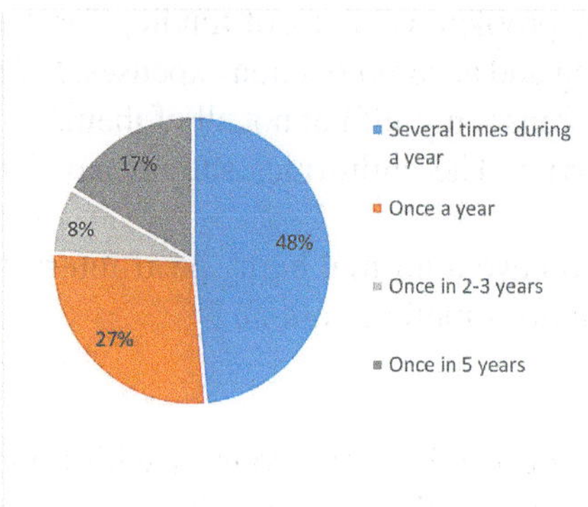

Figure 11.2. Teachers' views on frequency of CPD

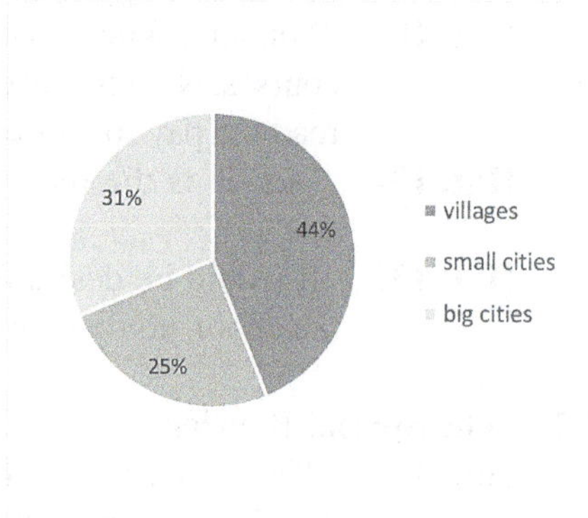

Figure 11.3. Distribution of teachers reluctant to have frequent CPD sessions.

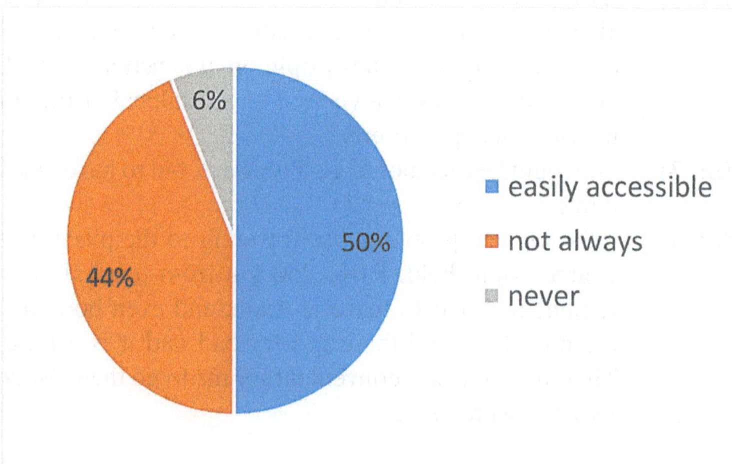

Figure 11.4. Teachers' views on access to CPD.

study everything yourself. Read and search for it. Practice. All we need is practice. Show me how to use it... She knows that we are working with other textbooks, but she used a textbook that she was comfortable with and that only one school in our city actually uses."

[Int. 2] "I was more interested in practical training, but I didn't get it. The courses were too theoretical and boring."

[Int. 8] "We must learn some practical stuff and gain experience. Not the theory they teach us. If they introduce new textbooks or manuals to us, we should apply them immediately, not just talk about them. We have to face up to the reality of school life and it is different from what we are told about during the courses. We were given only the theory, and when I got home, I went to YouTube and watched how other teachers do it practically. Let them show me the real teachers and class that work well and I'll say "Cool! I'll do this!" "Perhaps the material [of CPD courses] was interesting, but the presentation made me reluctant to learn it… No one asked us what exactly we needed for our professional development."

[Int. 11] "It was superfluous that we were shown a lot of innovations that can only be implemented at the tertiary level. But we cannot introduce them in our institutions due to insufficient training and age of students. So it was a sheer disappointment."

[Int. 13] "We studied a lot. There were, for example, lectures on history. But I don't need such lectures. This is useless. I would like to have only specialized subjects."

4.4. The Psychological Barrier

[Int. 5] "We don't really see any commitment from either the students or the school [administration]. Therefore, the desire to do something just disappears. There is no understanding on the part of parents and on the part of the school administration."

"It is highly desirable that these [core] subjects be taught in English. So that we have practice and are not afraid to speak English... We often just don't understand their pronunciation and are afraid to talk to them."

[Int. 6] "We were afraid of being judged by the facilitator. I felt that it was easier for me to sit out all the courses at the last desk and be left alone."

[Int. 9] "We are afraid that we may be laughed at. We have a fear of talking and debating. There is also a fear of speaking English."

5. Discussion

The sources of unequal access to CPD activities are complex. There is an extensive potential for educational policies and practices to allay them. We cannot define solutions for EFL teachers. But still, there are some recommendations that appear from the analysis of EFL teachers' access to CPD activities in Ukraine.

International experts have identified four main directions for professional development of teachers (OECD, 2014). They advise education policy makers in all countries to encourage schools to offer formal primary professional socialization measures for teachers who start their teaching career and encourage teachers to participate in different activities; support teachers' participation in mentoring programs at all levels of their teaching career; ensure that professional development programs are available; and eliminate possible obstacles to the professional development of teachers.

First research question was raised by the existence of the financial barriers to CPD activities. The findings are in agreement with the research of All-Ukrainian Monitoring Survey of Secondary School Teachers and Principals on TALIS methodology which showed a close connection between finances and access to CPD activities, lack of rewards or incentives for EFL teachers' professional development.

Due to the conducted research, financial barriers are based, on the one hand, on the formal and semi-formal mechanisms of payment for educational services that are common in the education system, and on the other—on the existing economic stratification in the country. Payment for educational services involves certain expenses. Two types of expenses could be defined concerning CPD. They are the following:

- direct expenses: monetary contributions in addition to the teacher's salary
- indirect expenses: transportation costs, costs for accommodation

In the frameworks of the "New Ukrainian School" reform, direct expenses are also supported by the fact that EFL teachers should pay for many CDP activities by themselves. Most respondents mentioned the necessity to pay for courses delivered by regional universities. These costs are not refunded by the state or school. The research results completely agree with UNESCO's definition on indirect expenses. The participants of our survey stated negative influence of transportation and accommodation costs on their ability to take part in the desired CPD activity. Additionally, some of the EFL teachers under research while discussing the financial side of the studied issue gave a hint at corruptive payment made to the trainers. For instance, some EFL teachers "thanked" in the mentioned way for the conducted workshops. Other EFL teachers were made to buy textbooks or other manuals authored by the trainers.

When explaining the financial barriers, we take into account the theory of rational choice, that is, educational decisions which are the result of individual considerations regarding costs, preferences, and probabilities of successful acquisition of educational (professional development) programs (Teran, 2007).

We would like to propose some recommendations on how to reduce financial constraints in order to ensure more equal access of EFL teachers to CPD activities. According to the research results, we suggest using two strategies: an alignment strategy and institutional transformation strategy (Vakhshtain, 2005). The first strategy states the need to overcome existing economic inequality by means of the educational system. The second strategy considers inequality in education, on the one hand, as a reflection of existing social stratification, and on the other—as a mechanism for its reproduction. Based on the abovementioned strategies, we propose the following steps on different levels:

- state should pay (refund) for EFL teachers' CPD activities, taking into consideration formal and informal education;
- educational institutions (schools) should inform EFL teachers about scholarships and grants aimed at professional development (including some activities abroad); and
- more amenability should be provided to the trainer for corruptive payments to avoid additional expenses.

Spatial barriers made it possible to put forward a second research question. The research results showed the difference between modes of EFL teachers' participation in CPD activities. This difference is seen in location. EFL teachers who live in rural areas prefer formal training. In contrast, townies and citizens choose an informal mode to develop professionally.

While speaking about spatial barriers, the distance from the place of residence to the educational institution provided training and the transport links are also taken into consideration. Most of the participants choose the regionally closest educational institution where they can be trained. Representatives of rural schools pointed out the opportunity to have face-to-face conversation with other teachers so that they could share their experiences or just have a breath of fresh air and take a break from housework. EFL teachers from towns and cities preferred not to spend their time on visiting institutions but to view on-demand webinars.

Another reason for rural EFL teachers choosing formal training is their lack of language skills and digital competences. In truth, some of the rural EFL teachers failed to open the form and fill it out while doing the Microsoft Forms survey. The interviewed EFL teachers noticed that the study program of their training included ICT classes. But very often the content of such workshops covered information on computer teaching tools or new programs helping in effective teaching. We strongly believe that insufficient knowledge of ICT usage reduces EFL teacher's access to CPD activities (i.e., webinars). Another obstacle for rural school EFL teachers to develop professionally is lack or absence of modern computer technologies and, in most cases, Wi-Fi connection. In our opinion, disability to use ICT properly and inappropriate Internet connection have also influenced the quality of teaching.

Some recommendations could be given to remove spatial barriers to EFL teachers' CPD activities:

- provide schools with innovative technologies (computers, projectors etc.).
- ensure available Internet or Wi-Fi connection to participate in distance CPD activities.

The third research question is connected with EFL teachers' opportunity to choose the content of CPD activities. Most of the research participants emphasized on the uselessness of lectures or workshops for them (i.e., Pedagogics). They also stressed on the lack of lectures on EFL Teaching Methods.

The results support the fact that educational institutions that provide training courses should examine EFL teachers' needs. The mentioned institutions should send out a short questionnaire on the subjects EFL teachers require for their professional growth and, as a result, for quality teaching.

The following recommendations we propose to reduce content barriers to CPD activities:

- give the EFL teachers the opportunity to choose the subjects or topics which are necessary for their professional development at the moment;
- provide workshops on ICT via modern computer technologies and online platforms to develop professionally and improve the quality of teaching;
- conduct trainings on new EFL Teaching Methods (i.e., proposed by British Council in Ukraine); and

- develop standard professional programs and plans for in-service training for every category of general secondary educator taking into account their competence levels.

The fourth RQ is based on psychological barriers. The research results show that most EFL teachers are afraid of participating in discussions or any other activities involving their speaking skills because of fear of making mistakes (grammar, pronunciation, etc.) and, as a consequence, to be negatively treated by their colleagues. For this reason, a differentiated approach to CPD activities should be used, as it takes into consideration the EFL teachers' level of professional competence.

We have identified various mentions of the EFL teacher trainer's personality features. Some teachers expressed their unwillingness to participate in CPD activities because of the instructor who conducted training sessions, and, as a result, they had no access to professional development in the institution chosen by their school. Native speakers should be involved (if possible) into specialty-oriented classes. It could also reduce EFL teachers' fear of speaking.

We suppose the most difficult barriers for teachers to overcome are the psychological ones as they totally depend on a teacher's personality. But we would like to give some recommendations on how to reduce them:

- organize summer schools or camps on improving language skills;
- invite native speakers (i.e., Peace Corps volunteers) to conduct workshops for teachers and students in order to foster spoken English skills; and
- use differentiated instruction while conducting trainings.

All the abovementioned recommendations could be implemented at the state and regional levels. We reveal them through different aspects.

1. State level.

Political aspect. It is essential to create a unit (units) that will be responsible for all types of support to EFL teachers in order to provide them with equal access to CPD. Such approach to institutional support, first of all, would help to coordinate the efforts of various managerial units, to concentrate the means of influence on the solution of a certain problem, to trace the process from the justification of the strategy for ensuring rights, and verification of its effectiveness.

Scientific and pedagogical (educational) aspect. It is necessary to present CPD issues through statistical data that should have open access. To do this, a special agency should be created that would collect information, process results, compile them, and publish. Methodology of educational research on teachers' professional development conducted in Ukraine should be in line with research conducted by world organizations (UN, EU, OECD, EUROSTAT, and UNESCO). This approach will allow managerial staff to draw parallels and identify the real state of affairs in Ukrainian CPD.

Scientific theses on teachers' professional development are of considerable interest (e.g., comparative education research). The authors of these theses offer practical recommendations for applying the experience of advanced countries in the teachers' training. Unfortunately, such recommendations are not taken into account and are not used at the state level. We consider it appropriate to create a working group of specialists who would study, systematize, and suggest using the recommendations when regulating the educational space of Ukraine.

The issue of diversification is relevant. It will make it possible to implement a diversification strategy that aims to create and provide a wide range of options for EFL teachers' training, which can be followed both at the state level and at the level of a single educational institution. In this regard, we consider it necessary to create a network of alternative educational institutions to provide options for transitioning from formal to non-formal and informal education. These provisions should be mentioned at the legislative level and receive state support in Ukraine.

Along with official authorities, public organizations (NGO) representing the EFL teachers' interests are widely and successfully functioning. In Ukraine, the mentioned organizations do not have the right to interfere with the state educational policy. We suggest creating mechanisms for their participation in decision-making process at the state level. In this case, we should mention, first off, the need to implement a strategy of civil support, which is based on active participation of public organizations in the discussion and making proposals for the organization of the teacher's professional training.

The *financial aspect* is reflected in the ensuring the financial support to schools, universities, as well as institutions of alternative, non-formal and informal education, aimed at providing CPD for EFL teachers.

2. Regional level

Social policy aspect. In Ukraine, there is a problem of insufficient information of EFL teachers about education institutions providing training on professional development, which reduces their chances of receiving quality training. We consider it necessary to direct/adjust the work of local authorities or educational institutions to raise awareness of EFL teachers about their opportunities to train. This can be achieved through cooperation with the media and educational institutions. Social networks could be also used to promote training by highlighting the advantages of professional development.

Aspect of educational institutions. Nowadays it is necessary for universities to receive a license to conduct professional training for EFL teachers. These universities could choose the content of the training courses and the specialists who will conduct them.

Harmonization of the teaching and learning structure is one of the pillars of quality training. We believe that it is necessary to provide teachers with the opportunity for lifelong learning to improve their skills not only in Ukraine, but also abroad and to learn foreign colleagues' experience. In the mentioned case the educational institutions (schools) could implement the strategy of internationalization of education, aimed at ensuring the teachers' mobility through the participation in the international educational programs or projects, which, in turn, can provide grants for training.

Taking into account the specifics of the professional training organization, it is needed to provide a link between EFL teachers' training and work that take place simultaneously. We believe it is necessary to create a mechanism at the level of educational institutions (schools) that would allow teachers to work and train at the same time or synchronously.

Psychological aspect. As our research stated, EFL teachers are afraid of being treated as non-professionals with low language proficiency. Psychological boost for overcoming personal difficulties in the process of transferring to competence education could be of great value.

3. Institutional level.

Guarantee high-quality pre-service teacher training. All university students should have access to high-quality education, including those with special needs, and regardless of their parents' income. Advanced teaching techniques should be used while conducting EFL Teaching Methods classes where teacher is a teacher-coordinator, facilitator, and moderator.

Ensure that future EFL teachers pass all the stages of training. To be competent in core subjects, pre-service EFL teachers need a smooth transition from on-the-job training to real world context, i.e., school work environment. All the mentioned stages should be logically interconnected allowing trainees to develop vocational skills, and use them giving high-quality classes.

Stay close (equal) to the age-related gaps in achievement. Basically, the conducted research showed no inequality in access to CPD activities of the different age-related groups of EFL teachers. In this context, the attention should be paid to the stereotypes that "old teachers" do not need any professional development as a result of approaching retirement.

6. Conclusions

This research is a work in progress. In light of the research results presented in this paper, equality in access to continuous professional development of EFL teachers in Ukraine is not often a priority. We consider equal access not only to the educational institutions that provided training, but also to the content of training. Despite visible equality, parity toward the freedom of choice of the place for training remains unclear.

The participants of the research expressed their attitude toward their experience of CPD, professional views on the nature of CPD, its providers and frequency. The results allowed us to define financial, spatial, content, and psychological barriers to equal access of EFL teachers to CPD in Ukraine. In the context of financial barrier our results almost coincide with the conclusions of All-Ukrainian Monitoring Survey of Secondary School Teachers and Principals (by the TALIS methodology) (Shchudlo *et al.* 2017).

What is still missing is a robust debate backed by research on how to reduce gaps in teacher professional development through appropriate and differentiated training courses based on their current needs. Such research should examine what counts as quality professional education for EFL teachers, how to expand and deepen teacher knowledge of innovative EFL teaching/learning methods and techniques, as well as use of educational technology to develop digital skills and competences. With new understanding of Ukrainian teachers' needs, we will be able to address with real confidence the inequality issues, and provide teachers with access to meaningful, personalized, inquiry-based teacher education, and life-long learning opportunities.

Appendix A Teachers' Views on CPD Practices Questionnaire

Thank you for agreeing to complete this survey to help analyze and improve continuing professional development (CPD) practices.

The survey is completely anonymous and your answers will remain confidential. The results will be available to public only in the form of summaries and generalized data. Please provide as much information and comments as you consider to be necessary.

Part 1

1. Your name (optional)
2. Your gender
   - male
   - female
   - other
3. Your country
4. The city / town / village etc. you work at
5. You work as a ...
   - teacher
   - administrator
   - other
6. The institution you work at
   - Primary school (state-owned or similar)
   - Lower secondary school (state-owned or similar)
   - Upper secondary school (state-owned or similar)
   - Private school / language courses
   - Vocational school
   - College
   - Education management body
   - other
7. What is your professional (teaching) experience?
   - Less than 5 years
   - 6–10 years
   - 11–15 years
   - 16–20 years
   - 21–25 years
   - 26–30 years
   - 31–35 years

36–40 years
More than 40 years

Part 2

8. When was the last time you were involved into CPD activities?
9. Who provided those CPD activities (institution / company / web portal etc.)?
10. Could you describe those CPD activities in brief?
11. Are you satisfied with those CPD activities?
    1 = not at all, 2 = Disagree, 3 = unsure, 4 = Agree, 5 = completely
12. How have those CPD activities influenced your professional competence?
    5 = My professional competence has changed a lot.
    4 = I've got a significant amount of new knowledge and/or skills.
    3 = I've got only some new knowledge and/or skills.
    2 = My professional competence has barely changed.
    1 = My professional competence hasn't changed at all.
13. Do you take part in teachers' associations and/or professional networks?
    3 = Yes
    2 = No, I don't have time for it.
    1 = No, I don't think it is important for my CPD.
14. Can you say that CPD activities are easily accessible for you?
    3 = Yes
    2 = Not always.
    1 = Never.

Part 3

15. What is the most appropriate form of CPD for teachers?

Choose more than one if appropriate

Formal
Non-formal
Informal

16. Who should provide CPD for teachers?

Choose more than one if appropriate

Specialized educational institutions (state-owned)
Specialized educational institutions (private and commercial)

Specialized non-government and public organizations
Universities
Teachers' associations and professional networks

17. How often should teachers take part in CPD activities?
    5 = Never. Everyday work is enough for personal CPD.
    4 = Once in 5 years
    3 = Once in 2–3 years
    2 = Once a year
    1 = Several times during a year
18. Should teachers be able to choose forms and topics of their CPD?
    4 = Yes, both forms and topics.
    3 = Yes, but only forms, not topics
    2 = Yes, but only topics, not forms.
    1 = No, forms and topics should be predefined by leading professionals in the area.
19. Are you and your colleagues ready to build personal CPD schemes and choose the ways to implement them?
    1 = not ready at all
    2 = somewhat ready
    3 = partially ready
    4 = mostly ready
    5 = completely ready

Appendix B Interview questions

Question Pool 1: Experience of CPD

What were your expectations before the last CPD activities you attended (took part in)?
Did you worry about anything before the last CPD activities you attended (took part in)?
What did you do at those activities? Did they meet your expectations?

Question Pool 2: Difficulties and Barriers

What would you like to change in your usual CPD practices?
Have you experienced any problems while having CPD?
Are there any factors that make the process of CPD difficult for you? Is it possible to eliminate them?

References

Abramova, I. (2013). Grappling with language barriers: Implications for the professional development of immigrant teachers. *Multicultural Perspectives, 15*(3), 152-157. doi: 10.1080/15210960.2013.809305)

Anderson, C. (2017). Ethics in qualitative language education research. In: S.-A. Mirhosseini (Ed.), *Reflections on qualitative research in language and literacy education*. Educational Linguistics, vol 29 (pp. 59-73). Cham: Springer International Publishing AG.

Ansalone, G. (2009). *Exploring Unequal Achievement in Schools*. New York: Lexington Books.

Antoninis, M., Deprato, M. & Benavot, A. (2016) Inequality in education: the challenge of measurement. In *World social science report, 2016: Challenging inequalities, pathways to a just world* (pp. 63-67). Paris: UNESCO/ISSC.

Apple, M. W. (2011). Democratic Education in Neoliberal and Neoconservative Times. International *Studies in Sociology of Education 21* (1), 21–31.

Badri, M., Alnuaimi, A., Mohaidat, J., Yang, G., & Al Rashedi, A. (2016). Perception of teachers' professional development needs, impacts, and barriers: The Abu Dhabi case. *SAGE Open*. doi: 10.1177/2158244016662901

Batruch, A., Autin, F., Bataillard, F. & Butera, F. (2018). School Selection and the Social Class Divide: How Tracking Contributes to the Reproduction of Inequalities. *Personality and Social Psychology Bulletin, Vol. 45* (3), 477-490.

Bhandari, Medani P. (2020) Second Edition- Green Web-II: Standards and Perspectives from the IUCN, Policy Development in Environment Conservation Domain with reference to India, Pakistan, Nepal, and Bangladesh, River Publishers, Denmark / the Netherlands. ISBN: 9788770221924 e-ISBN: 9788770221917

Bhandari, Medani P. (2020), Getting the Climate Science Facts Right: The Role of the IPCC (Forthcoming), River Publishers, Denmark / the Netherlands- ISBN: 9788770221863 e-ISBN: 9788770221856

Bhandari, Medani P. and Shvindina Hanna (2019) Reducing Inequalities Towards Sustainable Development Goals: Multilevel Approach, River Publishers, Denmark / the Netherlands- ISBN: Print: 978-87-7022-126-9 E-book: 978-87-7022-125-2

Blanden, J., & MacMillan, L. (2016). Educational Inequality, Educational Expansion and Intergenerational Mobility. *Journal of Social Policy, 45*(04), 589–614. doi:10.1017/s004727941600026x

Bredeson, P. (2002). The architecture of professional development: materials, messages and meaning. *International journal of educational research, 37*(8), 661–675.

Breen, R. & Jonsson J. (2005). Inequality of opportunity in comparative perspective: Recent Research on Educational Attainment and Social Mobility. *Annual Review of Sociology, 31*, 223–243 doi: 10.1146/annurev.soc.31.041304.122232

Breen, R., Luijkx, R., Muller, W & Pollak, R. (2010). Long-term Trends in Educational Inequality in Europe: Class Inequalities and Gender Differences. *European Sociological Review, 26*(1), 31-48.

Broad, J. (2015). So many worlds, so much to do: Identifying barriers to engagement with continued professional development for teachers in the further education and training sector. *London Review of Education. 13*, 16-30. doi: 10.18546/LRE.13.1.03

Bubbico,R.&Freytag,L.(2018).*InequalityinEurope*.EuropeanInvestmentBank.Retrieved from https://www.eib.org/attachments/efs/econ_inequality_in_europe_en.pdf

Burke, P., Heideman, R. & Heideman, C. (1990). *Programming for staff development: fanning the flame*. London; New York: Falmer Press.

Chzhen, Y., Gromada, A., Rees, G., Cuesta, J & Bruckauf, Zl. (2018). An unfair start: Inequality in children's education in rich countries. *Innocenti Report Card no. 15*. UNICEF Office of Research - Innocenti, Florence. Retrieved from: https://www.unicef-irc.org/publications/pdf/an-unfair-start-inequality-children-education_37049-RC15-EN-WEB.pdf

Cohen, L., Manion, L., & Morrison, K. (2007). *Research methods in education*. New York: Routledge.

Crimmins, G., Oprescu, F. & Nash, G. (2017). Three pathways to support the professional and career development of casual academics, *International Journal for Academic Development, 22*:2, 144-156, doi: 10.1080/1360144X.2016.1263962

Dailey-Hebert, A., Mandernach, B., Donnelli-Sallee, E. & Norris, V. (2014). Expectations, motivations, and barriers to professional development: Perspectives from adjunct instructors teaching online. *Journal of Faculty Development, 28* (1), 67-82.

Day, Ch. & Sachs, J. (2008). Professionalism, performativity and empowerment: discourses in the politics, policies and purposes of continuing professional development. In Ch. Day & J. Sachs (Eds.), *International handbook on the continuing professional development of teachers*. Maidenhead: Open University Press.

Del Carpio, X., Kupets, O., Muller, N. & Olefir, A. (2017). *Skills for a Modern Ukraine*. Directions in Development. Washington, DC: World Bank. Retrieved from: https://openknowledge.worldbank.org/handle/10986/25741

Duru-Bellat, M. (2015). Social inequality and schooling. In J. D. Wright (Ed.), *International Encyclopedia of the Social & Behavioral Sciences* (2nd ed.), Volume 22 (pp. 325–330). doi:10.1016/b978-0-08-097086-8.92015-4

Feist, L. (2003). Removing Barriers to Professional Development. *T.H.E. Journal, 30*(11). Retrieved from https://www.learntechlib.org/p/97481/

Fullan, M. (2012). *Change forces: Probing the depths of educational reform*. Hoboken: Taylor and Francis.

Guclu, N. & Onder, E. (2014). Analysis of Inequalities among Schools in Primary Education. *International Journal of Academic Research* 6(6), 170–177.

Hall, M. (2012). *Inequality and higher education: Marketplace or social justice?* Leadership Foundation for Higher Education. Retrieved from https://www.

salford.ac.uk/__data/assets/pdf_file/0015/76110/Inequality-and-Higher-Education-published-Jan-2012.pdf

Hargreaves, D. (2014) A self-improving school system and its potential for reducing inequality. *Oxford Review of Education, 40*(6), 696-714, doi: 10.1080/03054985.2014.979014;

Herrera, L. (2007). Equity, equality and equivalence – a contribution in search for conceptual definitions and a comparative methodology. *Revista Espanola de Educación Comparada, 13*, 319-340.

Hippe, R., Araújo, L. & Dinis da Costa, P. (2016). Equity in Education in Europe; Luxembourg (Luxembourg): Publications Office of the European Union; EUR 28285 EN. doi:10.2791/255948

Howe, K. (1994). Standards, assessment, and equality of educational opportunity. *Educational Researcher. Vol. 23*(8), 27-33.

Institute of Educational Analytics (2018). *Osvita v Ukraini: bazovi indykatory. Informatsiino-statystychnyi biuleten rezultativ diialnosti haluzi osvity u 2017/2018 n.r.* [Education in Ukraine: Basic indicators. Information and statistical bulletin of education in 2017/2018]. Retrieved from: https://mon.gov.ua/storage/app/media/nova-ukrainska-shkola/1serpkonf-informatsiyniy-byuleten.pdf

Jacob, J. C. (1981). Theories of Social and Educational Inequality: from dichotomy to typology. *British Journal of Sociology of Education, 2*(1), 71-89.

Joyce, B. & Showers, B. (2002). *Student achievement: through staff development.* Alexandria, VA: ASCD.

Kennedy, A. (2011). Collaborative continuing professional development (CPD) for teachers in Scotland: aspirations, opportunities and barriers. *European Journal of Teacher Education, 34*(1), 25-41. doi: 10.1080/02619768.2010.534980

Klasnić, I. (2019). Teaching practice in the Republic of Croatia – student perspective. *Journal of Educational Sciences & Psychology, Vol. IX (LXXI)* 2, 41-52.

Kopcha, T. (2012). Teachers' perceptions of the barriers to technology integration and practices with technology under situated professional development. *Computers & Education. 59*, 1109–1121. doi: 10.1016/j.compedu.2012.05.014.

Lee, K.-w. (2000). English teachers' barriers to the use of computer-assisted language learning. *The Internet TESL Journal, Vol. VI* (12). Retrieved from: http://iteslj.org/Articles/Lee-CALLbarriers.html

Levine, A. & Sun, J. (2002). Barriers to distance education. Distributed education: Challenges, Choices, and a New Environment, Sixth in a Series. Washington: American Council on Education.

Lindeman, E. (2013). *The meaning of adult education.* London: Windham Press Classic Reprints.

Lukina, T. (2019). Zagrozy realizaciyi derzhavnoyi polit`ky protydiyi obmezhennyu rivnogo dostupu do osvity v Ukrayini [Threats to the implementation of the state policy of countering the restriction of equal access to education in Ukraine]. *Visnyk pislyadyplomnoyi osvity* [*Bulletin of Postgraduate Education*]. *9*(38). doi: 10.32405/2522-9958-9(38)-10-33. Retrieved from:

http://umo.edu.ua/images/content/nashi_vydanya/visnyk_PO/9_38_2019/Bulletin_9_38__Management_and_administration_Tetyana_Lukina_UA.pdf

Macià, M. & Garcia, I. (2016). Informal online communities and networks as a source of teacher professional development: A review. *Teaching and Teacher Education. 55*. 291-307. doi: 10.1016/j.tate.2016.01.021.

McConnell, T.J., Parker, J.M., Eberhardt, J., Koehler, M. & Lundeberg, M. (2013). Virtual professional learning communities: Teachers' perceptions of virtual versus face-to-face professional development. *Journal of Science Education Technology, 22*(3), 267–277. doi: 10.1007/s10956-012-9391-y

Mehan, H. (1992). Understanding Inequality in Schools: The Contribution of Interpretive Studies. *Sociology of Education, 65*(1), 1-20.

Meyer, M. J. (1986). Types of Explanation in the Sociology of Education. In J. Richardson (Ed.), *Handbook of Theory and Research for the Sociology of Education* (pp. 341-59). Westport: Greenwood Press

Meuret, D. (2011). A system of equity indicators for educational system. In W. Hutmacher, D. Cochrane, N. Botanni (Eds.), *In pursuit of equity in education: using international indicators to compare equity policies*. Dordrecht; London: Springer.

Ministry of Education and Science of Ukraine (2019). National qualifications framework [Web-page]. Retrieved from https://mon.gov.ua/eng/osvita/nacionalna-ramka-kvalifikacij

Mongkolhutthi, P. (2019). Inequality and imbalance of professional development opportunities: The case of a higher educational institution in Southeast Asia. *Journal of Applied Research in Higher Education. 11*. doi: 10.1108/JARHE-01-2018-0010)

Montt, G., & Kelly, S. (2015). Educational Inequality. In J. Stone, R. Dennis, P. Rizova, and X. Hou (Eds.), *The Wiley Blackwell Encyclopedia of Race, Ethnicity, and Nationalism*. doi:10.1002/9781118663202.wberen367

Mukan, N., Yaremko, H., Kozlovskiy, Yu., Ortynskiy, V., & Isayeva, O. (2019). Teachers' continuous professional development: Australian experience. *Advanced Education, 12*, 105-113. doi: 10.20535/2410-8286.166606

Noble, H., & Smith, J. (2015). Issues of validity and reliability in qualitative research. *Evidence Based Nursing, 18* (2), 34-35. doi: 10.1136/eb-2015-102054

Pachamama Alliance (n.d.). *Social Inequality*. Retrieved 2020, January 12, from https://www.pachamama.org/social-justice/social-inequality

Pallant, J. (2016). *SPSS survival manual: a step by step guide to data analysis using SPSS*. New York, NY: McGraw-Hill Education.

Pfeffer, F. T. (2008). Persistent Inequality in Educational Attainment and Its Institutional Context. *European Sociological Review 24* (5), 543–565.

Razumkov Center (2002). Systema osvity v Ukraini: Stan ta perspektyvy rozvytku. Analitychna dovidka tsentru Razumkova [Ukrainian educational system: Current situation and development prospective. The analytical report of Razumkov Center]. *Natsionalna bezpeka i oborona* [National Security and

Defence], *4 (28)*, 2-35. Retrieved from: http://razumkov.org.ua/uploads/journal/ukr/NSD28_2002_ukr.pdf

Somel, R. (2019). *A relational approach to educational inequality: Theoretical reflections and empirical analysis of a primary education school in Istanbul.* Wiesbaden: Springer Fachmedien. doi: 10.1007/978-3-658-26615-8.

OECD (2014). *TALIS 2013 Results: An International Perspective on Teaching and Learning*. TALIS. Paris: OECD Publishing. doi: 10.1787/9789264196261-en

OECD (2017). *Educational opportunity for all: Overcoming inequality throughout the life course*. Educational Research and Innovation. Paris: OECD Publishing. doi: 10.1787/9789264287457-en.

Owens, A. (2018). Income Segregation between School Districts and Inequality in Students' Achievement. *Sociology of Education, 91*(1), 1–27.

Postanova Kabinetu Ministriv Ukrainy "Deiaki pytannia pidvyshchennia kvalifikatsiyi pedahohichnykh i naukovo-pedahohichnykh pratsivnykiv" [Resolution of the Cabinet of Ministers of Ukraine "Some issues of professional development of pedagogical and scientific-pedagogical workers"] № 800 (2019)

Rodríguez-Pose, A., & Tselios, V. (2009). Education and income inequality in the regions of the European Union. *Journal of Regional Science, 49*(3), 411-437

Rubenson, K., & Desjardins, R. (2009). The impact of welfare state regimes on barriers to participation in adult education. *Adult Education Quarterly, 59*(3), 187-207.

Schindler, S. & Lörz, M. (2012). Mechanisms of Social Inequality Development: Primary and Secondary Effects in the Transition to Tertiary Education Between 1976 and 2005. *European Sociological Review 28*(5), 647–60.

Shchudlo, S., Zabolotna, O. & Lisova, T. (2017). *Ukrainian Teachers and the Learning Environment. Results of All-Ukrainian Monitoring Survey of Secondary School Teachers and Principals (by the TALIS methodology)*. Drohobych: UERA - Trek LTD.

Sondergaard, L. (2018). *Ukraine education policy note: Introducing the New Ukrainian School in a fiscally sustainable manner*. Washington, DC: World Bank. Retrieved from: http://documents.worldbank.org/curated/en/322641535692262866/Ukraine-Education-Policy-Note-Introducing-the-New-Ukrainian-School-in-a-Fiscally-Sustainable-Manner

Teran, C. (2007). Financial barriers to higher education [Project Report]. Retrieved from: https://www.pdffiller.com/jsfiller-desk13/?projectId=447462259#4710a2c8f54d085231ad9b0883f411a7

Tezikova, S. & Plotnikov, Y. (2019). Detsentralizatsiia v osviti: novi pidkhody do profesiinoho rozvytku vchyteliv [Decentralization in education: new approaches to professional development of teachers]. In T. Finikov & R. Sucharski (Eds.), *Innovatsiinyi universytet i liderstvo: proekt i mikroproekty – III* [Innovative university and leadership: Projects and micro projects - III] (pp. 329-344). Warsaw: Faculty "Artes Liberales", Warsaw University.

UNESCO (2009). *Overcoming inequality: why governance matters. EFA Global Monitoring Report*. Paris: UNESCO. Retrieved from: https://unesdoc.unesco.org/ark:/48223/pf0000177683

UNESCO (2017). *Education for Sustainable Development Goals: Learning Objectives*. Paris: UNESCO. Retrieved from: https://www.unesco.de/sites/default/files/2018-08/unesco_education_for_sustainable_development_goals.pdf

Vakhshtain, V. (2005). Strategii obespechenija ravenstva obrazovatel'nyh vozmozhnostej v stranah OJESR: sravnitel'nyj analiz [Strategies of ensuring equality of educational opportunities in OECD countries: Comparative analysis]. In M. Larionova (Ed.), *Aktual'nye voprosy razvitija obrazovanija v stranah OJESR* [Current issues of education development in OECD countries]. (pp. 111-123). M.: Izdatel'skij dom GU VShe,

Villegas-Reimers, E. (2003). *Teacher professional development: an international review of literature*. Paris: UNESCO.

World Bank (2010). *Equal opportunities, better lives. Gender in Africa: Using knowledge to reduce gender inequality through World Bank activities*. Washington DC: The World Bank.

World Bank (2011). *Ukraine - Equal Access to Quality Education in Ukraine Project*. Washington, DC: World Bank. Retrieved from: http://documents.worldbank.org/curated/en/858531468143397357/Ukraine-Equal-Access-to-Quality-Education-in-Ukraine-Project

Yeaxlee, B. (1929). *Lifelong education*. London: Cassell.

Zakon Ukrainy "Pro Osvitu" [Law of Ukraine "On Education"] № 2145-VIII (2017)

Chapter 12

Socio-Cultural Dimensions of Socio-Economic Inequality in Household Food Security in Nepal

Prem B. Bhandari[1*], *Madhu Sudhan Atteraya*[2], *Medani P. Bhandari*[34]
[1]University of Michigan, USA
[2]Keimyung University, South Korea
[3]Akamai University, Hilo, Hawaii, USA
[4]Sumy State University, Ukraine

ABSTRACT
The chapter examines the association between socio inequalities and food security in Nepal. Specifically, we investigate the associations between caste/ethnicity and household wealth status on household-level food security, after controlling for several other demographic, geographic, and other factors. For this purpose, we used nationally representative data from the 2011 Nepal Demographic Health Surveys (NDHS 2011). Results from multilevel regression (binary logistic and OLS regressions) show that various household livelihood capitals are important in shaping household food security status of a household. These findings provide an implication in implementing "rights for food" as mandated by the Article 36 of the Constitution of Nepal.
Keywords: food security, social inequality, caste/ethnicity, Nepal

* Correspondence: Prem B. Bhandari, Institute for Social Research, University of Michigan, USA. E-Mail: prembh@umich.edu

1. Introduction

Food security is one of the global challenges. Nepal is not an exception and is one of the most food-insecure countries in the world, ranking 157 out of 187 countries in terms of food security (UNDP 2011; Joshi et al. 2012). In 2015, about one quarter of the Nepali population was still below poverty line (GoN 2015). In 2010/2011, of the Nepal's 75 districts, 38 were characterized as food insecure districts (GoN 2012). Among them, 2 districts were self-sufficient for less than 3 months, 3 districts were food secure for 3–6 months, 14 were food secure for 6–9 months, and 19 districts were food secure for 9–12 months. The Government of Nepal (GoN) aims at eradicating poverty, ending hunger and all forms of malnutrition by ensuring access to safe, nutritious, and sufficient food all year round for all people by 2030 through its sustainable development goals programs (Government of Nepal 2015). However, the bleak picture stated earlier shows that Nepal requires concerted efforts to meet the challenge of the food insecurity problem by 2030.

As a whole, Nepal is one among the most food insecure countries. However, there are inequalities in the rates of poverty and food security across social and economic groups and geographical areas (see, for example, GoN 2018; Goli et al. 2019; Bhandari 2019; Bhandari et al. 2020). Unless we identify socio-culturally and economically vulnerable groups of people within the country, addressing the problem of food insecurity in the country becomes an unaddressable challenge.

Previous studies on food security in Nepal examined several factors contributing to food security in Nepal. These factors range from geographical variation (Bhandari 2019), household livelihood capitals (Bhandari et al. 2020), and more (CITES). Some other studies explored the associations between remittances and food security (CITES). However, less is known about the associations between social-structural and economic inequality and household food insecurity in Nepal. Understanding the dire need, this chapter attempts to investigate the extent to which household food security is associated with social-cultural structure and economic status of households in Nepal. In this chapter, first we define the concept and measurement of food security. Second, we provide a theoretical link between social-structural (specifically, caste/ethnicity) and economic (household wealth) and food security. Third, because we provide an empirically-based evidence of the associations among social-structural and economic inequality and food security, we discuss about data, measures, and analytic methods. This section is followed by results and conclusions and implications. We believe the information and the evidence presented in this chapter will be an important guide in designing and perusing food security policies.

2. The Concept and Measurement of Food Security

Food security is a complex, and multidimensional concept. Thus, measuring the concept *per se* is not straightforward. D. G. Maxwell (1996) pointed out that there are already over 200 definitions of food security. Since then, several definitions and concepts on food security have been developed over time. Of the various definitions, the most commonly used definition of food security is the one provided by the World Food Summit (1996). The Summit defined this concept as: "*Food security exists when all people, at all times, have physical and economic access to sufficient, safe and nutritious food that meets their dietary needs and food preferences for an active and healthy life.*" This definition has acquired the broadest acceptance (Smith et al. 1993).

There are four dimensions of food security: availability, accessibility, utilization, and stability (Napoli 2010/11; FAO 2006). The first three dimensions (i.e., food availability, accessibility, and utilization) are the physical dimensions and stability is the temporal dimension (Napoli 2010/11). The nutritional status, the ultimate outcome, is determined by food utilization, which depends upon accessibility of food. The accessibility of food is determined by availability. However, the accessibility of food does not guarantee its proper utilization. The sustainable utilization, accessibility, and availability of food are greatly influenced by factors such as weather, conflict, unemployment, and diseases that may influence the stability dimension. Thus, stability dimension influences any or all of the other three components of the food insecurity framework over time.

Although there are several ways of measuring food security, in this study, we use the *Household Food Insecurity Access Scale (HFIAS)* developed by Food and Nutrition Technical Assistance (FANTA) of USAID (Coates et al. 2006; 2007). This scale assesses whether a household experienced the problem of food "access" during the past reference period such as the past 30 days or past 12 months. The measurement instrument consists of nine items which measure the occurrence and frequency of food access. These items collect information about food access experienced by a household as a result of limited resources.

In HFIAS, there are three domains of food insecurity access: (a) anxiety and uncertainty about household food supply, (b) insufficient quality that includes variety and preferences of food types, and (c) insufficient food intake. The first domain is *anxiety and uncertainty about household food supply,* also known as the "anxiety domain." This domain is measured by asking "Did you worry that your household would not have enough food?" to measure whether a household experienced uncertainty and anxiety about acquiring food during the period-specific in the item†. The second domain is *insufficient quality* which measures whether a household experienced having limited choices in the type of food that a household eats during a specific period. The following three items are asked to measure this domain.

- Were you or any household member not able to eat the kinds of foods you preferred because of a lack of resources?
- Did you or any household member have to eat a limited variety of foods due to a lack of resources?
- Did you or any household member have to eat some foods that you really did not want to eat because of a lack of resources to obtain other types of food?

The third dimension is *insufficient quantity,* which measures whether a household had to cut the amount of food during a specific period due to a lack of resources. The followings five items measure this domain.

- Did you or any household member have to eat a smaller meal than you felt you needed because there was not enough food?
- Did you or any household member have to eat fewer meals in a day because there was not enough food?
- Was there ever no food to eat of any kind in your household because of a lack of resources to get food?
- Did you or any household member go to sleep at night hungry because there was not enough food?
- Did you or any household member go a whole day and night without eating anything because there was not enough food?

† Recall period may vary – "in the past 30 days" or "in the past 12 month" depending upon the context.

These items are measured in a scale of 0–3 (0 = never, 1 = rarely, 2 = sometimes and 3 = often). The type and number of occurrence indicators and time frames may be modified depending upon the context. Using the aforementioned indicators (9 items or 7 items depending upon context), the HFIA scale is calculated. The scale is the sum of the frequency of occurrence of aforementioned items during the past reference period. The summated scale ranges from 0–27 (if 9 items are included) and 0–21 (if only 7 items are included). The higher number in the scale refers to a greater level of food insecurity and *vice versa.* This scale may be also used for food security groupings such as food secure, mildly food insecure, moderately food insecure, and severely food insecure households (Coates et al. 2007; NDHS 2011).

3. Nepal's Socio-economic Context and Household Food Security

This section provides the theoretical reviews, empirical evidence, possible theoretical explanations, and proposed hypothesis of the relationships between a household's caste/ethnicity and wealth status and food insecurity.

Nepali society is highly structured and stratified based on Hindu religion-based caste and ethnicity. Household members ascribe membership in a particular caste/ethnicity by birth or by marriage (for females). These societies exhibit a distinct pattern of social inequality based on caste/ethnicity which is uncommon in other societies globally. In general, the caste hierarchy involves four major domains: Brahmin (also known as Bahun), Chhetri, Vaishya, and Shudra (Bista, 1991). Bahun and Chhetri who are also known as the Tagadhari (sacred-thread wearing caste with the highest purity) caste groups are at the top of the caste hierarchy. This is followed by other groups such as Newar and Hill Janajati, for example, who belong to Vaishya (also termed as Matawali for whom alcohol is acceptable, with middle purity). Shudras are the impure caste groups who are at the bottom. Dalits also called untouchables (water unacceptable, also officially known as Dalit or traditionally skilled or occupational castes) are at the very bottom of the caste hierarchy. This caste hierarchy also forms the basis of social-economic hierarchy in Nepali society. Evidence shows that caste/ethnicity-based differentials in almost every sphere of life are prevalent. This is primarily associated with caste/ethnicity-based socio-economic hierarchy and associated disadvantages with it.

Although the discriminatory practices based on caste/ethnicity were legally abolished in 1962 in Nepal, discrimination continues to be prevalent. People are often considered advantaged/disadvantaged based on their caste/ethnicity (Bennet, Dahal and Govindasamy, 2008; Gellner, 2007; Pradhan, 2005). Their membership in a particular caste/ethnicity enables or hinders them the access to assets and services through social relations (Allison and Ellis 2001). The high caste Hindu (Bahun and Chhetri), and the ethnic Newar people are among the historically advantaged groups. These caste/ethnic groups are considered to have the most access to various economic and non-economic resources and opportunities (Bennet et al., 2008; DFID/World Bank, 2006). Bahun, Chhetri, and Newar are considered to be socio-culturally, economically, and politically advantaged as compared to other castes/ethnic groups.

Evidence supports that there is a close association between caste/ethnicity and household wealth status (Bhandari and Chan 2017). For example, NDHS (2011) shows that among Dalits, slightly over half (51 percent) of them belong to the lowest wealth quintile as compared to only 29 percent Bahun/Chhetri and only 4 percent Newar. On the other hand, nearly 46 percent of Newar and 23 percent of Bahun/Chhetri belong to the highest wealth quintile as compared to only 2 percent of Terai Dalit, 4 percent of Hill Dalit, and 9 percent of Terai Janajati. The lower castes (or Dalits) and other ethnic minority groups (Hill and Tarai ethnic groups) are disproportionately affected by widespread poverty and are relatively disadvantaged in terms of educational achievement, income, life expectancy, and have lower overall Human Development Index scores (Norwegian Refugee Council/Global IDP Project, 2003; NESAC, 1998; Bennet, Dahal & Govindasamy, 2008). In a study in rural Nepal, Maharjan and Khatri-Chhetri (2006) reported a difference in food security status by caste/ethnicity favoring privileged caste groups. Thus, *we hypothesize that households that belong to the higher or privileged caste/ethnic groups such as Brahmin/Chhetri and Newar are relatively food secure as compared to those at the lower level of caste/ethnic hierarchy (H1).*

The *household wealth* includes homeownership or ownership of other assets by the household, which is an important indicator of financial/economic capital. In the United States, homeowners were found to be more likely to report food secure than individuals still making mortgage payments and renters (Shobe et al. 2018). Similarly, Regmi et al. (2015) also found that earned incomes were used to improve households' food security in Bangladesh. In Nepal, according to Nepal Demographic Health Survey (2011), while only 18.1 percent of the households in the lowest (poorest) wealth quintile expressed that they were food secure as compared to 82 percent of those who were in the highest wealth quintile. Thus, we *hypothesize that households with a higher level of wealth status are more likely to be food secure as compared to those at the lower level of wealth status (H2).* In addition, as caste/ethnicity and household wealth status are closely intertwined, we also argue that *a large part of caste/ethnicity-based differences in food security is mediated by household wealth status (H3).*

4. Other Factors (Controls) Theoretically Associated with Food Security

We adjust for a number of factors that are theoretically known to influence both the outcome and explanatory variables. Below we discuss our controls. *Household size* influences food security in several ways. From a consumption perspective, the size of a household is the direct determinant of household food consumption. Thus, other things remaining the same, the household size is expected to have a negative influence on household food security. Garret and Ruel (1999) show evidence of a negative association between household size and food security in Mozambique. In Nepal, Maharjan and Joshi (2011) found a positive association between family size and food insecurity. From a production perspective, the availability of labor in a household is a building block for acquiring livelihood objectives and sustaining livelihood outcomes—here, food security.

Other controls include *sex and age of household head.* According to the Population Census of Nepal, nearly 26 percent of households in Nepal are headed or managed by females (GoN 2011), which increased from 15 percent in 2001. Women's involvement in agricultural production is significant in developing countries (Boserup 1971; Rahman, 2008). Women spend much longer hours in farming than their male counterparts (Acharya and Bennet 1981; Kumar and Hotchkiss 1988). This scenario holds true in Nepal. As nearly 60 percent of the households are engaged in agriculture, the gender of a household manager may have important influence on household food security.

Similarly, *age of the household head* could be another important factor to have an effect on household food security. Because younger individuals are less committed to farming occupations and are more likely to change occupations toward non-farm work (Mahesh 2002; Ogena and De Jong 1999) and elderly managers may be less likely to engage in production enhancing activities using modern farm inputs (Bhandari 2006; 2017). Thus, age of the manager may have a negative influence on household level food security. Other way round, elderly managers could acquire knowledge and experience to solve food insecurity problem (Beyene and Muche 2010).

Education enhances the knowledge and skills of individuals. Evidence shows that education is negatively associated with food insecurity, malnutrition, and hunger (Iftikhar 2017; Burchi 2006; Beyene and Muche 2010). Sseguya (2009) found that education was positively associated with household food security in Uganda.

Land holding and land ownership. The access to and ownership of farm land is crucial for sustaining livelihoods. The access to land provides employment and income to farmers. In addition, the ownership of land is an important criterion in defining one's position in the socio-politico-economic class hierarchy (De Janvry 1981; Blaikie, Cameron and Seddon 2002; Sugden 2009). The ownership of this vital resource increases control over other resources such as earned income from land, political power, and access to other institutions, for example, banks. The access and ownership of land provide food production and incomes to farmers thus helping to reduce food insecurity problem. In Nepal, the average size of operational holdings (actual area cultivated) is only 0.5 hectares. Ninety-three percent of operational holdings are operated by small farmers (<2 hectares) covering 69 percent of the cultivated area (Thapa 2009). Subedi and Dhital (2007) noted that small-holdings, marginal land for agricultural production, weather-dependent cultivation, and poor technology/technical know-how of farmers are some of the key factors influencing food insecurity in the country. Similarly, Maharjan and Khatri-Chhetri (2006) found that food-secure households have a significantly higher amount of total land, and have higher percentage of irrigated land as compared to food-insecure households.

Moreover, *livestock keeping* is an integral activity of farm households in Nepal. Animals provide food (meat, milk, egg, and other food products), employment, and income to farm families. Animals are also an important source of economic capital. Farm households may sell animals and earn incomes. Therefore, the ownership of livestock is a key to address the food insecurity problem of households (FAO 2011). Evidence suggests that households with large livestock size are less vulnerable to food insecurity especially in times of drought when crops fail (Little et al. 2006; FAO 2011). According to FAO (2011), animals can be a source of financial, economic as well as social capital in times of need and therefore, are important elements of food security assets. Maharjan and Khatri-Chhetri (2006) found that food security increases with the increase in a number of total livestock holdings.

A household's access to these basic community services may influence a household's food security status. Using this data, Bhandari (2019) finds that the *geographic proximity* of a household to the urban center may be one of the indicators of access to various resources. For example, in urban areas, most basic amenities and services including transportation and markets are accessible as compared to rural areas. Similarly, in Nepal, households in the Terai region or those in the eastern or central region have relatively better access to such services as compared to those in the Hill and Mountain regions or in the Western or Far-western regions. We control geographic variables in each model of our analysis (please refer to Bhandari 2019 for details).

5. Methods

5.1. Data

We used the nationally representative Nepal Demographic Health Survey (NDHS) data collected in 2010–11 to investigate household-level food security. The NDHS was designed to collect information and provide estimates of key population and health indicators for the country as a whole using a representative sample. The research design itself is multi-level in nature, and therefore, used a multi-stage sampling procedure to identify samples. The detailed procedure for sampling and data collection is provided in the NDHS 2011 report (Ministry of Health and Population (MOHP) [Nepal], New ERA, and ICF International Inc. 2012). Altogether, in the first stage, a total of 289 EAs (194 rural and 95 urban) were selected. In the second stage, upon a complete household listing of all EAs, 35 households from each rural cluster and 40 households from each urban cluster were randomly selected. Three questionnaires were administered—the Household Questionnaire, the Women's Questionnaire, and the Men's Questionnaire. The data utilized in this study comes from the household questionnaire that collected household-level information including the food security status of a household from 10,826 households. Of particular interest to this study, this data contains detailed information on household-level food security reported by a household member.

5.2. Measures

Below we discuss the measurement of outcome measure, explanatory measures, and controls.

Outcome measures—measures of food insecurity. As indicated earlier, HFIAS includes three domains of food insecurity access: (a) anxiety and uncertainty about household food supply, (b) insufficient quality that includes variety and preferences of food types, and (c) insufficient food intake. In a standard format, there are nine items. However, these items are context-specific. The NDHS (2011) included only seven measures. Each item was measured by asking specific questions. Each item was measured whether a household experienced that item in the past 12 months. Each item was measured in an ordinal scale with three responses "rarely, sometimes and often."

In this study, we used the household food security scale, the measurement of which follows. We calculated the food insecurity index which is the sum of the frequency of occurrence during the past 12 months. As each of these items is measured on a scale of 0–3 (0 = never, 1 = rarely, 2 = sometimes, and 3 = often), the summated index or scale ranged from 0–21. A higher number in the scale represents a greater level of food insecurity.

Explanatory measures. Caste/ethnicity and wealth status of a household are the two major explanatory variables used to explain food security.

Caste/ethnicity of the household head is grouped into eight categories. These categories are coded as: Brahmin/Chhetri, Terai Others, Hill Dalit, Newar, Hill Janajati, Terai Janajati, and Muslim. In the analysis, the most advantaged caste group Brahmin/Chhetri is used as the reference category.

A h*ousehold's wealth index, which is already calculated as a composite measure* (see, for example, NDHS 2011) is measured in quintiles and is coded as: 0 = lowest quintile, 1 = second quintile, 2 = third (middle) quintile, 3 = fourth quintile, and 5 = fifth (highest) quintile. The lowest quintile is used as the reference category.

Controls. We used several households and geographic controls that are theoretically known to influence a household's food security. *Household size* is used as the number of persons in a household. *Age of the household head* is measured in years. *Gender of household head* is coded as male = 1 and females = 0. *Education* of a household head is measured as 0 = no education, 1 = incomplete primary (grade 1–4), completed primary (grade 5), incomplete secondary (grades 6–9), completed secondary (10), and higher (11 plus). Household heads with no education are considered as the reference category.

Land ownership is the measure of a household's access to and ownership of land at the time of the survey. It is measured as whether a household owned any land (coded 1) and otherwise 0. *Livestock holding* includes the ownership of cattle/buffalo, cow/bulls, goats, chicken, and pigs each coded as yes (=1) or no (=0) owned by a household at the time of the survey.

Geographic controls include a *households' location by ecological regions, and* are grouped into three categories—Mountain, Hill and the Terai. As the Terai region is considered as the "bread basket" or the "granary" of the country, this region is used as the reference category. *Rural–urban location of a household* is another geographic region used in the analysis and rural location is considered as the reference category. Similarly, *location of a household by province* is also used as a control. There are seven provinces. Province 6 lies in the far-western–northern region of the country which has among the lowest socio-economic development indicators. Thus, this province is considered as the reference category.

5.3. Analytic Strategy

First, we calculated descriptive statistics of the measures used in this study (Table 12.1). Next, as the data was multilevel, we estimated multivariate models using multilevel modeling (hierarchical linear modeling) techniques to examine the associations between explanatory and outcome measures. The outcome measure household food security scale is a ratio measure, and thus, we estimated multilevel OLS regression models using the PROC MIXED SAS procedure. This SAS procedure takes into account of clustering of households by geographic clusters (Garson 2013).

We presented four models. Model 1 presents the association between caste/ethnicity and food security score. Model 2 presents the association between household wealth status and food security. Model 3 presents the associations between caste/ethnicity and wealth status adjusting for each other and other geographic measures. This model verifies whether caste/ethnicity and wealth status have independent effects on food security. It is believed that caste/ethnicity and household wealth are closely related. In the final model (Model 4), we controlled for additional household-level factors in addition to geographic measures.

6. Results

We provide descriptive statistics of the measures used in the analysis. Table 12.1 shows that the average household-level food insecurity scale (score), which is 3.77 on a scale that ranged from 0–21 (0 being food secure and 21 being highly food insecure). This result suggests that the distribution is highly skewed. About one-half of the households (48% with 0 score) were food secure.

Table 12.1. Descriptive statistics of measures used in the analysis (N = 8882 households).

Measures	Descriptive Statistics		
	Mean	SD	Min-Max
Outcome measures			
Overall food insecurity(score)	3.77	4.77	0-21
Anxiety and uncertainty about food supply (Yes = 1)	0.47	0.50	0-1
Insufficient quality (Yes = 1)	0.48	0.50	0-1
Insufficient food intake (Yes = 1)	0.19	0.39	0-1
Explanatory measures			
Caste/ethnicity (Brahmin/Chhetri = 0)			
Terai Others	0.06	0.23	0-1
Hill Dalit	0.12	0.32	0-1
Terai Dalit	0.03	0.16	0-1
Newar	0.04	0.20	0-1
Hill Janajati	0.24	0.43	0-1
Terai Janajati	0.08	0.28	0-1
Muslim	0.02	0.25	0-1
Household wealth status			
Wealth index (lowest quintile = 0)	0.21	0.41	0-1
Second quintile	0.19	0.32	0-1
Third (middle) quintile	0.18	0.38	0-1
Fourth quintile	0.19	0.39	0-1
Fifth (highest) quintile	0.24	0.43	0-1
Household size (numbers)	5.00	2.19	1-31
Gender of household head (Male = 1)	0.71	0.45	0-1
Age of household head (years)	43.13	14.02	15-95
Education of household head (Ref: No education)			
Incomplete primary (1–4)	0.16	0.37	0-1
Completed primary (5)	0.07	0.25	0-1
Incomplete secondary (6–8)	0.19	0.39	0-1
Completed secondary (9–10)	0.09	0.29	0-1

Measures	Descriptive Statistics		
	Mean	**SD**	**Min-Max**
Higher (11 plus)	0.09	0.28	0-1
Land ownership (owns = 1)	0.70	0.46	0-1
Animals[i]: Owns cattle/buffalo (=1)	0.33	0.47	0-1
Owns cow/bulls (=1)	0.44	0.50	0-1
Owns goats (=1)	0.48	0.50	0-1
Owns chicken (=1)	0.43	0.49	0-1
Owns pigs (=1)	0.12	0.32	0-1
Ecological regions: Terai (Reference)	0.43	0.50	0-1
Mountain	0.17	0.37	0-1
Hill	0.40	0.50	0-1
Rural-urban: Urban (Reference)	0.28	0.45	0-1
Rural	0.72	0.45	0-1
Federal provinces: Province 6 (Reference)	0.07	0.26	0-1
Province 1	0.23	0.42	0-1
Province 2	0.08	0.27	0-1
Province 3	0.17	0.38	0-1
Province 4	0.11	0.31	0-1
Province 5	0.18	0.38	0-1
Province 7	0.16	0.37	0-1

7. Caste/Ethnicity and Household Food Security

By caste/ethnicity, the largest proportion (40 percent) of the households belonged to Brahmin/Chhetri, which is followed by Hill Janajati (24 percent), Hill Dalit (12 percent), Terai Janajati (8 percent), Terai others (6 percent), Newar (4 percent), and Muslim (2 percent).

It is evident from Table 12.2 that there is a variation in food insecurity level of households by caste-ethnicity. Newar, Hill Bahun, and Terai Bahun are among the most food-secure households. For instance, in a HFIAS scale that ranges from 0–21, Newar had an average food security score of 1.69, Hill Bahun had an average score of 1.96, and Terai Bahun/Chhetri had an

Table 12.2. Household-level food insecurity by caste/ethnicity in Nepal (N = 8516 households*).

Caste/Ethnicity	Anxiety (worry) (%)	Insufficient food Quality (%)	Insufficient food Quantity (%)	Overall food insecurity score (mean/std. dev.)
Hill Bahun	28.8	31.6	8.2	1.96 (3.35)
Hill Chhetri	48.3	48.6	15.6	3.63 (4.63)
Terai Bahun/Chhetri	28.6	32.7	13.3	2.40 (4.18)
Other Terai caste	38.0	37.2	20.2	3.02 (4.50)
Hill Dalit	76.1	76.5	41.7	7.24 (5.47)
Terai Dalit	71.2	73.5	54.0	8.10 (6.41)
Newar	23.7	29.2	5.8	1.69 (3.19)
Hill Janajati	46.2	46.5	13.7	3.37 (4.17)
Terai Janajati	50.1	54.3	25.2	3.94 (4.54)
Muslim	61.1	59.1	44.8	6.16 (5.92)

Source: Bhandari et al. (2020).

* Total number of households in NDHS sample is 10,226. Because caste/ethnicity variable was not available at the household level and this variable was brought in from individual data, we could match 8516 cases. However, the results with both numbers of cases are very similar.

average score of 2.40. On the contrary, Terai Dalit, Hill Dalit, and Muslim are the most food-insecure households with a score of 8.10, 7.24, and 6.16, respectively. Similarly, there is a variation in response in food insecurity for each domain. A higher proportion of Hill and Terai Dalit and Muslim households expressed insecurity in all three aspects of food security measures as compared to other caste groups.

Below we discuss results from multilevel multivariate analysis provided in Table 12.3. The results in model 1, model 2, and model 3 are adjusted for geographic variations. The results in model 1 show that there are statistically significant differences in overall food security of households by caste/ethnicity. Adjusting for macro-level geographic regions (i.e., ecological regions, rural and urban areas, and provinces), the households that belong to the Hill Janajati, Terai Janajati, Hill Dalit, Terai Dalit, Terai Others, and Muslim reported a statistically significantly higher level of food insecurity score than Brahmin/Chhetri castes. However, those who belonged to Newar caste had lower level of food insecurity but were not significantly different as

Table 12.3. Multilevel models estimating the influence of capital assets (including caste) on food insecurity controlling for geographic differentials in Nepal, 2011 (N=8882 households).

Measures	**Overall food insecurity (score) (Model 1)**	**Overall food insecurity (score) (Model 2)**	**Overall food insecurity (score) (Model 3)**	**Overall food insecurity (score) (Model 4)**
Explanatory variables				
Caste/ethnicity (Brahmin/Chhetri=0)				
Terai Others	0.88*		0.06	-0.52*
Hill Dalit	3.46***	-	2.22***	1.65***
Terai Dalit	5.38***	-	3.47***	2.07***
Newar	-0.06	-	0.18	-0.13
Hill Janajati	0.72***	-	0.09	-0.27*
Terai Janajati	1.61***	-	0.25	-0.25
Muslim	3.38***	-	2.27***	1.13***
Household wealth index				
Lowest quintile=0)				
Second quintile	-	-2.15***	-1.98***	-1.81***
Third (middle) quintile	-	-4.20***	-3.88***	-3.55***
Fourth quintile	-	-6.04***	-5.52***	-5.03***
Fifth (highest) quintile	-	-7.99***	-7.23***	-6.49***

Table 12.3 (*Continued*)

Measures	**Overall food insecurity (score) (Model 1)**	**Overall food insecurity (score) (Model 2)**	**Overall food insecurity (score) (Model 3)**	**Overall food insecurity (score) (Model 4)**
Household size (numbers)				0.12^{***}
Gender of household head (Male=1)	-	-	-	0.43^{***}
Age of household head (years)	-	-	-	-0.01^{***}
HH head education (Ref: No education)				
Incomplete primary (1–4)	-	-	-	-0.44^{***}
Completed primary (5)	-	-	-	-0.96^{***}
Incomplete secondary (6–9)	-	-	-	-1.25^{***}
Completed secondary (9–10)	-	-	-	-1.49^{***}
Higher (11 plus)	-	-	-	-1.96^{***}
Land ownership (owns=1)	-	-	-	-1.04^{***}
Animals[!]: Owns cattle/buffalo (=1)	-	-	-	-0.52^{***}
Owns cow/bulls (=1)	-	-	-	-0.41^{***}
Owns goats (=1)	-	-	-	-0.37^{***}
Owns chicken (=1)	-	-	-	-0.04
Owns pigs (=1)	-	-	-	-0.20
Ecological regions: Terai (Reference)				
Mountain	0.50	-2.50^{***}	-2.15^{***}	-1.53^{***}
Hill	0.87^{**}	-1.27^{***}	-0.99^{***}	-0.52^{*}

Rural-urban: Urban (Reference)				
Rural	1.48***	-0.88***	-0.72***	-0.23
Provinces: Province 6 (Reference)				
Province 1	-4.28***	-3.51***	-3.24***	-3.07***
Province 2	-3.32***	-2.63***	-2.83***	-2.67***
Province 3	-3.85***	-2.18***	-1.91***	-2.09***
Province 4	-4.54***	-2.81***	-2.69***	-2.68***
Province 5	-3.15***	-2.99***	-2.76***	-2.62***
Province 7	-2.23***	-2.71***	-2.43***	-2.17***
Intercept	4.63***	12.21***	10.90***	11.70***
Deviance (-2 R log LL)	49929.5 (Null = 50782.2)	48810.7 (Null = 50782.2)	48445.6 (Null = 50782.2)	48082.3 (Null = 50782.2)

+p<.10, *p<.05, **p<.01, ***p<.001.

[1]Only a small fraction of households own horses, sheep, ducks, and yak. Therefore, these are excluded from analysis.

Note: Total number of households in NDHS 2011 is 10,226. Because caste/ethnicity variable was not available at the household level and this variable was brought in from individual data, we could match 8882 cases. Other results except caste/ethnicity were very similar suggesting there is not much influence of missing cases.

compared to Brahmin/Chhetri caste. When we controlled for the household wealth status along with geographic controls, some of the results of caste/ethnicity remained but other results were wiped out or changed with meaningful variation in the magnitude of the association. For example, the results for households that belong to the (a) Hill Dalit, Terai Dalit, and Muslim remained with slight reduction in the magnitude, remained statistically significantly higher level of food insecurity score, and (b) Hill Janajati, Terai Janajati, and Terai Others are still positive but turned out to be statistically not significant, and (c) those of Newar household turned out to be positive but still significantly not different as compared to Brahmin/Chhetri. As expected, much of this association between caste/ethnicity and household food insecurity was mediated by a household's wealth status (as shown in model 3, Table 12.3).

In model 4 (Table 12.3), the results present after adjusting all control variables in the analysis, including geographic controls, household wealth status, and other important household-level factors such as household size, age, and education of the household head and ownership of land and animals. The results show that caste/ethnicity was still independently associated with the household level food insecurity. According to Table 12.2, the households that belonged to the Hill Dalit, Terai Dalit, and Muslims reported a higher amount of food insecurity score as compared to the Brahmin/Chhetri households. However, interestingly, the households that belonged to Terai Others, Newar, Hill Janajati, and Terai Janajati reported lower food insecurity score as compared to the Brahmin/Chhetri households. However, only the results for Terai Others and Hill Janajati were statistically significant. These results clearly show that not all the households that belong to other castes/ethnic groups are relatively food insecure as compared to the so-called advantaged caste group Brahmin/Chhetri. In fact, the households that belonged to Terai Others and Hill Janajati are statistically significantly more food secure than Brahmin/Chhetri.

8. Household Wealth Status and Food Security

From Table 12.1, we know that one-fifth (21 percent) of the households belonged to the lowest wealth quintile, whereas 24 percent of the households belonged to the highest wealth quintile. According to the Nepal Demographic Health Survey (2011), while 18.1 percent of the households in the lowest wealth quintile expressed that they were food secure, 82 percent of those were in the highest wealth quintile (Figure 12.1). Similarly, nearly 70 percent of households from the lowest wealth quintile reported that they were

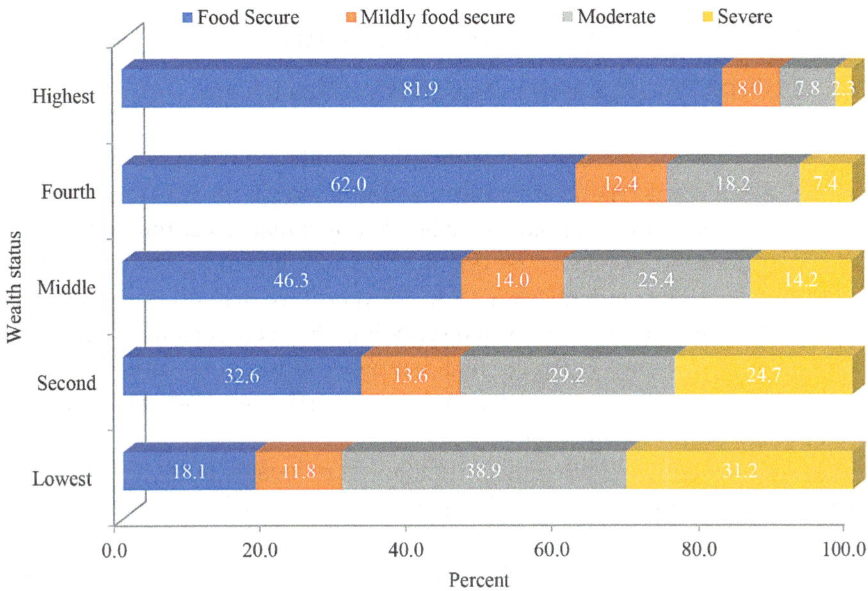

Figure 12.1. Food security by households' wealth status, 2011, Nepal.
Source: Bhandari et al. (2020).

moderately or severely food insecure; whereas, this proportion was only about 10 percent among wealthy households.

Our multivariate results in Table 12.3 show that there is a statistically significant association between wealth and food insecurity status of a household. As expected, the households that are at the higher level of wealth index scale are statistically significantly less food insecure as compared to the households that are at the bottom of wealth index hierarchy (lowest quintile). For example, the households that belonged to the second wealth quintile had 2.15 points lower score (regression coefficient =-2.15; p<.001, model 2, Table 12.3) than those that are at the lowest wealth quintile, adjusting for geographic controls. When caste/ethnicity was simultaneously included in the analysis of model 2, the results in model 3 remained with a slight reduction in the magnitude of the association. In model 4, when other household level controls were included along with controls used in model 3, the association still holds (regression coefficient =-1.81; p<.001, model 4, Table 12.3) than those that are in the lowest level of wealth status. Similarly, the households that belonged to the fifth (the highest) wealth quintile had 6.49 points lower score than those that are at the lowest wealth quintile. Interestingly, food insecurity score systematically reduced with the increase of wealth status.

9. Other Factors Associated with Household Food Security Status

Table 12.1 that presents the distribution of other variables shows that on average, a household had five family members which ranged from 1 to 31. Seventy-one percent households were headed by a male as compared to only 29 percent households headed by females. The average age of the household head was 43 years (ranging from 15 to 95 years). Nearly 40 percent of the household heads did not have formal schooling. Sixteen percent of them had 1–4 years of schooling, 7 percent had completed primary education, 19 percent had 6–8 years of schooling, 9 percent had 9–10 years, and only 9 percent had 11 years or more education.

Geographically, 43 percent households were located in the Terai region, 40 percent were from the hills, and 17 percent were from the Mountain region. Seventy-two percent households were from the rural areas. By province, the highest proportions of households were from Province 1 (23 percent) and the lowest proportions were from Province 6 (7 percent).

Seventy percent of households owned some farmland. In addition, 33 percent households owned cattle/buffalo, 44 percent owned cow/bulls, 48 percent owned goats, 43 percent owned chicken, and 12 percent owned pigs.

Now, let's turn to the multivariate results of our controls in Table 12.3—the associations between various controls used in the analysis and the outcome—household food security or insecurity scores. In general, the results of our controls are as expected, net of all other factors. For example, the large household size as associated with high food insecurity, which is expected. A household with an older head was less food insecure as compared to a household head with a younger household head. Similarly, an increase in education of a household head was negatively associated with the food insecurity score. Those households that did not own land and/or animals reported being more food insecure as compared to those that owned these assets. Both these results suggest that land and livestock ownerships are important factors to reduce food insecurity in Nepal.

If a household head is a male, the household reported more food insecurity as compared to those households with a female household head. This finding seems plausible. In developing countries, women spend much time in agriculture and contribute to production. In addition, males may be outside of home due to out-migration and the households with a female head may have been working hard to reduce food insecurity or it could be due to the remittances received by migrant-sending households. However, the evidence from Brazil shows that this is the opposite. The likelihood of food insecurity in female-headed households was reported to be greater than those of male-headed households (Felker-Kantor & Wood 2012). While this is not the focus of this paper but it may be an important agenda for future research. This analysis of the controls allows us the examination of the internal validity of our controls and provides confidence on the external validity or generalizability of our results.

10. Conclusions and Policy Implications

This chapter examined the associations between caste/ethnic and household wealth status and household level food security using the nationally representative DHS (2011) data from Nepal. Our findings show that both caste/ethnicity and household wealth status are independently associated with household food security, adjusting for geographic and household level factors known to influence food security.

First, the food security status of a household is found to be significantly associated with a household's food insecurity status. As we argued earlier, evidence revealed that the food security status of households varied by caste/ethnicity favoring advantaged caste/ethnic groups. This evidence further suggests that caste/ethnicity requires attention while formulating food policies in Nepal. The results clearly showed that not all the households that belong to other castes/ethnic groups are relatively food insecure as compared to so-called advantaged caste group such as Brahmin and Chhetri caste. In fact, the households that belonged to Terai Others and Hill Janajati are more food secure than Brahmin/Chhetri. However, the households that belonged to the Hill Dalit, Terai Dalit, and Muslims had significantly higher amount of food insecurity score as compared to the Brahmin/Chhetri households.

Second, as per the common perception that caste/ethnicity and household wealth status are intertwined and if either caste/ethnicity or household's wealth status is taken into consideration for policy, that will resolve the food security issue. However, because both of these explanatory factors are independently associated with household food security status, this assumption is ruled out. However, there is quite a large mediation effect which can be shown by the reduction of the magnitude and a change in the direction of the relationships of caste/ethnicity after controlling for household wealth status.

Third, a household's wealth status is also independently associated with household food security status. The households with lower wealth status were significantly more likely to report greater level of food insecurity, adjusting for all other factors.

Fourth, as per the common assumption that caste/ethnicity and household wealth status are intertwined and if either caste/ethnicity or household's wealth status is taken into consideration for policy purposes, that will resolve the food security issue. However, because both of these explanatory factors are independently associated with household food security status, this assumption is ruled out. The evidence helps us conclude that both of these factors should be considered in policy to solve the food insecurity problem in the country. However, there is quite a large mediation effect which can be shown by the reduction of the magnitude and a change in the direction of relationship of caste/ethnicity after controlling for household wealth status suggesting for the support of mediation effect as well. The theoretically expected results of our controls increase our confidence on the generalizability of our results.

Fifth, the strength of the paper is that it included all socio-economic, demographic, and geographical factors in examining the association between social inequalities factors and food security. The findings are robust, generalizable to the entire nation, as well as it can be an important reference for food security policy design and policy implementation. This paper makes important contribution to framing food security policies at the time when the constitution of Nepal 2015 has expressed state's commitment to food security enshrining the right to food as fundamental rights of every individual. However, the paper is not free from some limitations. One limitation is its cross-sectional data and our results are rather associations. Second, the evidence is at the country level and provinces and geographic niches in Nepal have social-economic and cultural variations.

Sixth, in conclusion, social inequalities factors should be considered while designing food security for all. In the context of Nepal, a person belonging to the lowest social gradient (i.e., belonging to underprivileged castes/ethnic groups, poor economic status, and lower level of educational attainment) were less likely to be food secured than compared to their counterpart. Moreover, provincial differences in food security status have also been found. Therefore, it is imperative to reach the lower social gradients and to provide food security to these sections of populations as the Article 36 of the 2015 Constitution of Nepal commits— the right to food as fundamental rights for every individual.

Finally, philosophically, inequality is a global problem boasted by cultures, religions, political, economic, and social structures of the society. This problem is as old as human civilization; therefore, to overcome this deep-rooted problem it is necessary to evaluate all societal norms and values and societies need to be capable to change those aspects which divide the society and help to increase inequality. We need to change societal behavioral pattern and the egocentric mentality of the society by boasting the principles of Bashudhaiva Kutumbakam (the whole of humanity is one family. We all are related and equal in this planet) and Live and Let Other Live (Bhandari 2019a, 2019b; 2020).

Conflict of Interest: The authors do not have any conflict of interest.

References

Acharya, M., & Bennet, L. 1981. The status of women in Nepal: The rural women of Nepal. Kathmandu, Nepal: Tribhuvan University, Centre for Economic Development and Administration.

Allison, E. H. & Ellis, F. (2001). The livelihoods approach and management of small-scale fisheries. Marine Policy 25, 377-388.

Bennet, L., Dahal, D.R. & Govindasamy, P. (2008). Caste, ethnic and regional identity in Nepal: further analysis of the 2006 Demographic and Health Surveys. Calverton, Maryland, USA: Macro International Inc.

Beyene, F. & Muche, M. (2010). Determinants of food security among rural households of Central Ethiopia: An empirical analysis. Quarterly Journal of International Agriculture, 49 (4), 299–318.

Bhandari Medani P. (2019)a. "BashudaivaKutumbakkam"- The entire world is our home and all living beings are our relatives. Why we need to worry about climate change, with reference to pollution problems in the major cities of India, Nepal, Bangladesh and Pakistan. Adv Agr Environ Sci. (2019);2(1): 8–35. DOI: 10.30881/aaeoa.00019 (second part) http://ologyjournals.com/aaeoa/aaeoa_00019.pdf

Bhandari, Medani P, (2019)b. Live and let other live- the harmony with nature /living beings-in reference to sustainable development (SD)- is contemporary world's economic and social phenomena is favorable for the sustainability of the planet in reference to India, Nepal, Bangladesh, and Pakistan? Adv Agr Environ Sci. (2019);2(1): 37–57. DOI: 10.30881/aaeoa.00020 http://ologyjournals.com/aaeoa/aaeoa_00020.pdf

Bhandari, Medani P. (2020), Getting the Climate Science Facts Right: The Role of the IPCC (Forthcoming), River Publishers, Denmark / the Netherlands- ISBN: 9788770221863 e-ISBN: 9788770221856

Bhandari, Prem. (2019). Regional Variation in Food Security in Nepal. Dhaulagiri Journal of Sociology and Anthropology, 12: 1-10. https://doi.org/10.3126/dsaj.v12i0.22174.

Bhandari, P. B., Karki, L. B., & Rasali, D. P. (2020). Food security and sustainable livelihood framework. In D. P. Rasali, P. B. Bhandari, U. Karki, M. N. Parajulee, R. Acharya & R. Adhikari (Eds.), Principles and Practices of Food Security: Sustainable, Sufficient, and Safe Food for Healthy Living in Nepal (pp. 9–35). USA: Association of Nepalese Agricultural Professionals of Americas (NAPA).

Bhandari, P. B. (2013). Rural livelihood change? Household capital, community resources and livelihood transition." Journal of Rural Studies, 32:126-136. PMCID: PMC3772533.

Bhandari, P. (2016). Remittance receipt by households in rural Nepal: Does migrant's destination make a difference? Dhaulagiri Journal of Sociology and Anthropology, 10, 1–35.

Bhandari, P. B. (2006). Technology use in agriculture and occupational mobility of farm households in Nepal: Demographic and socioeconomic correlates (Unpublished doctoral dissertation). The Pennsylvania State University, Department of Agricultural Economics and Rural Sociology, University Park.

Blaikie, P. M., Cameron, J. & Seddon, J. (2000). The Struggle for basic needs in Nepal. Delhi: Adroit Publishers.

Boserup, E. (1971). Women's Role in Economic Development. London, England: George Allen and Unwin.

Burchi, F. (2006). Education, human development, and food security in rural areas: Assessing causalities. The Human Development and Capability Approach. Groningen.

Coates, J., Frongillo, E. A., Rogers, B. L., Webb, P., Wilde, P. E., & Houser, R. (2006). Commonalities in the experience of household food insecurity across cultures: What are measures missing? Journal of Nutrition, 136(5), 1438S–1448S.

Coates, J., Swindale, A. & Bilinsky. P. (2007). Household food insecurity access scale (HFIAS) for measurement of food access: Indicator guide. Version 3. FANTA Food and Technical Assistance, USAID, Washington, DC. https://www.fantaproject.org/sites/default/files/resources/HFIAS_ENG_v3_Aug07.pdf

De Janvry, A. (1981). The agrarian question and reformism in Latin America. Baltimore: Johns Hopkins University Press.

DFID and World Bank (2006). Unequal citizens: Gender, caste and ethnic exclusion in Nepal. A Co-publication of the World Bank and the Department for International Development, U.K., Kathmandu, Nepal.

FAO (2006). Food security. Policy Brief. Issue 2, June 2006. http://www.fao.org/forestry/13128-0e6f36f27e0091055bec28ebe830f46b3.pdf

FAO (2011). World livestock 2011: Livestock in food security. Food and Agriculture Organization of the United Nations. Rome. http://www.fao.org/docrep/014/i2373e/i2373e.pdf

Felker-Kantor, E. & Wood, C.H. Food Sec. (2012) 4: 607. https://doi.org/10.1007/s12571-012-0215-y.

Garson, G. D. (2013). Fundamentals of hierarchical linear and multilevel modeling. Hierarchical linear modeling: Guide and applications by Garson. Los Angeles, London, New Delhi, Singapore, Washington DC: Sage Publications Inc, pp. 3–26.

Goli, S., N.K. Maurya, Moradhvaj, Prem Bhandari. 2019. Regional Differentials in Multidimensional Poverty in Nepal: Rethinking Dimensions and Method of Computation. Sage Open, 1-18. https://doi.org/10.1177/2158244019837458.

GoN (2012). Agriculture atlas of Nepal. National Planning Commission Secretariat, Government of Nepal, Kathmandu.

GoN (2015). Sustainable development goals 2016–2010. National (Preliminary) Report. National Planning Commission. Nepal.

GoN (2018). Multidimensional poverty index 2018. National Planning Commission. Government of Nepal, Nepal. https://www.npc.gov.np/images/category/Nepal_MPI.pdf Retrieved on August 22, 2019.

Iftikhar, S. (2017). Human capital development and food security nexus: An empirical appraisal from districts of Punjab province. J Food Drug Res., 1(1), 3–9.

Joshi, K. D., Conroy, C. & Witcombe, J. R. (2012). Agriculture, seed, and innovation in Nepal: Industry and policy issues for the future. International Food Policy Research Institute (IFPRI). Project Paper, December 2012. Washington, D.C.

Kumar, S. K., & Hotchkiss, D. (1988). Consequences of deforestation for women's time allocation, agricultural production, and nutrition in hill areas of Nepal (Research Report No. 69). Washington, DC: International Food Policy Research Institute.

Little, P.D., Stone, M.R., Mogues, T., Castro, A.P. & Negatu, W. (2006). Moving in place: drought and poverty dynamics in south Wollo, Ethiopia. Journal of Development Studies, 42(2), 200–225.

Maharjan, K. L. & Khatri-Chhetri, A. (2006). Household food security in rural areas of Nepal: Relationship between socio-economic characteristics and food security status. Poster paper prepared for presentation at the International Association of Agricultural Economics Conference, Gold Coast, Australia August 12–18, 2006.

Maharjan, K. L. & Joshi, N. P. (2011). Determinants of household food security in Nepal: A binary logistic regression analysis. Journal of Mountain Science, 8(3), 403–413.

Mahesh, R. (2002). Labour mobility in rural areas: A village-level study. Discussion Paper No. 48. Kerala Research Programme on Local Level Development, Center for Development Studies, Thiruvananthapuram, India.

Maxwell, D.G. (1996). Measuring food insecurity: The frequency and severity of coping strategies. Food Policy, 21(3), 291–303.

Napoli, M. (2010/11). Towards a food insecurity multidimensional index (FIMI). Master in Human Development and Food Security (2010/2011). nitroPDFprofessional.

NDHS (2011). Nepal demographic and health survey 2011. Ministry of Health and Population (MOHP) [Nepal], New ERA, and ICF International Inc. (2012). Kathmandu, Nepal: Ministry of Health and Population, New ERA, and ICF International, Calverton, Maryland.

NESAC (1998). Nepal human development report 1998. Nepal South Asia Centre, Nepal.

Ogena, N. B. & De Jong, G. F. (1999). Internal migration and occupational mobility in Thailand. Asian and Pacific Migration Journal 8(4), 419-46.

Rahman, S. A. (2008). Women's involvement in agriculture in northern and southern Kaduna State, Nigeria. *Journal of Gender Studies*, *17*(1), 17-26.

Regmi, M., Paudel, K. P. & Mishra, A. K. (2015). Impact of remittance on food security in Bangladesh. Selected Paper prepared for presentation at the Southern Agricultural Economics Association (SAEA) Annual Meeting, Atlanta, Georgia, February 1-4.

Shobe, M. A., Narcisse, M. & Christy. K. (2018). Household financial capital and food security. Journal of Poverty 22 (1), 1-22.

Smith, M., Pointing, J. & Maxwell, S. (1993). Household's food security, concepts and definitions; An annotated bibliography.

Sseguya, H. (2009). Impact of social capital on food security in southeast Uganda (unpublished Dissertations). 10747. http://lib.dr.iastate.edu/etd/10747.

Thapa, G. (2009). Smallholder farming in transforming economies of Asia and the Pacific: Challenges and opportunities. A discussion paper prepared for the side event organized during the thirty third session of IFAD's Governing Council, 18 February 2009.

UNDP (2011). Human development report 2011 sustainability and equity: A better future for all. UNDP, Washington, D.C.

World Food Summit (1996). Declaration on world food security. World Food Summit, 13-17 November, 1996, Rome. http://www.fao.org/docrep/003/w3613e/w3613e00.htm.

Chapter 13

Conclusion: Social Dimensions of Inequality

Medani P. Bhandari

Inequality can be seen as a global communicable disease of human civilization. Social inequalities are seen in the service sectors—access to healthcare, access to education, social protection, access to housing systems, childcare, elderly cares etc. Another neglected but deep-rooted issue is cultural inequality, which segregates people from the mainstream. The minority groups face recognition problems of their social status, language, religion, customs, norms etc. The chapters of this book try to unveil the real picture of these issues with exemplary case studies from various countries.

It is true that primarily the world is divided between haves and have-nots; and social strata are created as comfort zones to those who dominate the social, economic, and political systems. Such unequal strata can be grounded on innumerable factors—such as social, cultural, political, geographical, or due to environmental (anthropogenic or natural) catastrophe. On the basis of various case studies, this book encourages to rethink societal development through the lens of growing inequalities and disparities. The book presents new insights for evaluating the progress on social development, with the coverage of major inequalities due to gender, age, origin, ethnicity, disability, sexual orientation, class, and religion.

The rules and regulations, social norms and values, are always created and implemented by the social elite; therefore, the marginalized groups have no or very minimal stake in the society, and they are always the victims. There is a need for a complete social, economic, and political reform to minimize the designed division. Until or unless the marginalized group is empowered, the inequality issue cannot be solved or even minimized.

Social inequalities are grounded and boasted by the religious and cultural systems, therefore, social phenomena automatically creates the socially biased inequal circumstances particularly to the women. In another words, inequality in the society is created by the elite groups for their benefit, it is present in each society; however, most dangerously, the gender inequality is maintained by the males of each houses. The social ethics and cultures are prepared by the male; therefore, it is omnipresent. Another neglected but deep-rooted issue is cultural inequality, which segregates people from mainstream. The minority groups face recognition problems of their social status, language, religion, customs, norms etc.

Social, cultural, and religious discriminations are often shadowed in the society, because the show-up face is made up by the elite males who hold the knowledge of made-up stories; therefore, actual discriminated face of the society is rarely understood. Society functions through the economic, social, religious, cultural, and political (the pillars of the society) systems. Therefore, even if one of the pillars becomes dysfunctional, it can disbalance the entire social order.

"This book is the result of collective work of scholars around the world. At the present time when we talk of inequality, we immediately focus our vision to economic inequality. But there are other major areas where human has discriminated their fellow members in terms of culture, languages, ethnicity, sex, social status and others" (forward).

"Achieving social equity is an important issue everywhere. Despite the vast improvements in human well-being the world has seen in the past, social inequality remains a stubborn problem in the human society. Whether it is about access to positions of power, employment opportunities, health and education services, availability of decent housing or other amenities, there is a gap between what is available for the more privileged and marginalized groups in societies.

The social inequalities can be driven by various factors such as economic, racial, and ethnic class, and sexual orientation or gender differences. For example, in most societies, minorities and women often face discrimination in workplace, political arena and social life in general. To preserve and protect their power and advantageous positions, the elites and groups belonging to the higher echelon in the society may create policies, laws, regulations and protocols to secure their position of privilege in the system. The legal and policy regimes could seem fair, but there might be a tacit agreement among many in the advantageous situation to help preserve the status quo" (Forward).

"One must recognise the stark reality that this would not be attained in post-pandemic era when the world has been witnessing states' failing strategies in containing the spread of a contagious disease, simply called COVID-19. This situation is heavily impacting on peoples' lives and psychic in an eroding environment of public health care system and rapidly deteriorating quality of life, coupled with vanishing small and medium scale economic enterprises and loss of employment opportunities. Recovery and transformation from this dual-edge shocks (health and economy), created by the global pandemic, is unknown and faltering, and future is uncertain. Multilateralism, in a multi-polar world tension, faces serious challenges in addressing growing socio-economic inequalities. We need transformative re-thinking whether traditional measures are still relevant for new challenges if we are serious about finding pragmatic approaches to equitable equilibrium level of wellbeing for all, and thereby reducing deep disparities and inter-social inequalities. Although this may not sound new, but the landscape has altered significantly warranting re-definition of practical approaches and realistic measures for reducing inequalities for creating a semblance of equity equilibrium. Time to contemplate differently and deep than in the past. I leave this question to scholars, academics, economists, sociologists, and political pundits to drill down for a workable answer" (Forward).

This book unveils societal inequalities through regional and country-specific cases, mostly with the social principles. Each chapter uses and strictly follows the prescribed scholarly standards of scientific writing and provides new insights; therefore, each chapter is unique. The chapters in this book provide in-depth problems and consequences of social inequality, with the case studies of various countries of the world. Each chapter is unique, complete, theoretically grounded, and supported by the factual data.

Whereas the first chapter by *Medani P. Bhandari,* provides the theoretical framework of inequality; with reference of various scholarly work by the social scientists as well as the organizations; and other stakeholders (Barker 2004; Beall, Guha-Khasnobis and Kanbur, 2012; Bhandari 2018; 2019, 2020; Bhandari and Shivindina 2019; Brockmann, and Delhey 2010; Carson, 1962; Cheng, 2018; Colyvas, and.Powell 2006; Crossan, and Bedrow, 2003; Daly, 2007; DESA 2013; Freistein, and. Mahlert, 2015; Holvino, and Merrill-Sands 2004; IISD 2005; IUCN, UNDP & WWF 1991; Jackson, 2009; Jan-Peter et al 2007; Harris 2000; Lélé 1991; Marshall 1961; Marx 1953ff; Meadows, 1998; Meadows, et al 1972; Norgaard, 1988; North, 1981; Purvis et al 2018; Rockström 2009; Scott, 2003; Soares, et al 2014; Sterling, 2010; United Nations 1972;1997; 2002; 2010; 2012, 2015, 2015 a, b, c, d; UNEP 2010; UNESCO 2013; Vander-Merwe and Van-der-Merwe 1999; Vos 2007;

Wals 2009; World Bank 2015; WCED 1987; WID 2018; Wright 1994; World Economic Outlook 2007; World Bank. 2011; 2020; United Nations 2020; World Economic Forum 2020; ILO 2020; UNDP 2020; United Nations 1972;1997; 2002; 2010; 2012, 2015, 2015 a, b, c, d; UNDP 2019; Blau, and Kahn 2017; Goldin and Rouse 2000; Neumark, Bank, and Nort, 1996; Olivetti, and Petrongolo, 2008; Ortiz-Ospina 2018; Jayachandran 2015; Christiansen and Jensen 2019; Bhandari and Shvindina 2019; United Nations 2020a, b; United Nations 2020; Dehm, 2018; Chhabria 2016; Freistein and Mahlert 2015; United Nations 2020). Chapter paves the theoretical grounding of inequality as well as unveils the causes of incremental inequalities. Chapter shows the practical examples (through pictural stories) of how, discrimination due to gender, age, origin, ethnicity, disability, sexual orientation, class, and religion has been playing a major role in increasing social inequality, particularly in the developing countries.

Chapter 2 by G.B. Sahqani Sahqan, provides the Inequality, a Historical Perspective—With Reference to Society, Politics, and the Economy. This chapter unveils the causes inequality, and how inequality has been persistent for centuries given multidisciplinary factors. Further, chapter suggests the possible action to reduce inequality.

Similarly, Chapter 3, by Salvin Paul and Maheema Rai presents a real picture of the Religion and Gender Inequality in India, whereas they conclude that, the gender inequality is structured in the religious and societal customs appears to exist all over the world. These inequalities that are sanctioned through the norms and values of cultures and religions have become universal phenomenon where women themselves support such practices to make them officially invisible. In this backdrop, how women are unable to grasp and comprehend in their daily life the understanding on equality and its of late discourses in a religiously driven society like in India where secular understanding of equality has become major issue of contemporary debate.

Chapter 4, by Tetyana Vasilyeva, Anna Buriak, Yana Kryvych, and Anna Lasukova, titled Behavioral View on Inequality Issue: Bibliometric Analysis, uses the concept of behavioral economics as basis for future policy initiatives to promote equal opportunities in society. It is agreed that nowadays policy of reducing inequality should extend beyond standard manifestations of inequality like income, gender, social status and focus on processes and determinants that generate inequality in society. As the most widespread research view is about the consequences and causes of inequality on macroeconomic levels, this chapter encompasses conceptual framework for understanding behavior and decision-making of economic units under economic and social inequality in the country.

In Chapter 5, Oleksii Demikhov examines the Public Health Policy Communication with Other Policies in The Context of Inequality. Chapter explores the health sector of the EU and compares it with Ukraine in order to formulate proposals for mitigating health inequalities in access to health services, as well as developing new standards and to have an integrated approach to work out an effective public health policy. Author uses a multidisciplinary and systematic approach in research as a baseline, methods of analysis, synthesis, generalization, comparison, and economic-statistical method.

Chapter 6, by Mirjana Radovic-Markovic, Milos Vucekovic, and Aidin Salamzadeh, investigates Employment Discrimination and Social Exclusion, in Serbia, whereas they analyze the situation of particular groups in society—women, people over 55 years old, Roma people, and persons with disabilities. In some cases, a comparison with Slovenia is made in order to explain the prevalence of discriminatory experiences and their relevance for the study of social policy against discrimination. Chapter shows that the integrated approaches and strategies on all levels of society are needed to improve the social inclusion of marginalized groups.

Likewise, Chapter 7 by Sotnyk Iryna, Oleksandr Kubatko, Tetiana Kurbatova, Leohid Melnyk, Yevhen Kovalenko, and Almashaqbeh Ismail Yousef Ali, presents the Impact of Subsidies on Social Equality and Energy Efficiency Development, with the case of Ukraine. Given the low incomes of the majority of population and high prices for utilities, the issue of reducing social inequality in Ukraine is extremely urgent. Today, it is being addressed by the government through the operation of a large-scale state program of utilities subsidies, a significant part of which is paid for energy. At the same time, the national economy has a high level of energy intensity and the housing and utilities sector is the main contributor to this outcome. Authors have proposed improvement for the state mechanism of utilities subsidies, which should (1) be combined with the existing state programs of energy-efficient development, in particular, the program of "warm" credits and feed-in tariff for renewable energy objects in households, (2) enhance the expansion of cooperation with international financial organizations to implement energy-efficient measures by Ukrainian households and utilities suppliers and (3) allow comprehensive solutions to eradicate the energy poverty, ensure the growth of energy efficiency in the national economy, and provide social equality among society members.

Chapter 8 by Oksana Zamora, and Svitlana Lutsenko, titled Combating Inequality Via an Intercultural Strategy of The City: A Case Study of a Ukrainian City also presents the case study of Ukraine, where they examine how inequality is increasing in city environment. Chapter is devoted to the analysis of a comprehensive approach of equality promotion via designing and implementing the intercultural strategy of the city. A multicultural city of Sumy at the North-Eastern part of Ukraine is used as a case study. Currently the city has just adopted the intercultural strategy incorporated within the city development strategy. It presents the best Ukrainian practices and challenges of the solution search for the issues of unequal access to the city infrastructure and municipal services in key spheres of the community life of the city visitors, migrants, and minorities of all kinds. Chapter presents the empirical analysis of the cultural and educational components which appear to be essential in building the proper mentality and understanding of the need for the equality within the local community. There are a number of outlined strategic solutions based on the best world practices of the intercultural cities and adapted to the reality of a Ukrainian provincial city.

Chapter 9 authored by Tetiana Semenenko, Volodymyr Domrachev, and Vita Hordiienko, analyzes the Causes and Ways to Overcome Socio-Economic Inequality in Ukraine. Chapter proposes, that, the main tools of state regulation of social inequality should be: legal support of the social sphere; direct state payments from budgets of different levels to finance the social sphere; social transfers in the form of social subsidies; establishment of social guarantees (minimum wage, subsistence level, consumer basket, minimum size of old-age pensions, scholarships, etc.); state regulation of prices for basic necessities; social insurance; implementation of state programs to solve certain social problems; establishment of social and environmental norms and standards, and control over their observance.

Chapter 10 by Shapoval Vladyslav, Medani P. Bhandari, and Shvindina Hanna, titled, Education as A Tool for Prevention Discrimination Against LGBTQ+ People, provides a unique picture of LGBTQ+ People, how they are discriminated in socio-political and economic arena. In general, inequality can be seen in every spheres of political, social, economic, cultural, religious systems; however, LGBTQ+ people's struggles are relatively new and painful. This chapter outlines the general history of such struggles and the current situation of LGBTQ+ people. Ukraine remains on the process of developing Human Rights for LGBTQ+ and creating an inclusive society. It became possible after the collapse of the USSR and the Revolution of Dignity that boosted the Human Rights activities. We believe education is an essential tool for fighting with discrimination and harassment against the LGBTQ+ community. Results showed that students that have an LGBTQ+ curriculum at school are less likely to become wrongdoers and LGBTQ+ students suffer less bullying. That is why it is so vital to have governmental support and assistance that may provide not only a legal basis but comprehensive protection and support on different levels.

The chapter 11 investigates the issue of access to the continuing professional development (CPD) of in-service English as foreign language (EFL) teachers in Ukraine in the context of current educational reform. Chapter reveals the barriers and challenges that teachers with different social and professional background face while trying to get involved into the high quality CPD activities. In addition, the chapter examines the existing ways of in-service EFL teachers training and the mismatch between the provided activities and those needed by teachers. The study utilizes holistic description, analysis, and surveys to address teachers' experience in CPD activities, as well as their beliefs on forms and methods of such training and provided an in-depth exploration of teachers' professional needs and the role of certain factors influencing the access of different teacher groups to quality professional development. Poor technological facilities, unclear goals, and practices, as well as financial factors, such as funding to support travel or cost of attendance, were identified as key barriers to CPD access.

Chapter 12 by Prem B. Bhandari, Madhu Sudhan Atteraya, and Medani P. Bhandari, titled, Socio-Cultural Dimensions of Socio-Economic Inequality in Household Food Security in Nepal, examines the association between socio inequalities and food security in Nepal. Specifically, we investigate the associations between caste/ethnicity and household wealth status on household-level food security, after controlling for several other demographic, geographic, and other factors. For this purpose, we used nationally representative data from the 2011 Nepal Demographic Health Surveys (NDHS 2011). Results from multilevel regression (binary logistic and OLS regressions) show that various household livelihood capitals are important in shaping household food security status of a household. These findings provide an implication in implementing 'rights for food' as mandated by the Article 36 of the Constitution of Nepal.

The book also contents the forwards from prominent scholars of social sciences—academia, former staff of United Nations Agencies, high ranking government officials who argue that; there is a need transformative re-thinking whether traditional measures are still relevant for new challenges if we are serious about finding pragmatic approaches to equitable equilibrium level of wellbeing for all, and thereby reducing deep disparities and inter-social inequalities. Although this may not sound new, but the landscape has altered significantly warranting re-definition of practical approaches and realistic measures for reducing inequalities for creating a semblance of equity equilibrium. Time to contemplate differently and deep than in the past. There are questions to scholars, academics, economists, sociologists, and political pundits to drill down for a workable answer.

Reference

Barbier, E. B., (1987), 'The concept of sustainable economic development,' Environmental Conservation, Vol. 14, No. 2 (1987), pp. 101–110.

Barker, Chris (2004), The SAGE Dictionary of Cultural Studies, SAGE Publications, London / Thousand Oaks / New Delhi https://zodml.org/sites/default/files/%5BDr_Chris_Barker%5D_The_SAGE_Dictionary_of_Cultural__0.pdf

Berg, Andrew, and Jonathan Ostry (2011). Inequality and unsustainable growth: two sides of the same coin? IMF Staff Discussion Note. SDN/11/08. Washington, D.C.: International Monetary Fund. 8 April.

Bertelsmann Stiftung and Sustainable Development Solutions Network (2018), SDG Index and Dashboards Report 2018-Global Responsibilities, Implementing the Goals, G20 and Large Countries Edition. www.pica-publishing.com, http://www.sdgindex.org/assets/files/2018/00%20SDGS%202018%20G20%20EDITION%20WEB%20V7%20180718.pdf

Bhandari, Medani P. (2019), Inequalities with reference to Sustainable Development Goals in Bhandari, Medani P. and Shvindina Hanna (edits) Reducing Inequalities Towards Sustainable Development Goals: Multilevel Approach, River Publishers, Denmark / the Netherlands- ISBN: Print: 978-87-7022-126-9 E-book: 978-87-7022-125-2

Bhandari Medani P. and Shvindina Hanna (2019), The Problems and consequences of Sustainable Development Goals, in Bhandari, Medani P. and Shvindina Hanna (edits) Reducing Inequalities Towards Sustainable Development Goals: Multilevel Approach, River Publishers, Denmark / the Netherlands- ISBN: Print: 978-87-7022-126-9 E-book: 978-87-7022-125-2

Bhandari, Medani P (2017), Climate change science: a historical outline. Adv Agr Environ Sci. 1(1) 1-8: 00002. http://ologyjournals.com/aaeoa/aaeoa_00002.pdf

Bhandari, Medani P. (2018), Green Web-II: Standards and Perspectives from the IUCN, Published, sold and distributed by: River Publishers, Denmark / the Netherlands ISBN: 978-87-70220-12-5 (Hardback) 978-87-70220-11-8 (eBook),

Blau, Francine D., and Lawrence M. Kahn. (2017). "The Gender Wage Gap: Extent, Trends, and Explanations." Journal of Economic Literature, 55(3): 789-865.

Brockmann, Hilke, and Jan Delhey (2010), "The Dynamics of Happiness". Social Indicators Research 97, no.1 (2010): 387–405.

Callanan, Laura and Anders Ferguson (2015), A New Pilar of Sustainability, Philantopic-Creativity, Foundation Center, New York, https://pndblog.typepad.com/pndblog/2015/10/creativity-a-new-pillar-of-sustainability.html

Carson, Rachel (1962), Silent Spring, A Mariner Book, Houghton M1fflin Company, Boston, New York

Chhabria, Sheetal (2016), Inequality in an Era of Convergence: Using Global, Histories to Challenge Globalization Discourse, World History Connected Vol. 13, Issue 2.

Christiansen C.O., Jensen S.L.B. (2019) Histories of Global Inequality: Introduction. In: Christiansen C., Jensen S. (eds) Histories of Global Inequality. Palgrave Macmillan, Cham. https://doi.org/10.1007/978-3-030-19163-4_1

Clark, W. C., and R. E. Munn (Eds.) (1986), Sustainable Development of the Biosphere, Cambridge: Cambridge University Press

Daly, H. E (2007), Ecological Economics and Sustainable Development, Selected Essays of Herman Daly, Advances in Ecological Economics, MPG Books Ltd, Bodmin, Cornwall http://library.uniteddiversity.coop/Measuring_Progress_and_Eco_Footprinting/Ecological_Economics_and_Sustainable_Development-Selected_Essays_of_Herman_Daly.pdf

Fischer, D.; Jenssen, S.; Tappeser, V. Getting (2015) an Empirical Hold of the sustainable University: A Comparative Analysis of Evaluation Frameworks across 12 Contemporary Sustainability Assessment Tools. Assess. Eval. High. Educ. 40, 785–800

Fobes (2019), America's Wealth Inequality Is At Roaring Twenties Levels, Contributor-Jesse Colombo, Forbes (https://www.forbes.com/sites/jessecolombo/2019/02/28/americas-wealth-inequality-is-at-roaring-twenties-levels/#62f244642a9c).

Freistein, K., and. Mahlert, B. (2015), The Role of Inequality in the Sustainable Development Goals, Conference Paper, University of Duisburg-Essen

See discussions, stats, and author profiles for this publication at: https://www.researchgate.net/publication/301675130

Galbraith, James K. (2012). Inequality and Instability: The Study of the World Economy Just before the Great Crisis. Oxford: Oxford University Press

Girard, Luigi Fusco (2010), Sustainability, creativity, resilience: toward new development strategies of port areas through evaluation processes, Int. J. Sustainable Development, Vol. 13, Nos. 1/2, 2010 161

Goldin, C., & Rouse, C. (2000). Orchestrating impartiality: The impact of" blind" auditions on female musicians. American Economic Review, 90(4), 715–741

Guidetti, Rehbein (2014), Theoretical Approaches to Inequality in Economics and Sociology, Transcience, Vol. 5, Issue 1 ISSN 2191-1150 https://www2.hu-berlin.de/transcience/Vol5_No1_2014_1_15.pdf

Håvard Mokleiv Nygård (2017), Achieving the sustainable development agenda: The governance – conflict nexus, International Area Studies Review, Vol. 20(1) 3–18

Holvino, E., Ferdman, B. M., & Merrill-Sands, D. (2004), Creating and sustaining diversity and inclusion in organizations: Strategies and approaches. In M. S. Stockdale & F. J. Crosby (Eds.), The psychology and management of workplace diversity (pp. 245-276). Malden, Blackwell Publishing.

https://wir2018.wid.world/files/download/wir2018-full-report-english.pdf

IISD, (2005), Indicators. Proposals for a way forward. Prepared L. Pinter, P. Hardi & P. Bartelmus. International Institute for Sustainable Development Sustainable Development, Canada Retrieved January 8, 2015, from https://www.iisd.org/pdf/2005/measure_indicators_sd_way_forward.pdf.

IISD, (2013), The Future of Sustainable Development: Rethinking sustainable development after Rio+20 and implications for UNEP. International Institute for Sustainable Development Retrieved November 5, 2015, from http://www.iisd.org/pdf/2013/future_rethinking_sd.pdf

ILO (2020). Gender Statistics, International Labor Organization (ILO), 1990-2016

IUCN (1980), World Conservation Strategy: Living Resource Conservation for Sustainable Development. Retrieved November 7, 2015, from https://portals.iucn.org/library/efiles/documents/WCS-004.pdf.

IUCN, UNDP & WWF, (1991), Caring for the Earth. A Strategy for Sustainable Living. International Union for Conservation of Nature and Natural Resources, United Nations Environmental Program & World Wildlife Fund Retrieved November 8, 2015, from https://portals.iucn.org/library/efiles/documents/CFE-003.pdf

IUCN, UNDP & WWF, (1991), Caring for the Earth. International Union for Conservation of Nature and Natural Resources, United Nations Environmental Program & World Wildlife Fund

Jackson, Tim (2009). Prosperity without Growth: Economics for a Finite planet. Abingdon, United Kingdom: Earthscan.

Jan-Peter Voß, Jens Newig, Britta Kastens, Jochen Monstadt† & Benjamin No¨ Lting (2007), Steering for Sustainable Development: a Typology of Problems and Strategies with respect to Ambivalence, Uncertainty and Distributed Power, Journal of Environmental Policy & Planning Vol. 9, Nos. 3-4, September–December 2007, 193–212 https://www.researchgate.net/profile/Jochen_Monstadt/publication/233049753_Steering_for_Sustainable_Development_A_Typology_of_Problems_and_Strategies_with_Respect_to_Ambivalence_Uncertainty_and_Distributed_Power/links/577ff29608ae5f367d370a97/Steering-for-Sustainable-Development-A-Typology-of-Problems-and-Strategies-with-Respect-to-Ambivalence-Uncertainty-and-Distributed-Power.pdf

Jayachandran S. (2015), The Roots of Gender Inequality in Developing Countries, Annu. Rev. Econ. 2015.7:63-88. Https://faculty.wcas.northwestern.edu/~sjv340/roots_of_gender_inequality.pdf Downloaded from www.annualreviews.org

Jonathan M. Harris (2000), Basic Principles of Sustainable Development, GLOBAL DevelopmentAndEnvironmentInstitute,WorkingPaper00-04,GlobalDevelopment and Environment Institute, Tufts University, https://tind-customer-agecon.s3.amazonaws.com/11dc38b4-a3e2-44d0-b8c8-3265a796a4cf?response-content-disposition=inline%3B%20filename%2A%3DUTF-8%27%27wp000004.pdf&response-content-type=application%2Fpdf&AWSAccessKeyId=AKI-

AXL7W7Q3XHXDVDQYS&Expires=1560578358&Signature=T%2Bp-MgFFZjQvmVL8EHsy74Ds%2FKAM%3D

Julia Dehm, (2018), 'Highlighting Inequalities in the Histories of Human Rights: Contestations Over Justice, Needs and Rights in the 1970s' https://doi.org/10.1017/S0922156518000456 (published online 19 September 2018).

Lélé, Sharachchandra M. (1991), Sustainable development: A critical review. World Development, Vol 19, No 6, 607-621 https://edisciplinas.usp.br/pluginfile.php/209043/mod_resource/content/1/Texto_1_lele.pdf

Mair, Simon, Aled Jones, Jonathan Ward, Ian Christie, Angela Druckman, and Fergus Lyon (2017), A Critical Review of the Role of Indicators in Implementing the Sustainable Development Goals in the Handbook of Sustainability Science in Leal, Walter (Edit.) https://www.researchgate.net/publication/313444041_A_Critical_Review_of_the_Role_of_Indicators_in_Implementing_the_Sustainable_Development_Goals

Meadows, D.H. (1998), Indicators and Information Systems for Sustainable Development. A report to the Balaton Group 1998. The Sustainability Institute.

Meadows, D.H., Meadows, D.L., Randers, J. & Behrens III, W.W. (1972), The Limits of Growth. A report for the Club of Rome's project on the predicament of mankind. Retrieved September 20, 2015, from http://collections.dartmouth.edu/published-derivatives/meadows/pdf/meadows_ltg-001.pdf.

Neumark, D., Bank, R. J., & Van Nort, K. D. (1996). Sex discrimination in restaurant hiring: An audit study. The Quarterly Journal of Economics, 111(3), 915–941.

Norgaard, R. B., (1988), 'Sustainable development: A coevolutionary view,» Futures, Vol. 20, No. 6 pp. 606–620.

Olivetti, C., & Petrongolo, B. (2008). Unequal pay or unequal employment? A cross-country analysis of gender gaps. Journal of Labor Economics, 26(4), 621–654.

PAP/RAC, (1999), Carrying capacity assessment for tourism development, Priority Actions Program, in framework of Regional Activity Centre Mediterranean Action Plan Coastal Area Management Program (CAMP) Fuka-Matrouh – Egypt, Split: Regional Activity Centre

Purvis, Ben, Yong Mao and Darren Robinson (2018), Three pillars of sustainability: in search of conceptual origins, Sustainability Science, Springer, 23https://doi.org/10.1007/s11625-018-0627-5r15%20-low%20res%2020100615%20-.pdf

Soares, Maria Clara Couto, Mario Scerri and Rasigan Maharajh -edits (2014), Inequality and Development Challenges, Routledge, https://prd-idrc.azureedge.net/sites/default/files/openebooks/032-9/

UN, United Nations (1972), Report of the United Nations Conference on the Human Environment. Stockholm. Retrieved September 20, 2015, from http://www.un-documents.net/aconf48-14r1.pdf.

UN, United Nations (1997), Earth Summit: Resolution adopted by the General Assembly at its nineteenth special session. Retrieved November 4, 2015, from http://www.un.org/esa/earthsummit/index.html.

UN, United Nations (2002), Report of the World Summit on Sustainable Development, Johannesburg; Rio +10. Retrieved November 4, 2015, from http://www.unmillenniumproject.org/documents/131302_wssd_report_reissued.pdf.

UN, United Nations (2010), The Millennium Development Goals Report. Retrieved September 20, 2015, from http://www.un.org/millenniumgoals/pdf/MDG%20 Report%202010%20En%20

UN, United Nations (2012). Resolution „The future we want ". Retrieved November 5, 2015, from http://daccess-dds-ny.un.org/doc/UNDOC/GEN/N11/476/10/PDF/N1147610.pdf? .

UN, United Nations (2015). Retrieved September 21, 2015, from http://www.un.org/en/index.html.

UN, United Nations (2015b), 70 years, 70 documents. Retrieved September 21, 2015, from http://research.un.org/en/UN70/about.

UN, United Nations (2015c), Resolution „Transforming our world: the 2030 Agenda for Sustainable Development. Retrieved November 5, 2015, from http://www.un.org/ga/search/view_doc.asp?symbol=A/RES/70/1&Lang=E.

UN, United Nations (2015d), The Millennium Development Goals Report 2015. Retrieved November5,2015,from http://www.un.org/millenniumgoals/2015_MDG_Report/pdf/MDG%20

UNDESA-DSD –(2002). United Nations Department of Economic and Social Affairs Division for Sustainable Development, 2002. Plan of Implementation of the World Summit on Sustainable Development: The Johannesburg Conference. New York. UNESCO – United Nations Educational, Scientific and Cultural Organization, 2005. International Implementation Scheme. United Nations Decade of Education for Sustainable Development (2005-2014), Paris.

UNDP (2020). UN Human Development Report, United Nations Development Program http://hdr.undp.org/en/data# http://data.worldbank.org/data-catalog/world-development-indicators
http://reports.weforum.org/global-gender-gap-report-2020/dataexplorer/
http://www.ilo.org/ilostat/

UNEP (2010), Background paper for XVII Meeting of the Forum of Ministers of Environment of Latin America and the Caribbean, Panamá City, Panamá, 26 -30 April 2010, UNEP/LAC-IG.XVII/4, UNEP, Nairobi, Kenya http://www.unep.org/greeneconomy/AboutGEI/WhatisGEI/tabid/29784/Default.aspx.

UNESCO (2013), UNESCO's Medium-The Contribution of Creativity to Sustainable Development Term Strategy for 2014-2021, http://www.unesco.org/new/fileadmin/MULTIMEDIA/HQ/CLT/images/CreativityFinalENG.pdf

United Nations (2015), Transforming our world: the 2030 agenda for sustainable development. New York (NY): United Nations; 2015 (https://sustainabledevelopment.un.org/post2015/transformingourworld, accessed 5 October 2015).

United Nations (2020), World Social Report 2020, Inequality in A Rapidly Changing World,

United Nations (2020). – Gender Statistics, United Nations (Minimum Set of Gender Indicators, as agreed by the United Nations Statistical Commission in its 44th Session in 2013) https://genderstats.un.org

United Nations (2020a), Inequality – Bridging the Divide- Shaping our future together- United Nations 75-2020 and Beyond, United Nations, New York https://www.un.org/en/un75/inequality-bridging-divide

United Nations General Assembly. (1987), Report of the world commission on environment and development: Our common future. Oslo, Norway: United Nations General Assembly, Development and International Co-operation: Environment.

Vander-Merwe, I. & Van-der-Merwe, J. (1999), Sustainable development at the local level: An introduction to local agenda 21. Pretoria: Department of environmental affairs and

Vos, Robert O. (2007), Perspective Defining sustainability: a conceptual orientation, Journal of Chemical Technology and Biotechnology, l 82:334–339 (2007)

WB, The World Bank (2015), World Development Indicators. Retrieved September 2, 2015, from http://data.worldbank.org/data-catalog/world-development-indicators.

WCED (1987), Our Common Future World Commission on Environment and Development New York: Oxford University Press

WID (2018), World Inequality Report 2018, The Paris School of Economics, Inequality Lab, WID.world,

World Bank (2020). – World Development Indicators, World Bank

World Bank (2020). Gender Statistics- World Bank – https://datacatalog.worldbank.org/dataset/gender-statistics

World Bank. (2011). Gender Equality and Development, World Bank http://siteresources.worldbank.org/INTWDR2012/Resources/7778105-1299699968583/7786210-1315936222006/Complete-Report.pdf

World Economic Forum (2020). – Global Gender Gap Report, World Economic Forum

World Economic Outlook (2007), Globalization and Inequality, World Economic Forum, DC

Wright, E. O., (1994), Interrogating Inequalities. New York: Verso

Wu,SOS, Jianguo (Jingle) (2012), Sustainability Indicators Sustainability Measures: Local-Level SDIs494/598–http://leml.asu.edu/Wu-SIs2015F/LECTURES+READINGS/Topic_08-Pyramid%20Method/Lecture-The%20Pyramid.pdf

Yarime, M.; Tanaka, Y. (2012), The Issues and Methodologies in Sustainability Assessment Tools for Higher Education Institutions—A Review of Recent Trends and Future Challenges. J. Educ. Sustain. Dev. 6, 63–77.

Word Phrase Definitions and Abbreviation: Social Inequality as a Global Challenge

Abbreviation

COVID-19	Infectious disease caused by the strain of coronavirus SARS-CoV-2
CERF	UN Office for the Coordination of Humanitarian Affairs
EU	European Union
FAO	Food and Agriculture Organization
FIVIMS	Food Insecurity and Vulnerability Information and Mapping Systems
HIV	Human Immunodeficiency Virus
IBRD	International Bank for Reconstruction and Development (World Bank Group)
ISSC	International Social Science Council
IISD	International Institute for Sustainable Development
ILO	International Labor Organization
IPCC	Intergovernmental Panel on Climate Change
IUCN	International Union for Conservation of Nature and Natural Resources
LGBTQ+	Lesbian, Gay, Bisexual, Transgender, and Queer and/or Questioning
MDG	Millennium Development Goal
MEA	Millennium Ecosystem Assessment
OCHA	UN Office for the Coordination of Humanitarian Affairs
ODI	Overseas Development Institute
PAP	Priority Actions Program
RAC	Regional Activity Centre

SA	South Asia
SD21	Sustainable Development in the 21st Century
SDGs	Sustainable Development Goals
UN	United Nations
UN DESA	United Nations Department of Economic and Social Affairs
UN ESCAP	Economic and Social Commission for Asia and the Pacific
UNAIDS	Joint United Nations Programme on HIV/AIDS
UNCCD	United Nations Convention to Combat Desertification
UNDP	United Nations Environmental Program
UNEP	United Nations Environment Programme
UNESCO	United Nations Educational, Scientific and Cultural Organization
UNHCR	Office of the United Nations High Commissioner for Refugees
UNHCR	UN High Commissioner for Refugees
UNICEF	United Nations Children's Fund
Vasudhaiva Kutumbakam	The entire world is my home and all living being are my relatives
WB	World Bank
WFP	World Food Program
WHO	World Health Organization
WWF	World Wildlife Fund

Word Phrase Definitions

Climate Change (a) The Inter-governmental Panel on Climate Change (IPCC) defines climate change as "a change in the state of the climate that can be identified (e.g., by using statistical tests) by changes in the mean and/or the variability of its properties, and that persists for an extended period, typically decades or longer. Climate change may be due to natural internal processes or external forcing, or to persistent anthropogenic changes in the composition of the atmosphere or in land use" (IPCC).

Climate Change Includes both global warming and its effects, such as changes to precipitation, rising sea levels, and impacts that differ by region (IPCC).

COVID-19 Infectious disease caused by the strain of coronavirus SARS-CoV-2 discovered in December 2019. Coronaviruses are a large family of viruses which may cause illness in animals or humans. In humans, several coronaviruses are known to cause respiratory infections ranging from the common cold to more severe diseases such as Middle East Respiratory Syndrome (MERS) and Severe Acute Respiratory Syndrome (SARS). The most recently discovered coronavirus causes coronavirus disease COVID-19 (WHO, 2020).

Cultural Inequality: discriminations based on gender, ethnicity and race, religion, disability, and other group identities. (ISSC, IDS and UNESCO 2016)

Economic Inequality: differences between levels of incomes, assets, wealth, and capital, living standards and employment. (ISSC, IDS and UNESCO 2016)

Empowerment Refers to increasing the personal, political, social, or economic strength of individuals and communities. Empowerment of women and girls concerns women and girls gaining power and control over their own lives. It involves awareness-raising, building self-confidence, expansion of choices, increased access to and control over resources and actions to transform the structures and institutions which reinforce and perpetuate gender discrimination and inequality (UNICEF 2017).

Environmental Inequality: unevenness in access to natural resources and benefits from their exploitation; exposure to pollution and risks; and differences in the agency needed to adapt to such threats; (ISSC, IDS and UNESCO 2016).

Food Insecurity A situation that exists when people lack secure access to enough safe and nutritious food for normal growth and development and an active and healthy life. It may be caused by the unavailability of food, insufficient purchasing power, inappropriate distribution, or inadequate use of food at the household level. Food insecurity, poor conditions of health and sanitation, and inappropriate care and feeding practices are the major causes of poor nutritional status. Food insecurity may be chronic, seasonal, or transitory. (FIVIMS).

Food Security A situation that exists when all people, at all times, have physical, social and economic access to sufficient, safe and nutritious food that meets their dietary needs and food preferences for an active and healthy life. (FIVIMS)

Gender A social and cultural construct, which distinguishes differences in the attributes of men and women, girls, and boys, and accordingly refers to the roles and responsibilities of men and women. Gender-based roles and other attributes, therefore, change over time and vary with different cultural contexts. The concept of gender includes the expectations held about the characteristics, aptitudes and likely behaviors of both women and men (femininity and masculinity). This concept is useful in analyzing how commonly shared practices legitimize discrepancies between sexes (UNICEF 2017).

Gender A social and cultural construct, which distinguishes differences in the attributes of men and women, girls, and boys, and accordingly refers to the roles and responsibilities of men and women. Gender-based roles and other attributes, therefore, change over time and vary with different cultural contexts. The concept of gender includes the expectations held about the characteristics, aptitudes and likely behaviors of both women and men (femininity and masculinity). This concept is useful in analyzing how commonly shared practices legitimize discrepancies between sexes (UNICEF 2017).

Gender Analysis A critical examination of how differences in gender roles, activities, needs, opportunities, and rights/entitlements affect men, women, girls and boys in certain situations or contexts. Gender analysis examines the relationships between females and males and their access to and control of resources and the constraints they face relative to each other (UNICEF 2017).

Gender Analysis A critical examination of how differences in gender roles, activities, needs, opportunities, and rights/entitlements affect men, women, girls and boys in certain situations or contexts. Gender analysis examines the relationships between females and males and their access to and control of resources and the constraints they face relative to each other (UNICEF 2017).

Gender and Development (GAD) Gender and Development (GAD) came into being as a response to the perceived shortcomings of women in development (WID) programs. GAD-centered approaches are essentially based on three premises: 1) Gender relations are fundamentally power relations; 2) Gender is a socio-cultural construction rather than a biological given; and 3) Structural changes in gender roles and relations are possible. Central to GAD is the belief that transforming unequal power relations between men and women is a prerequisite for achieving sustainable improvements in women's lives. The onus is on women and men to address and re-shape the problematic aspects of gender relations. The conceptual shift from "women" to "gender" created an opportunity to include a focus on men and boys (UNICEF 2017).

Gender Balance -This is a human resource issue calling for equal participation of women and men in all areas of work (international and national staff at all levels, including at senior positions) and in programs that agencies initiate or support (e.g., food distribution programs). Achieving a balance in staffing patterns and creating a working environment that is conducive to a diverse workforce improves the overall effectiveness of our policies and programs and will enhance agencies' capacity to better serve the entire population (UNICEF 2017).

Gender Empowerment Measure (GEM)-Developed by the United Nations system in 1995, GEM measures inequalities between men's and women's opportunities in a country. An annually updated tool, it is used in formulating and applying gender equality indicators in programs. It provides a trends-tracking mechanism for comparison between countries, as well as for one country over time (UNICEF 2017).

Gender Equality The concept that women and men, girls and boys have equal conditions, treatment, and opportunities for realizing their full potential, human rights, and dignity, and for contributing to (and benefitting from) economic, social, cultural and political development. Gender equality is, therefore, the equal valuing by society of the similarities and the differences of men and women, and the roles they play. It is based on women and men being full partners in the home, community, and society. Equality does not mean that women and men will become the same but that women's and men's rights, responsibilities and opportunities will not depend on whether they are born male or female (UNICEF 2017).

Gender Gap Disproportionate difference between men and women and boys and girls, particularly as reflected in attainment of development goals, access to resources and levels of participation. A gender gap indicates gender inequality (UNICEF 2017).

Gender Inequality Index A composite measure which shows the loss in human development due to inequality between female and male achievements in three dimensions: reproductive health, empowerment, and the labor market. The index ranges from zero, which indicates that women and men fare equally, to one, which indicates that women fare as poorly as possible in all measured dimensions (Gender Inequality Index (GII), UNDP, 2015).

Human Immunodeficiency Virus The virus infects cells of the immune system, destroying or impairing their function. Infection with the virus results in progressive deterioration of the immune system, leading to "immune deficiency." The immune system is considered deficient when it can no longer fulfill its role of fighting infection and disease. Infections associated with severe immunodeficiency are known as "opportunistic infections", because they take advantage of a weakened immune system (World Health Organization, WHO, 2017).

Human Rights Agreed international standards that recognize and protect the inherent dignity and the equal and inalienable rights of every individual, without any distinction as to race, color, sex, language, religion, political or other opinion, national or social origins, property, birth, or other status (The Universal Declaration of Human Rights).

Human Rights Agreed international standards that recognize and protect the inherent dignity and the equal and inalienable rights of every individual, without any distinction as to race, color, sex, language, religion, political or other opinion, national or social origins, property, birth, or other status (The Universal Declaration of Human Rights).

Inequality Unfair situation in society when some people have more opportunities, money, etc. than other people. (Cambridge Dictionary, and FAO, 2014).

Inequality-adjusted Human Development Index (IHDI) The IHDI combines a country's average achievements in health, education, and income with how those achievements are distributed among country's population by "discounting" each dimension's average value according to its level of inequality. Thus, the IHDI is distribution-sensitive average level of human development. Two countries with different distributions of achievements can have the same average HDI value. Under perfect equality the IHDI is equal to the HDI but falls below the HDI when inequality rises (UNDP).

Knowledge-based Inequality differences in access and contribution to different sources and types of knowledge, as well as the consequences of these disparities. (ISSC, IDS and UNESCO 2016)

Labor Binding labor law instruments include the 1973 Minimum Age Convention (No. 138), the 1999 Worst Forms of Child Labor Convention (No. 182), the 2011 Domestic Workers Convention (No. 189) of the International Labor Organization and the Protocol of 2014 to the Forced Labor Convention, 1930 (UNICEF 2017).

LGBTQ+ Umbrella term for all persons who have a non-normative gender or sexuality. LGBTQ stands for lesbian, gay, bisexual, transgender, and queer and/or questioning. Sometimes a + at the end is added to be more inclusive. A UNICEF position paper, "Eliminating Discrimination Against Children and Parents Based on Sexual Orientation and/or Sexual Identity (November 2014)," states all children, irrespective of their actual or perceived sexual orientation or gender identity, have the right to a safe and healthy childhood that is free from discrimination (UNICEF 2020).

Maternal Mortality Death of a woman while pregnant or within 42 days of termination of pregnancy, irrespective of the duration and site of the pregnancy, from any cause related to or aggravated by the pregnancy or its management but not from accidental or incidental causes. World Health Organization (UN, 2014).

Political Inequality the differentiated capacity for individuals and groups to influence political decision-making processes and to benefit from those decisions, and to enter into political action. (ISSC, IDS and UNESCO 2016)

Sexual Exploitation Any actual or attempted abuse of a position of vulnerability, differential power, or trust, for sexual purposes. (UN, 2014). Sexual violence under international law encompasses rape, sexual slavery, enforced prostitution, forced pregnancy, enforced sterilization, trafficking and any other form of sexual violence of comparable gravity, which may, depending on the circumstances, include situations of indecent assault, trafficking, inappropriate medical examinations, and strip searches (ONU, 2014).

Social Inequality differences between the social status of different population groups and imbalances in the functioning of education, health, justice, and social protection systems (ISSC, IDS and UNESCO 2016).

Social Protection, or Social Security is a human right and is defined as the set of policies and programs designed to reduce and prevent poverty and vulnerability throughout the life cycle. Social protection includes benefits for children and families, maternity, unemployment, employment injury, sickness, old age, disability, survivors, as well as health protection. Social protection systems address all these policy areas by a mix of contributory schemes (International Labor Organization, ILO 2020).

Son Preference-The practice of preferring male offspring over female offspring, most often in poor communities, that view girl children as liabilities and boy children as assets to the family (UNICEF 2017).

Spatial Inequality: spatial and regional disparities between centers and peripheries, urban and rural areas, and regions with more or less diverse resources. (ISSC, IDS and UNESCO 2016)

Substantive Equality This focuses on the outcomes and impacts of laws and policies. Substantive equality goes far beyond creating formal legal equality for women (where all are equal under the law) and means that governments are responsible for the impact of laws. This requires governments to tailor legislation to respond to the realities of women's lives. Striving for substantive equality also places a responsibility on governments to implement laws, through gender-responsive governance and functioning justice systems that meet women's needs. Substantive equality is a concept expressed in the Convention on the Elimination of All Forms of Discrimination against Women (CEDAW). It recognizes that because of historic discrimination, women do not start on an equal footing to men (UNICEF 2017).

United Nations Girls' Education Initiative (UNGEI) A multi-stakeholder partnership committed to improving the quality and availability of girls' education and contributing to the empowerment of girls and women through education. The UNGEI Secretariat is hosted by UNICEF in New York City (UNICEF 2017).

Women in Development (WID)-A Women in Development (WID) approach is based on the concept that women are marginalized in development-oriented interventions, with the result that women are often excluded from the benefits of development. Hence, the overall objective is to ensure that resources and interventions for development are used to improve the condition and position of women (UNICEF 2017).

Index

Bibliometric 15, 71, 73–74, 77, 80, 82, 84–87

Economic Growth 4, 13, 15, 36, 43, 45, 73, 79, 84–85, 150, 190–191, 194, 206–207

H

Q

R

Systematic Approach 90–91

U

Authors Biographies: Social Inequality as a Global Challenge

Ambika P. Adhikari is a Principal Planner managing the long-range planning division at City of Tempe in Arizona, USA. He has over 30 years of professional and academic experience in urban and environmental planning and international development in several countries including Nepal, India, USA, Canada, Mexico, Kenya, and Fiji.

Ambika was a Village Planner and Project Manager at the City of Phoenix, and a Senior Planner at SRPMIC in Scottsdale, Arizona. He then joined Arizona State University as a Program and Portfolio Manager for Research. Prior to that, Ambika worked as a Director of international programs at DPRA Inc. in Toronto and Washington DC, where he managed international urban and environmental and economic development programs and projects in Canada, USA, Mexico, Nepal, and India.

In Nepal, he served as Nepal's Country Representative of the Switzerland based IUCN—International Union for Conservation of Nature, leading Nepal's national and some Asia regional programs in several environmental and conservation fields. He was also a member of the Government of Nepal's National Water and Energy Commission, the highest policy-making national body in this sector. Ambika was an Associate Professor of Architecture and Planning at the Institute of Engineering, Tribhuvan University, Nepal.

Ambika has authored one book, co-edited six books, published numerous reports and articles and refereed articles in journals. He regularly writes and lectures on sustainability and urban planning, international development, climate change policy, and international environmental policies and programs.

Ambika received his Doctor of Design (DDes.) degree in Urban Planning and Design from Graduate School of Design, Harvard University. He obtained his post-graduate fellowship from Massachusetts Institute of technology (MIT), M. Arch. from University of Hawaii, and B. Arch from M. S, University of Baroda.

He is a member of American Institute of Certified Planners (AICP) and holds the designations of Project Management Professional (PMP) through Project Management Institute (PMI), and LEED AP (ND) through the US Green Building Council (USGBC). He is Fellow of American Society of Nepalese Engineers, where he also serves in the Advisory Council.

Ambika has been active in serving the Nepali community in the US and Canada by volunteering for various community organizations. He was the president of the NRNA National Coordination Council of US and Canada, NRNA Regional Coordinator for North America, Advisor to NRNA International Coordination Council (ICC), and Patron of NRNA ICC. He is the Chair of Board of Trustees for the Association of Nepalese in the Americas (ANA), Board Member of Asta-Ja USA, and Executive Member of Global Policy Forum Nepal (GPFN).

Almashaqbeh Ismail Yousef Ali is Ph.D. student of the Economics, Entrepreneurship and Business Administration Department, Sumy State University. Ismail Almashaqbeh has a master's degree in engineering, graduated Aircraft faculty Kiev city, specialist in development construction and economic sector in Jordan and the Middle East. Almashaqbeh Ismail has provided many lectures and seminars about how to develop construction works and its impacts on the economics sectors. Also, he is a PhD student in Sumy State University on the Economics program.

Medani P. Bhandari holds M.A. Anthropology (Tribhuwan University, Nepal), M.Sc. Environmental System Monitoring and Analysis (ITC-The University of Twente, the Netherlands), M.A. Sustainable International Development (Brandeis University, Massachusetts, USA), M.A. and Ph.D. in Sociology (Syracuse University, NY, USA).

Prof. Bhandari is a well-known humanitarian, author, editor, co-editor of several books and author of hundreds of scholarly papers on social and environmental sciences: a poet, essayist, environment, and social activist, etc. He is a true educator, who advocates that education should not be only for education shake and for employment (get position, earn money, etc.) but should bring meaningful change in students' lives, which could provide the inner peace in life and be able to face any kind of challenges. Prof. Bhandari is also a motivational speaker—life skill coach (how to remain in peace and calm when situation is not in our control, with application of Basudhaiva Kutumbakam principles)—and in writings and teaching, he exemplifies how studies of scientific theories can be enjoyable and applicable in day-to-day life. As such the epistemology of knowledge which is the basis of theory building is considered not fun subject for everybody; however, Prof. Bhandari's works show that, there is always lucid and attractive side in difficult subject.

Dr. Medani P. Bhandari has been serving as a Professor of inter-disciplinary Department of Natural Resource & Environment / Sustainability Studies, at the Akamai University, USA and Professor at the Department of Finance, Innovation and Entrepreneurship, Sumy State University (SSU), Ukraine. Dr. Bhandari is also serving as Editor in Chief of International Journal—The Strategic Planning for Energy and the Environment (SPEE), Denmark, The Netherlands, USA- ISSN: 1546-0126 (Online Version), ISSN: 1048-4236 (Print Version) https://www.journal.riverpublishers.com/index.php/SPEE/about He is also serving as managing editor at the Asia Environment Daily, (the largest environmental journal of the world)- https://asiaenvdaily.com/index.php/editorials-2/2-uncategorised/29760-about-asia-environmental-daily

Dr. Bhandari has spent most of his career focusing on the Social Innovation, Sociological Theories; Environmental Sustainability; Social Inclusion, Climate Change Mitigation and Adaptation; Environmental Health Hazard; Environmental Management; Social Innovation; Developing along the way expertise in Global and International Environmental Politics, Environmental Institutions and Natural Resources Governance; Climate Change Policy and Implementation, Environmental Justice, Sustainable Development; Theory of Natural Resources Governance; Impact Evaluation of Rural Livelihood; International Organizations; Public/ Social Policy; The Non-Profit Sector; Low Carbon Mechanism; Good Governance; Climate Adaptation; REDD Plus; Carbon Financing; Green Economy and Renewable Energy; Nature, Culture and Power.

Dr. Bhandari's major teaching and research specialties include: Sociological Theories and Practices; Environmental Health; Research Methods; Social and Environmental innovation; Social and Environmental policies; Climate Change Mitigation and Adaptation; International environmental governance; Green Economy; Sustainability and assessment of the economic, social and environmental impacts on society and nature. In brief, Prof. Bhandari has sound theoretical and practical knowledge in social science and environment science. His field experience spans across Asia, Africa, the North America, Western Europe, Australia, Japan, and the Middle East.

Dr. Bhandari has published several books, and scholarly papers in international scientific journals. His recent books are *Green Web-II: Standards and Perspectives from the IUCN (2018); 2nd Edition 2020; Getting the Climate Science Facts Right: The Role of the IPCC; Reducing Inequalities Towards Sustainable Development Goals: Multilevel Approach; and Educational Transformation, Economic Inequality—Trends, Traps and Trade-offs; Social Inequality as a Global Challenge, Inequality -the Unbeatable Challenges, Perspectives on Sociological Theories, Methodological Debates and Organizational Sociology.* Additionally, in creative writing, Prof. Bhandari has published 100s of poems, essays as well as published two volumes of poetry with Prajita Bhandari.

Prem B. Bhandari is a Social Researcher (Demographer). He has a Ph.D. in Rural Sociology and Demography from the Pennsylvania State University, USA. He completed his M.Sc. in Rural Development Planning from the Asian Institute of Technology, Thailand and earned a B.Sc. degree in Agricultural Economics from Tribhuvan University, Nepal. Currently, Bhandari is working as the Evaluation Coordinator/Data Analyst for a School Feeding Program Evaluation in Cambodia and Haiti through KonTerra Research Group, Washington, D.C. In addition, he is the Managing Director of South Asia Research Consult, Inc. He is also an Adjunct Professor at Agriculture and Forestry University in Nepal. Prior to this, he worked as a researcher (Co-Investigator) and Assistant Research Scientist at the Institute for Social Research, University of Michigan. He has over 20 years of experience in designing and managing large scale social (survey) research programs. In Nepal, he worked as the assistant professor of agricultural economics at Tribhuvan University, Nepal. His major research interests include social research methods (survey design, survey instrument design, pre-testing and finalization, survey data collection, survey management and quantitative analysis of survey data, and qualitative surveys); socio-economic and cultural inequalities and determinants of demographic behaviors (population health and migration); social inequalities; food security and nutrition; and overall rural social change. He has a strong background and experience in quantitative (and qualitative) analysis. Bhandari has published a number of articles in peer-reviewed journals and as book chapters and presented dozens of papers in national and international professional meetings. He is the editorial member of several journals and an occasional reviewer of over a dozen of peer-reviewed journals.

Keshav Bhattarai, born and raised in a rural area of Balkot (Current Chhatradev Village Palika), Arghakhanchi district of Nepal, Dr. Keshav

Bhattarai has contributed hugely to bridge Nepali and US education system.

Bhattarai family composed of four siblings (three brothers and one sister). The family lived on subsistence farming. Bhattarai completed High School (10th grade) from Mahendra Vidya Bodh High School, Balkot, Arghakhanchi district of Nepal. He earned Intermediate of Science (I. Sc.), Bachelor of Science (B. Sc.), and Bachelor of Arts (BA) from Tribhuvan University. He also earned Associate of Indian Forest College (AIFC) post-graduate Diploma from Indian Forest College, Dehradun, India, and Master of Science (M.Sc.) in Natural Resource Management from the University of Edinburgh, Scotland, U.K.

He completed Ph.D. from Indiana University, Bloomington, Indiana, US in 2000 in Geography. He later joined Eastern Kentucky University as a faculty member, and in 2002, he moved to the University of Central Missouri, Warrensburg, Missouri.

His career on academia began as an Associate Instructor at the Department of Geography, Indiana University, Bloomington, Indiana in 1996. He has since taught a variety of courses at Eastern Kentucky University and University of Central Missouri. Dr Bhattarai is a member of Nepal Foresters' Association since 1983 and Association of American Geographers since 2002. A recipient of more than a dozen grants, Dr. Bhattarai has also taken the responsibility of journal article reviewer and book reviewer.

Dr. Bhattarai has been working as a Professor/ Geography at the School of Geoscience, Physics, and Safety Sciences at the University of Central Missouri since 2011. Dr. Bhattarai has also worked as a Short-term Consultant to the World Bank to assess Hydropower potential in Nepal. He has also worked as an Interim Chair, Department of Geography at the University of Central Missouri from 2009 to 2011.

Dr. Bhattarai bridges between the University of Central Missouri (UCM) and various universities of Nepal. Nepali universities and UCM have started 2+2 and 3+1 programs with UCM. The 2+2 refers to completing two years of undergraduate studies in Nepal and remaining 2 years in the US to complete undergraduate studies and vice-versa. The 3+1 program refers to completing three years education in Nepal and transferring to the US and vice-versa to complete an undergraduate degree.

Before embarking for his higher studies in the USA, Dr. Bhattarai worked in various capacities under the Ministry of Forest and Soil Conservation in Nepal from 1983 to 1995. Bhattarai left the Government job due to repeated political illogical impositions on him on the settlements of political cadres on forest areas without any norms and land use planning.

Dr. Bhattarai's urbanization research reveals that Nepal's urbanization is moving too quickly without needed infrastructure for the urban dwellers.

The unplanned political ad-hoc decision to annex many rural areas into urban definition has increased vulnerabilities among urban dwellers since Nepal is located in between frequently moving Eurasian and Indian tectonic plates. Nepali urbanization has become *ruralopolis* (political decision to annex rural areas into municipalities). Such ad-hoc urbanization has increased the costs of lands and taxes on low-income people without giving any needed facilities to them. Dr. Bhattarai served as Fulbright Specialist at the Central Department of Geography (CDG) at Tribhuvan University, Kathmandu in the summer of 2018 to facilitate CDG to embark to active participation in urban planning through research.

Dr. Bhattarai is the recipient of Gorkha Dakhsina Bahu, award for civil servant decorated by the King Birendra for outstanding performance in forestry service and administration in 1993. He has also received Mahendra Vidyabhusan for his academic excellence in 2004. Moreover, he is also the recipient of All Round Forest Award by Indian Forest College, Dehradun, India in 1983, International Scholar Award by the University of Central Missouri in 2003 and Chairperson's award for Outstanding Academic Performance by the Department of Geography, Indiana University, Bloomington in 1999. He has also bagged half a dozen faculty research award for his academic excellence from 2007 to 2020.

He has published a number of academic papers, authored four books, co-authored one book and contributed many book chapters, journal articles, and presented his research at a number of professional meetings.

Dr. Mary Jo Bulbrook, her career expansion covers over 51 years primarily serving higher education in leadership positions in academia starting in community college education in nursing, then higher education at Texas Woman's University to University of Utah. Then she was guided to move out of country to Memorial University of St. John's Newfoundland, Canada, and from there to Edith Cohen University in Perth, Western Australia. On the side her role was a practitioner, educator, and researcher in a range of energy therapies starting with Therapeutic Touch, Healing Touch, Touch for Health, Transform Your Life through Energy Medicine (TYLEM) her own creation blending her profession as a psychiatric mental health clinical specialist with energy therapies. This expansive background of energy therapies was not only taught throughout USA and Canada but also taken to Australia, New Zealand, South Africa, and Peru where she worked with traditional healers while education professors, nurses, and other health care, and lay people. Her journey served to capture the worldwide perspective on education, health and healing to address local, national, and international ways to improve the human condition where she gained firsthand perspectives on worldwide needs, values, and challenges.

The issues of "what if," "why," and "how" dominated her search to understand the broader perspectives of world strategies, cultural norms, and expectation of how to prepare individuals, families, and communities for being in the world, surviving and thriving. We are not alone and need each other standing as equals became her trademark and motto in search of how to prepare educators in different cultures regarding the blending cultural needs with innovative strategies and techniques of national heroes of health and healing with native traditions. She stood alongside those she served as equals not as "I am better than" or "have all the answers" to a "we are one," "united by our humanness as equals," and "each with different perspective on the meaning of life," living in this world in a cooperative way with a win-win perspective as her personal guiding light which serves as the philosophy in her new role leading Akamai University as their first woman president and innovator holding the light for the faculty and students to shine their light.

Jacek P. Binda is a Recotr and a researcher at Bielsko-Biała School of Finance and Law, Poland (pol. Wyższa Szkoła Finansów i Prawa w Bielsku-Białej, Polska). Prof. Binda obtained his PhD at the Warsaw University of Technology and his Post-Doc (habilitation) at University of Zylina, Republik of Slovakia. His principal fields of research include economy and finance, problems of the high-risk financial instruments, banking, local and public finances, e-economy, local government, project management and controlling. Since 2005, he has coordinated several medium-size and large research projects focusing on border studies, supported by the EU's Framework Programmes, National Research and Development Centre in Warsaw, Poland. He is an expert of the National Research and Development Centre in Warsaw, European Commission, Scientific Grant Agency of the Ministry of Education of Slovak Republic and the Academy of Science. Author or co-author of numerous monographs, over 60 national and international publications in the field of finance and information technology. He is a member of the Scientific Boards and Editorial Boards, incl.: Strategic Planning for Energy and the Environment (Journal in River Publishers), MDPI Journals, Switzerland; Sumy National Agrarian University: Sumy, UA; Scientific Journal of Bielsko-Biala School of Finance and Law (Editor-in-Chief), PL. Awarded the Gold Cross of Merit for achievements in research and teaching by President of Poland.

Anna Buriak graduated from Ukrainian Academy of Banking of the National Bank of Ukraine (M.S in "Banking," Ph.D. in "Money, finance and credit"),

Visiting Lecturer at Technische Universität Bergakademie, Freiberg, Germany (2016), Visiting Scholar at Fachhochschule der Deutschen Bundesbank, "University of Applied Science," Hachenburg, Germany (2013). She is a leader and contributor in international projects, such as Jean Monnet projects (2014), grant projects such as "Development of Dialogue Between Banks and Civil Society in the Context of Ensuring Democratic Processes in Ukraine" (DAAD, 2016), Young Scientist Research Grant of the Ministry of Education and Science of Ukraine on "The Economic-mathematical modeling of the mechanism for restoring public trust in the financial sector: a guarantee for the economic security of Ukraine" (2017–2020), Development grant of The Economics Education and Research Consortium (EERC) on "The systemically important banks Ukraine" (2014). She is an educational project manager at the non-governmental organization «Council of Young Scientists» (from 2014). Currently works as Associate Professor at the Department of Finance, Banking and Insurance at Sumy State University (Ukraine). Scientific interests include public policy, regulation and supervision, behavioral economics, decision-making, households, and trust.

Oleksii Demikhov graduated from Sumy National Agrarian University (M.S. in "Accounting and audit," Ph.D. in "Public Administration"), won ERASMUS + KA1 (2019), Jean Monnet Summer School (2018), AAF/OMI (The American Austrian Foundation) Program (2019, 2020) scholarships. Currently works as a Senior Lecturer at the Department of Management at Sumy State University (Ukraine). He is expert of the National Agency for Quality in Higher Education (2019–2021). He is member of editorial boards of scientific journals (Ukraine). He is author of more than 30 papers. Scientific interests include Public Health and Health Economics, Public Administration.

Volodymyr Domrachev graduated from Moscow Institute of Physics and Technology (MSc. in "Computer science," PhD in "Computational Mathematics and Cybernetics"). Currently works as Associate Professor at the Faculty of Information Technology at Taras Shevchenko National University of Kyiv (Ukraine). He is a member of the Ukrainian Academy of informatics. He is the author of more than 240 papers. Scientific interests include Data Science, Scientific modelling, Ecology, and Monetary economics. He is an activist and is a CEO at NGO "Ecological medical Academy" (http://emacademy.in.ua), where he provides courses on Ecology and Medicine.

Douglass Lee Capogrossi has established and administered a variety of formal and non-formal education programs including distance learning colleges, trade apprenticeships, correctional education programs, work experience projects, on-the-job training ventures within industry and the human services, cross-border university affiliations, distance learning training programs for industry, and adult job training through center-based programs. He holds permanent teaching credentials in the United States in commerce and social studies and lifetime certification as a counseling teacher for the emotionally disturbed. Dr. Capogrossi has ten years' experience developing and delivering successful correctional education programs for adult inmates in Hawaii prison facilities with emphasis in parenting, cognitive development, transition to work and community, and adult basic education skills. Dr. Capogrossi is an experienced community service administrator, where his expertise rests primarily with creation of NGO corporate structures, program funding, excellence of Board operation, and implementation of a wide spectrum of government and nonprofit training programs and emergency services projects. Dr. Capogrossi has held top management positions in industry, serving as General Manager of Micrographic Systems, a medical camera-manufacturing firm in Silicon Valley California for a brief period in the early 1980s and he owned and operated America Builders, a successful licensed general contracting firm. Dr. Capogrossi has extensive community service experience on NGO boards and is highly experienced with founding activities for new nonprofit ventures, especially with antipoverty programs. He has served as a member of the Board of an international quality assurance agency in higher education. Dr. Capogrossi earned his bachelor's in business administration, his Master's in Curriculum and Instruction, and his Ph.D. in Adult and Continuing Education from Cornell University, USA, where he completed an extensive dissertation investigating the effectiveness of the American education system. Among his more recent published scholarly papers, The Assurance of Academic Excellence among Nontraditional Universities was published in the Journal of Higher Education in Europe. Dr. Capogrossi has dedicated his career in service to humanity, through his efforts to improve the human condition.

Durga D. Poudel, the Founder of Asta-Ja Framework and Professor in School of Geosciences at University of Louisiana at Lafayette, Lafayette, USA, was born and raised in Tanahu, Nepal. He spent his childhood in Tanahu and Lamjung districts, both in the Mid-Hills of Nepal. Dr. Poudel passed his S.L.C. exam from Nirmal Vocational High School, Damauli, Tanahu in 1977 and received an I.Sc. in Agriculture at Tribhuvan University at Rampur (IAAS), Chitwan, Nepal in 1980; a B.Sc. in Agriculture (Major: Agricultural Economics) at University of Agriculture, Faisalabad, Pakistan in 1987; a M. Sc. in Natural Resource Development and Management at Asian Institute of Technology, Bangkok, Thailand in 1991; and a Ph.D. in Soil Science, University of Georgia, Athens, GA, U.S.A. in 1998. Dr. Poudel is an expert in environmental science, climate change adaptation, and sustainable agricultural development; soil physical, chemical, and mineralogical characterization; soil classification; soil and water conservation, water quality, roadside vegetation management, and natural resources conservation and development. His recent research focuses on water quality monitoring and modeling, smallholder mixed-farming system, climate change adaptation, geohazards, wildflower, and environmental soil chemistry. Dr. Poudel has published over 100 peer-reviewed journal articles, conference proceedings papers, book chapters, research briefs, and scientific abstracts. He has published numerous popular articles. He has given over 100 scientific presentations to regional, national, and international conferences. Dr. Poudel has received more than $4.48 million in external funding as a PI and more than $1.76 million in external funding as a CO-PI. He has supervised more than two dozen graduate students in their Thesis/Dissertation research.

Dr. Poudel's professional experience consists of Research Fellow at Asian Vegetable Research and Development Center, Taiwan (1991–1994); Graduate Research Assistant in Sustainable Agricultural and Natural Resource Management Collaborative Research Support Program, University of Georgia (1994–1998); and Visiting Research Scholar, University of California Davis (1998–2000), USA. Dr. Poudel joined the University of Louisiana at Lafayette, USA, as an Assistant Professor of Soil Science in August 2000, and currently is a tenured Professor and Coordinator of Environmental Science Program of School of Geosciences, Director of Ag. Auxiliary Units (Model Sustainable Agriculture Complex (600-acre Cade Farm), Crawfish Research Center, and Ira Nelson Horticulture Center), and Board of Regents Professor in Applied Life Sciences at the University of Louisiana at Lafayette, Louisiana, USA. Dr. Poudel's professional affiliations include membership in Soil Science Society of America, American Society of Agronomy, Crop Science Society of America, Geological Society of America, American Geophysical Union, and Soil and Water Conservation Society, USA. Dr. Poudel is the life member of NRNA, Nepalese Association in Southeast America (NASeA), Asta-Ja Abhiyan Nepal, Asta-Ja USA, and Association of Nepalese Agricultural Professionals of Americas (NAPA). Dr. Poudel is the founding member of Asta-Ja Abhiyan Nepal, and the founding President of Asta-Ja Research and Development Center (Asta-Ja RDC), Kathmandu, Nepal, and Asta-Ja USA, Honolulu, Hawaii, USA. Dr. Poudel was the Chair of NRNA ICC Agriculture

Promotion Committee America Region, 2017–2019. Dr. Poudel served as a Board Member of the Bayou Vermillion Preservation Association (BVPA), Lafayette, Louisiana, USA, and an Advisor of the NASeA. Dr. Poudel is a Resource Person of Louisiana Organics, and a member of the Louisiana Technical Advisory Committee, USDA-NRCS, Louisiana, USA. In 2017, in recognition of the impact and quality of one of his research papers, Dr. Poudel and his co-authors were awarded for the "2017 Best Research Paper Award for Impact and Quality Honorable Mention" by the Journal of Soil and Water Conservation. Dr. Poudel believes on providing his services to others, promoting social inclusion, integrity, goodwill, and peace, and advancing on better understanding of the world.

Hemant Ojha works partly as Associate Professor for University of Canberra and partly as a Principal Advisor for Sydney-based research and development consulting firm *Institute for Study and Development Worldwide (IFSD)*. Dr Ojha's work cut across academic theorizing, social entrepreneurship, and public policy engagement in relation to social development and environmental sustainability. A PhD on environment and development from University of East Anglia, Dr Ojha is a well-recognized public intellectual on Nepal's community forestry which is seen as a global institutional innovation for tackling the twin challenges of environment and development. He has championed science-policy interface works in South Asia in the field of forestry, agriculture, and climate change. He founded and led various policy think tanks and education institutions in South Asia and Australia. He has taught at Australia's University of New South Wales and has conducted research in Asia and Africa on a range of environmental governance issues. His current research focuses on social entrepreneurship, public leadership, and science-policy interface in sustainable development. His Books have been published by Routledge and Cambridge and writes actively in both academic and public media.

Vita Hordiienko graduated from Sumy State Pedagogical Institute (Teacher of geography and biology), Ukrainian Academy of Banking of the National bank of Ukraine (Specialist in accounting and auditing, specialty "Accounting and auditing"), Ph.D. (Candidate of Sciences in specialty 08.00.06 "Economics of natural resources and environmental protection"). Currently works as Associate Professor at the Department of Management and the deputy head of the Center of Potential Personnel Development at Sumy State University (Ukraine). She is the author of more than 90 papers. Scientific interests include regional economy, state regulation of economy, decentralization of public power, investment potential of the region, and problems of life-long learning education.

Kedar Neupane holds M.A. (Economics) from Tribhuvan University of Nepal and a Rockefeller Fellow Scholar at the School of Economics of University of the Philippines, was trained as emergency preparedness and operations manager at the University of Wisconsin (USA). After retiring from work for over three decades of international service with the United Nations System Mr. Neupane currently resides in Geneva (Switzerland). He began international professional career starting from the UN Department of Technical Cooperation and Capital Development Fund, attached to the UNDP Sudan, prior to moving to UNHCR.

Mr. Neupane has worked for the United Nations High Commissioner for Refugees in various capacities rotating between several countries in the North-Africa, the Middle East, Africa, Asia, and Europe. With varied work experience acquired from several countries, Mr. Neupane has unique perspectives on developmental issues and operational challenges, to bottom-up approach to the people-oriented program, design, planning, and implementation with target beneficiaries at the center of operational complexity of development syndromes, financing, and strategies for sustainability, organizational change management, and reform.

Yevhen Kovalenko is Ph.D. (Econ.), Senior Lecturer at the Economics, Entrepreneurship and Business Administration Department, Sumy State University. Yevhen Kovalenko graduated from Sumy State University (Master in "Economics of Enterprise" 2002; Master in "Law" 2017). Ph.D. in "Economics and Management of National Economy"—National Aviation University, Kyiv (Ukraine), 2018. He is an author and co-author of more than 50 scientific and other publications and a contributor of more than 10 research projects. His current research interests include corporate social responsibility, energy efficiency, environmental entrepreneurship, sustainable development, and transaction cost management.

Krishna Prasad Oli, currently member of National Planning Commission, Government of Nepal comes from Tehrathum District, Province No.1 of Nepal. Dr. Oli has the experiences of working with National and International organizations in a wide range of development and environmental issues at policy–program-action continuum in particular to environment planning and management, ecosystem management, implementation of multilateral environmental treaties, developing regional cooperation mechanism, capacity building, trans-boundary landscape management, and policy coordination and networking with stakeholders at different levels. He is a member of Nepal Bar Association and before joining the Government he was a visiting faculty at TU, teaching Environmental Law, he was Adjunct Professor at Faculty of Social Development and Western China Development Studies in Sichuan University in China and Distinguished Professor at Xian Minzu University, China. He is also a visiting professor at Kathmandu School of Law and adjunct faculty at Akamai University, Hawaii, USA. He has extensively visited different parts of Nepal, China, South and South East Asia, Europe, Canada, North America, Australia, and Africa during his career.

Yana Kryvych graduated from the Ukrainian Academy of Banking of the National Bank of Ukraine (M.S in "Management in foreign economic activity", Ph.D. in "Money, finance and credit"), participant of the international training courses under the auspices of the Banking Institute Banking University (Praha, Czech Republic, 2017). She is a leader of social and educational projects in career guidance for youth, moderator of the independent local conference TEDxUABS (2016), a researcher at Young Scientist Research project of the Ministry of Education and Science of Ukraine on "The Economic-mathematical modeling of the mechanism for restoring public trust in the financial sector: a guarantee for the economic security of Ukraine" (2017–2020). Currently works as Associate Professor at the Department of Finance, Banking and Insurance at Sumy State University (Ukraine). Scientific interests include marketing, decision-making, financial markets, and customer relationships.

Oleksandr Kubatko, Dr.Sc. (Econ.) is Professor, Professor at the Economics, Entrepreneurship and Business Administration Department, Sumy State University. Oleksandr Kubatko has published more than 50 scientific papers, including fourteen papers in international peer-reviewed journals. He is a leader of three and a contributor of more than ten scientific and research projects, including international ones. He is a Deputy Editor-in-Chief of the International Scientific Journal "Mechanism of Economic Regulation" (Ukraine), and Editor of International Scientific Journal "Environment, Development and Sustainability" (Netherlands). The sphere of his scientific interests includes environmental economics, sustainable development, and health economics.

Tetiana Kurbatova is Ph.D. (Econ.), Senior Lecturer at the International Economic Relations Department, Sumy State University. Tetiana Kurbatova is a Senior Lecturer at the International Economic Relations Department of Sumy State University. Her scientific interests cover renewable energy, energy policy, energy economics, low-carbon economies, sustainable development, and environmental policy. She has been working on various scientific projects in the renewable energy field for 2012–2020 and has published more than 60 scientific papers, including papers in international peer-reviewed journals.

Anna Lasukova graduated from Ukrainian Academy of Banking of the National Bank of Ukraine (M.S in "Banking," Ph.D. in "Money, finance and credit"), participant of social projects like "STEP IN: Step into the Dialogue" dedicated to exploring modern communication tools in the multicultural environment (Tallinn, Estonia) and "TRANSFORMERS: Migrants' Transition to Participation" focused on the issue of internally displaced migrants (Tbilisi, Georgia). She is a contributor in the Jean Monnet educational project—"Financial aspects of the European integration projects: summer/winter schools for young researchers and post-graduate students" (2014), Young Scientist Research Grant of the Ministry of Education and Science of Ukraine on "The Economic-mathematical modeling of the mechanism for restoring public trust in the financial sector: a guarantee for the economic security of Ukraine" (2017–2020), as a project manager at the NGO "Council of Young Scientists" (2015–2017). Currently works as Lecturer at the Department of Finance, Banking, and Insurance at Sumy State University (Ukraine). Scientific interests include behavioral finance, decision-making, financial markets, social responsibility, and trust.

Svitlana Lutsenko graduated from Sumy State Pedagogical University (specialist in History and Social Sciences; Pedagogy, M.S.in Higher School Pedagogy), Zaporizhia Classical Private University (Ph.D. in Public Administration "Mechanisms of Public Administration"), a graduate of international programs: "Intercultural Education for Democracy," "Leaders of Ukrainian education" (Poland, 2006–2008), "Teaching tolerance" (Israel, Netherland, Germany, France, 2006–2009), Educational Reform Program in Ukraine "Democratic School" (2017–2018). Currently works as an Associate Professor at the Department of Pedagogy, Special Education and Management of the Sumy Regional Institute of Postgraduate Pedagogical Education and Associate Professor at the Department of Management at Sumy State University (Ukraine). Scientific interests include Innovation management, Informatization of public administration, Civic education, Public service, Management technologies. She is the author of more than 80 scientific and scientific-methodical publications. She is a co-founder of the NGO "Center for Corporate Education AKME," which provides postgraduate education for teachers of Sumy region, a regional Internet project with international participation "Sumy Chronicles"—a collective hypertext wiki-encyclopedia of history and modernity of Sumy region.

Madhu S. Atteraya, Ph.D., is an assistant professor at Keimyung University, South Korea. His research areas include family welfare, equity in health care, social justice, migration, and international maternal and child health. Atteraya serves as a Focal Point for the partnership for research on maternal, newborn, & child health at PMNCH, WHO. Atteraya also serves at the BMC Women's Health journal and Research on Social Work Practice journal as an editorial board member.

Mirjana Radovic-Markovic is a full professor of Entrepreneurship, Doctor of Science. She holds B.Sc, M. Sc. and Ph.D. in Economics; she conducted Post-Doctoral Studies in Multidisciplinary Studies in Lomonosow (Russia). She has served as a professor at a number of international universities, foundations, and institutes. She is a founder and editor-in-chief of three Peer Journals as follows: "The Journal of Women's Entrepreneurship and Education" (JWE), "International Review" (IR), and "Journal of Entrepreneurship and Business Resilience" (JEBR). She has written more than thirty books and more than two hundred peers' journal articles. Her publications are published by top world publishing companies. Her article "Acceptance and use of lecture capture system (LCS) in executive business studies" published in Interactive Technology and Smart Education has been selected by the editorial team of Emeralds as an Outstanding Paper in the 2018 Emerald Literati Awards. Namely, her article was chosen as a winner as it is one of the most exceptional pieces of work the team has seen throughout 2017.

Maheema Rai, graduated in History (Honours) from Southfield College (Formerly Loreto College), under University of North Bengal, Darjeeling. Post Graduation from Sikkim University, Department of Peace and Conflict Studies and Management. Ph. D. in "Gender Justice and Uniform Civil Code in India: A Study of Darjeeling District in West Bengal," and MPhil in "Health security of women labourers living in slums of Darjeeling town" from Department of Peace and Conflict Studies and Management, Sikkim University. Recipient of Indian Council of Social Science Research (ICSSR) 2017–2018. Currently working as Guest Faculty in the Department of Peace and Conflict Studies and Management, Sikkim University. Area of interests are Gender Issue, Child rights, Traditional and Non-Traditional Security, and Human Security and written many articles on Women's security and Peace.

G.B. Sahqani is an International Trade and Tax Consultant; previously he has served in the Ministry of Finance, Pakistan. He was involved in the preparation of proposals for trade policy, tariffs and annual budgets, assessment of custom duty, and determining actual HS Code of import/export goods in line with Customs Tariff and WCO harmonized system of classification. Dealing with international trade involved interpretation and implementation of trade laws and policies. He is writing articles and research papers on tax, revenue, and socio-economic issues. He has done Master's in Economics. He lives in Karachi, Pakistan.

Aidin Salamzadeh is an assistant professor at the University of Tehran. His interests are startups, new venture creation, and entrepreneurship. Aidin serves as an associate editor in *Revista de Gestão* and *Innovation & Management Review* (Emerald) and an editorial advisory in *The Bottom Line* (Emerald). Besides, he is a reviewer in numerous distinguished international journals. Aidin is a member of the European SPES Forum, the Asian Academy of Management, and Ondokuz Mayis University.

Salvin Paul graduated from Mahatma Gandhi University, Kottayam, Kerala (M.A and M.Phil in Politics and International Relations, and Ph.D. in "Intellectual Property Rights and Biodiversity Associated Traditional Knowledge"). Currently, he teaches Peace and Conflict Studies in Sikkim University, India. He is the author of more than 60 papers. His areas of interest include non-traditional threats in security discourses, human rights of migrants, refugees, women, children, and elderly people, and nuclear security, and water governance.

Tetiana Semenenko graduated from Kharkiv Engineering and Economic Institute (now the Simon Kuznets Kharkiv National University of Economics) (MSc. in "Data science") and Sumy State University (Ph.D. in "Environmental Economics and Natural Resources Protection"). Currently, she works as Associate Professor at the Department of International Economic Relations and a Head of the Group of Information Management for Web Resources of Center for Benchmarking and Web Management at Sumy State University (Ukraine). She is a member of the Ukrainian Association of International Economics. She is the author of more than 40 papers. Scientific interests include Data Science, Inequality, and Behavioral economics.

Vladyslav Shapoval is a senior student at Sumy State University majoring in International Economic Relations. He is an alumnus of Governmental International Programs such as the U.S. Department of State "Global Undergraduate Exchange Program (Global UGRAD)" and the European Union "ERASMUS+." Mr. Shapoval is an active participant in many projects and initiatives that were mainly in the field of leadership and inclusiveness, and that helped him to build a strong interest in peace and inclusive society building. Another passion of Vladyslav is communication. He believes it is an essential tool that can help to raise awareness and combat inequality and discrimination around the world. Mr. Shapoval is also a trainer on leadership and communication for high schools' students in a framework of the All-Ukrainian Leadership Movement, mainly working in Sumy region and Sumy city with the City High School Students Organization "MOST."

Hanna Shvindina graduated from Sumy State University (M.S. in "Management of Organizations, Ph.D. in "Environmental Economics and Natural Resources Protection," Dr. Sci. in "Economics & Management at the Enterprises"). She was awarded several individual scholarships and was a Post-Doc researcher in France in the US (University of Montpellier, Purdue University) in "Business Studies and Management," Fulbright Alumna (2018–2019), won ERASMUS MUNDUS scholarships (2014, 2015). Currently, she works as Head of the department and Associate Professor at the Department of Management at Sumy State University (Ukraine). She is a member of CENA community, Researchers' excellence network (RENET), Ukrainian Association for Management Development and Business Education. She is a member of editorial boards of several scientific journals (Ukraine, Poland, and Switzerland), invited reviewer for the international conferences (Poland, Germany, and Ukraine). She is the author of more than 90 papers. Scientific interests include Coopetition Paradox, Strategies and Innovations, Change Management, Organizational Development, Civic Education and Leadership. She is an activist and is a CEO at NGO "Lifelong learning centre," where provides courses on Leadership and Communications, organizes the program on empowerment for the local community, and offers advisory services.

Iryna Sotnyk, Dr. Sc. (Econ.) is Professor, Professor at the Economics, Entrepreneurship and Business Administration Department, Sumy State University. Iryna Sotnyk graduated from Sumy State University (specialist in "Economics of Enterprise," Ph.D., Dr. Sc. in "Environmental Economics and Natural Resources Protection"), Fulbright Alumna (2019–2020), won the Cabinet of Ministers of Ukraine Prize for youth's contribution to the development of the state (2004) and scholarships of the Cabinet of Ministers

(2011–2012) and the Verkhovna Rada of Ukraine (2012–2013). Currently she works as Professor at the Department of Economics, Entrepreneurship and Business Administration at Sumy State University (Ukraine). Scientific interests include Economics of Energy and Resource Saving, Energy Economics, Social and Solidarity Economics. She is a Deputy Editor-in-Chief of the International Scientific Journal "Mechanism of Economic Regulation" (Sumy, Ukraine), Guest Editor of the International Scientific Journal "Global Energy Issues," a leader and a contributor of more than 25 research projects, including international ones. Under her leadership, 7 research projects including 4 grants of the President of Ukraine (2006, 2008, 2012 and 2016) were performed. Iryna Sotnyk had international training in Russia (2013), Estonia (2015), Lithuania (2016), Israel (2016), and USA (2019–2020).

Jagadish Timsina (Dr.) is a systems agronomist specializing on crop, soil, nutrient, and water management; conservation agriculture based sustainable intensification; agroforestry and community forestry systems; and crop and systems modelling. He has diverse experience in agronomic, economic, and environmental management of agro-ecologies ranging from levelled lands in Australia to hills and mountains in South and Southeast Asia. He has worked with University of Melbourne and Commonwealth Scientific and Industrial Research Organization (Australia), International Rice Research Institute (Bangladesh and Philippines), International Maize and Wheat Research Centre (Bangladesh and Mexico), and Tribhuvan University and Agricultural and Forestry University (Nepal) for about 40 years. Currently, he is an editor to Agricultural Systems journal, a Senior Fellow at the Global Evergreening Alliance (Melbourne), an organization dedicated to land and forest restoration and drawing down of greenhouse gases globally, and a Senior Adviser to the Institute for Study and Development Worldwide (Sydney).

Tetyana Vasilyeva is a Director of Academic and Research Institute of Business, Economics and Management, Professor (2010, Professor's degree of the Department of Management) and Doctor of Economics (2008, speciality 08.00.08—money, finance and credit). She is the author of more than 200 scientific papers, 70 monographs, and numerous materials of the conferences. She is Editor-in-chief of the scientific journal "Business Ethics and Leadership," member of the editorial board of the journals: "Strategy and Development Review," "Carbon Accounting and Business Innovation Journal." She is a mentor for PhD students and Post-Docs and is a member of the Specialized Academic Committee at Sumy State University; one of the founders and a member of NGO "Council of Young Scientists." She was the initiator of launching the Business Support Center in Sumy region, and now she is a member of this organization that gained the support of the European Bank for Reconstruction and Development. She is a team leader, an author, and participant of numerous scientific projects that gained the highest scores in Ukraine. Research interests include banking, investment, risk management, innovation activity, social responsibility, education. She is a leader and contributor in international projects, such as Jean Monnet projects (2014, 2015, 2017), grant projects such as "Development of Dialogue Between Banks and Civil Society in the Context of Ensuring Democratic Processes in Ukraine" (DAAD, 2016), "Enhancing Energy Security by Swiss–Ukrainian–Estonian Institutional Partnership" SCOPES IZ74Z0_160564 (Swiss National Science Foundation, 2015–2017) and many others. She is a contributor in many educational projects financed by Ministry of Economic Cooperation and Development of Germany, DVV International, DESPRO in Switzerland, American Council for Economic Education NCEE, International Foundation "Vidrodzennia" etc. Professor Vasilyeva initiated many scientific and educational events concerning the development of corporate social responsibility (e.g., winter/summer schools for youth "Economic and business values," "School of innovations and social entrepreneurship," "Management of socially responsible business," etc.)

Miloš Vučeković graduated Faculty of Informatics and Computing (FIR) at Singidunum University, where he acquired the title—bachelor–computer scientist. After completing his primary studies, he attended and completed a master's degree in engineering management at the same University with an average grade of 10.00, with the thesis "Comparative analysis of modern project management models on the example of business information systems implementation" which he defended in November 2019. He is currently attending the second year of doctoral studies in the same field, and at the same University.

He started his professional career in the private sector in 2004 as a freelance programmer, and then as an IT specialist for information technology, opening a company for IT consulting in 2007. Since 2009, he has been working as an IT project manager and IT consultant on software development projects and implementation of business information systems. At the beginning of 2018, he re-established a consulting company for innovative entrepreneurship, with a focus on the development of technology startups and innovative products, as well as the education of young entrepreneurs. Since 2019, he has been permanently employed as a Project Manager at NCR Corporation, where he leads business solution development and innovation projects in collaboration with the world's largest IT vendors such as Microsoft, Oracle, and QuickBase. He is also engaged in scientific research, with papers published in scientific journals on project management, digital transformation, and business information systems.

Oksana M. Zamora has graduated from Sumy National Agrarian University (Ukraine) with a master's degree in foreign economic affairs and junior specialist in law, Ph.D. in regional economics (Council for Productive Forces Study of NASU, Ukraine), later—a postgraduate education ("Management of projects funded by the EU within the financial perspective 2014–2020," Lodz University of Computer Science and Art, Poland). A research scholarship of the Polish Committee for UNESCO in 2013. Has experience in design/implementation of 13 grant-funded international projects funded by the EC, CzDA, FAO, CEI, and EBRD. Has participated in 30+ training courses/youth exchanges/mobilities/fellowships in Ukraine and abroad including Erasmus+, Youth in Action, USA FAP, British Council, EBRD, CEASC. Since January 2019, is a co-founder of the NGO "Association of Project Managers of Ukraine." She is a reviewer for four foreign scientific journals (Portugal, Poland), a member of two editorial boards (Czech Republic, Poland). She provides services as a non-formal education trainer and an international development coach. Currently, she works as an Associate Professor at the Department of International Economic Relations of Sumy State University.

Leohid Melnyk, Dr. Sc. (Econ.), Prof., Professor at the Economics, Entrepreneurship and Business Administration Department, Sumy State University, Ukraine melnyksumy@gmail.com. Leonid Melnyk is also a Director of the Institute for Development Economics of Ministry of Education & Science and The National Academy of Science of Ukraine, Sumy State University, Ukraine; Academician of the Academy of Business and Management of Ukraine; Academician of the International Academy of Information; Academician of the Academy of Engineering Sciences of Ukraine; member of International Society for Ecological Economics; the project leader of 100 research and educational projects on international, national and local levels including 30 international projects; author and co-author of more than 400 scientific and other publications, and 45 books. His research interests focus on development economics, the Third Industrial Revolution and Industry 4.0, social and economic EU studies, economics and business; ecological economics, economics of systems development, economics of enterprise, information economics, economics of sustainability.